Chemistry and Brain Development

ADVANCES IN EXPERIMENTAL MEDICINE AND BIOLOGY

Chemistry and Brain Development

Proceedings of the Advanced Study
Institute on "Chemistry of Brain
Development," held in Milan, Italy,
September 9–19, 1970

Edited by
Rodolfo Paoletti

Chairman
Institute of Pharmacology and Pharmacognosy
University of Milan, Italy

Alan N. Davison

Professor of Biochemistry
Department of Biochemistry
Charing Cross Hospital Medical School
University of London, U.K.

⊕ SPRINGER SCIENCE+BUSINESS MEDIA, LLC

Library of Congress Catalog Card Number 75-150494

ISBN 978-1-4684-7238-7   ISBN 978-1-4684-7236-3 (eBook)
DOI 10.1007/978-1-4684-7236-3

© *1971 Springer Science+Business Media New York*
Originally published by Plenum Press, New York 1971
Softcover reprint of the hardcover 1st edition 1971

ACKNOWLEDGEMENTS

Acknowledgement is gratefully made to the following organizations who sponsored this Advanced Study Institute:

NORTH ATLANTIC TREATY ORGANIZATION
Scientific Affairs Division
Bruxelles, Bèlgium

DREYFUS MEDICAL FOUNDATION
2 Broadway
New York, U.S.A.

INTERNATIONAL SOCIETY FOR
    BIOCHEMICAL PHARMACOLOGY
Via Vanvitelli 32
20129 Milan, Italy

and to the valuable assistance of the following individuals:

INTERNATIONAL ADVISORY COMMITTEE
A.N. Davison, London, England
J. Folch-Pi, Belmont, U.S.A.
R. Paoletti, Milan, Italy
E. Roberts, Duarte, U.S.A.

LOCAL ORGANIZING COMMITTEE
Bruno Berra
Francesco Clementi
Remo Fumagalli
Claudio Galli
Enrica Grossi Paoletti
Guido Tettamanti

SECRETARY TO THE ORGANIZING COMMITTEE
Hasel J. Prain

Professor Rodolfo Paoletti
Scientific Director

v

CONTENTS

SECTION II

BIOCHEMICAL AND MORPHOLOGICAL INTERRELATIONS

SECTION III

MEMBRANE FORMATION AND FUNCTION

SECTION IV

NUTRITION AND BRAIN DEVELOPMENT

INTRODUCTION

J. Folch-Pi

Director of Scientific Research,

McLean Hospital, Belmont, Mass., U.S.A.

The development of the central nervous system is
possibly the most significant aspect of the growth of a
mammal from embryo to adulthood. The central nervous
system is obviously the main repository not only of the
species' inherited functional characteristics but also
of the process of individuation. Whatever "engrams"
constitute the basis of individual characteristics are
laid down mainly in the central nervous system, and
especially the brain, during its growth.

The chemical aspect of this process is clearly of
great importance and the significance of its study should
be self evident. Nevertheless, it is only one aspect
of a parellel series of morphological, physiological,
biochemical and psychological events which take place as
an integrated process, the final result of which is the
transformation of the post-embryonic nervous system into
the functioning adult system.

It is imperative, therefore, that any study or
description of the chemical events during the development
of the CNS should be undertaken in full awareness of the
concomitant morphological, physiological and psychological
events. It is only against this multidisciplinary
informational framework that the chemical events during

1

development can be correctly interpreted and acquire
their full significance.

With this in mind, the introduction to this volume
may best serve its purpose by describing briefly the
morphological and physiological events that accompany
the chemical aspect of development.

During development the CNS presents a constantly
changing histological picture. Neurons, the conducting
elements, are the first to appear in large numbers and
the first to reach their eventual total number; since
they do not reproduce their number remains constant for
the rest of the life of the animal, except for whatever
losses occur through neuronal lysis. The neuroglia,
which retains its ability to reproduce, continues to
increase in numbers much later than neurons; in man,
for instance, neurons have reached their eventual number
before birth while neuroglia, by comparison, keeps
accumulating for many years after birth. These different
rates of accumulation do not describe the whole picture
which is complicated by the remarkable growth of neurons,
and by the acquisition of myelin sheaths by the naked
axons in the course of neuronal growth. Hence we see,
at birth in rat and mouse, a CNS containing mainly
neurons in the early stages of growth, some neuroglia,
but no myelin. As CNS increases in weight, neurons
increase in size and especially in the length of their
axons and in the number, size and increasing arborization
of their dendritic processes. This neuronal growth
increases the volume of each neuron several-fold, thus
changing the respective contribution of the different
parts of the neurons to the overall histological picture:
at the beginning neurons are mainly somas; in the later
stages, the somas become quantitatively less important,
and the processes come to represent the major part of
neuronal meterial (Fig. l)(l). At the same time, since
the number of neurons remains constant while the weight
of brain increases five or six-fold, the number of
neurons per unit volume actually decreases in inverse
ratio to the increase in weight of CNS (Fig. 2)(2).
At about the age of seven days after birth in mouse, and
ten days after birth in rat, myelin appears and starts

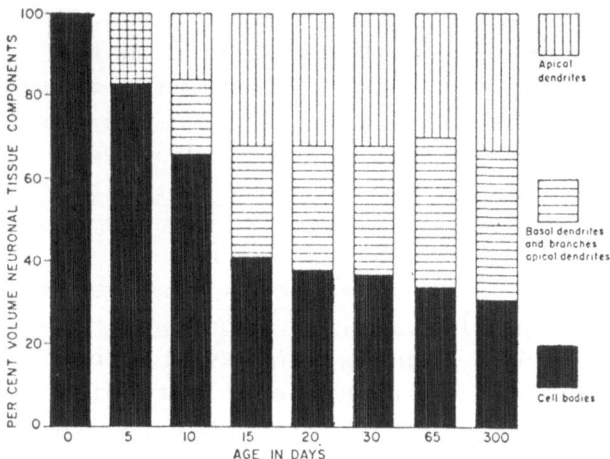

FIG. I. The percentage distribution in volume of nerve cell bodies and of apical and basal dendrites during postnatal development in area 2/3 of rabbit cortex.

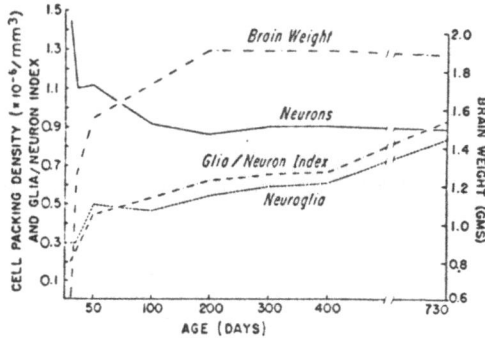

FIG. 2. Cell packing densities and glia/neuron index in fixed-stained tissues of rat cerebral cortex. Brain weight determined on fresh specimens.

to accumulate rapidly with a preceding and commensurate accumulation in the number of oligodendroglia. The rate of accumulation of myelin reaches a peak within a few days, and then decreases gradually to a very slow rate of growth at forty to fifty days of age in the mouse and somewhat later in the rat.

The protoplasmatic astrocytes increase in number in temporal relation to a marked increase in the vasculari-zation of CNS. In rat, relative vascularization, meas-ured as millimeters of capillaries per cubic millimeter of tissue, increases about three-fold between the ages of 10 and 20 days, the ages corresponding to the onset of myelination and the bulk of myelin accumulation. After 20 days it increases further but at a much slower rate (Fig. 3)(3). This increase in vascularization is reflected in a roughly parallel increase in the blood flow and in a marked increase in oxygen consumption which itself reflects both an increase in metabolism and a fast rate of biosynthesis incidental to the growth of CNS.

The physiological changes in the course of develop-ment are numerous and well documented. Even in summary form, a comprehensive description of these changes is outside the scope of this presentation. Only as an illustration let it be stated that spontaneous electrical

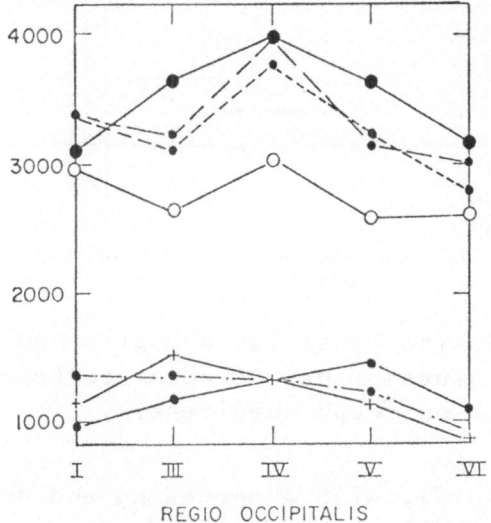

REGIO OCCIPITALIS

FIG. 3. Relative vascularity (in millimeters of capil-laries per cubic millimeter of fresh tissue) of the laninae of the associative part of the occipital cortex of rats at different ages. Birth ●—●; 5 days old +—+; 10 days old ●—··—●—·—●—·—●; 20 days old 0———0; 90 days old ●-----●; 140 days old ●----●; 390 days old ●——●.

activity appears at a certain stage in neuronal development, and that it slowly evolves into the pattern of the
adult EEG. As a general rule, the spontaneous electrical
activity precedes but not by much the onset of myelination; the adult EEG is reached way beyond the time of
completion of myelination.

   The chemical changes will be discussed in detail
by the contributors to this Study Course. Hence they
will be described here only in outline. These chemical
changes may be divided, for the purposes of simplicity
of presentation, into changes in structural components,
changes in metabolism and metabolites, and changes in
enzymes.

   The changes in structural components are possibly
the best known because of the availability of post-mortem
material for their study, and of methods for their
characterization and quantitation. In broad terms there
is, during development, a gradual desiccation of the
tissue, i.e., a relative loss of water and the exact
inverse gain in solids. This desiccation is due in no
small part to myelination and therefore it is much more
pronounced in white matter than in gray matter. The gain
in solids is correspondingly more marked in lipids,
especially the so-called myelin lipids, than in proteins,
although the increase in relative concentration with
development affects almost all structural components to
different degrees. In Table I (4) these general changes are
illustrated in the case of mouse brain. Myelination in
mice occurs on the 7th or 8th day after birth and reaches
its static or slow growth period by 40 days of age. As
can be seen, proteolipid protein and cerebrosides (which
include sulfatides) appear and accumulate in temporal
relationship with myelination. The other groups of components show different rates of change.

   In summarizing changes in metabolism and metabolites
it becomes necessary to discuss briefly the development
of the blood-brain barrier. Since the CNS has no lymphatic system, it can be postulated that the exchange
between central nervous tissue and the rest of the body
takes place at the level of the capillary walls, since

TABLE I.

COMPOSITION OF BRAIN OF MICE AT DIFFERENT AGES[a]

Age, days	Average weight of brain, mg	Water	Solids	Proteins	Trypsin-resistant protein residue	Strandin	Lipids (exclusive of strandin)	Cholesterol	Phosphatides	Cerebrosides	Proteolipid protein
1	90	76.9	12.3	7.81	—	0.206	—	0.38	1.94	<0.03	<0.03
	85	73.9	11.1	—	—	—	2.36	—	1.80	—	—
2	123	107.4	16.1	9.45	0.08	0.418	2.90	—	3.08	<0.04	<0.04
4	177	155.0	22.2	13.2	0.21	0.62	5.64	0.92	3.88	<0.05	<0.05
5	175	153.0	17.4	—	—	—	5.80	—	4.30	—	—
7	261	225.0	35.5	20.7	0.50	1.38	9.31	1.28	6.60	<0.07	<0.07
8	275	238.0	38.8	22.2	—	1.21	9.80	1.73	7.85	<0.07	—
10	318	268.0	49.6	28.6	—	—	14.3	2.61	10.6	0.48	0.29
13	360	296.0	64.5	36.0	—	2.09	18.5	3.56	13.5	0.65	—
16	376	307.0	69.5	39.1	1.35	2.26	21.3	3.83	16.5	0.94	0.565
19	379	306.0	73.1	41.4	1.51	2.24	23.8	3.90	16.0	1.33	0.760
22	384	304.0	80.6	43.4	—	2.26	26.2	5.11	18.1	2.92	—
25	402	316.0	86.0	45.9	—	—	30.4	6.25	19.8	3.06	0.805
30	396	310.0	86.4	44.0	2.60	2.26	29.9	5.80	19.6	3.84	—
	433	341.0	92.0	—	—	—	32.2	6.55	21.5	3.03	0.97
35	386	306.0	89.5	47.0	2.75	2.58	32.1	5.80	18.9	3.10	1.08
	431	339.0	93.0	—	—	—	32.7	7.03	21.9	3.46	1.12
40	414	324.0	89.5	47.6	—	2.58	32.1	5.80	18.9	3.10	1.36
50	410	318.0	91.5	47.4	—	2.26	31.6	6.15	20.5	4.10	—
75	444	344.0	101.0	53.4	—	2.48	35.6	7.65	24.0	4.80	1.60
90	445	342.0	103.0	53.9	—	2.58	37.5	7.90	24.6	5.2	1.80
180	424	322.0	101.0	51.4	—	2.26	37.4	8.50	22.0	5.55	2.06

[a] Average amount (in mg.) present in 1 brain.

in the case of most metabolites the possible contribution
of cerebrospinal fluid to this body-CNS exchange would
necessarily be negligible. It is known that in the adult
there is a control or regulation of the movement of many
metabolites from the blood stream into the CNS and vice
versa. This control usually results in keeping the net
passage of these metabolites in and out of the CNS at a
rate much lower than it would be by free diffusion. For
want of a better term, the name blood-brain barrier has
been used to describe this state of affairs. This is
unfortunate because instead of a barrier, which suggests
a passive hindrance to free passage, the phenomena
observed under the name of blood-brain barrier appear to
be the result of dynamic processes, not of passive
hindrances. Be as it may, the evidence available shows
that this blood-brain barrier is not operative or, at
least, not wholly operative in very young animals but
that it appears in the course of development. In rat,
the onset of a recognizable blood-brain barrier coincides
fairly closely with the onset of myelination.

The overall oxygen consumption of the CNS increases
markedly during development, reaching a peak and then
decreasing to a somewhat lower rate in the adult. The
increase in rate and the subsequent slight decrease exhibit many regional differences, being especially marked
in the cortex and less marked in the white matter. This
increase in oxygen consumption corresponds to the already
mentioned increase in vascularization which results, as
would be expected, in an increase in blood flow (Fig. 4)
(5).

At the same time that the rate of oxygen consumption of CNS increases, there is a marked decrease in the
resistance of CNS to anoxia and agents that interfere
with oxygen transport or utilization, such as, carbon
monoxide or cyanide. This decrease in resistance of CNS
to oxygen deprivation is very dramatic; the newborn rat
being able to survive total oxygen deprivation for as
long as one hour whereas the adult can survive for only
a few minutes. Obviously this tremendous decrease in
resistance to oxygen deprivation cannot be explained in

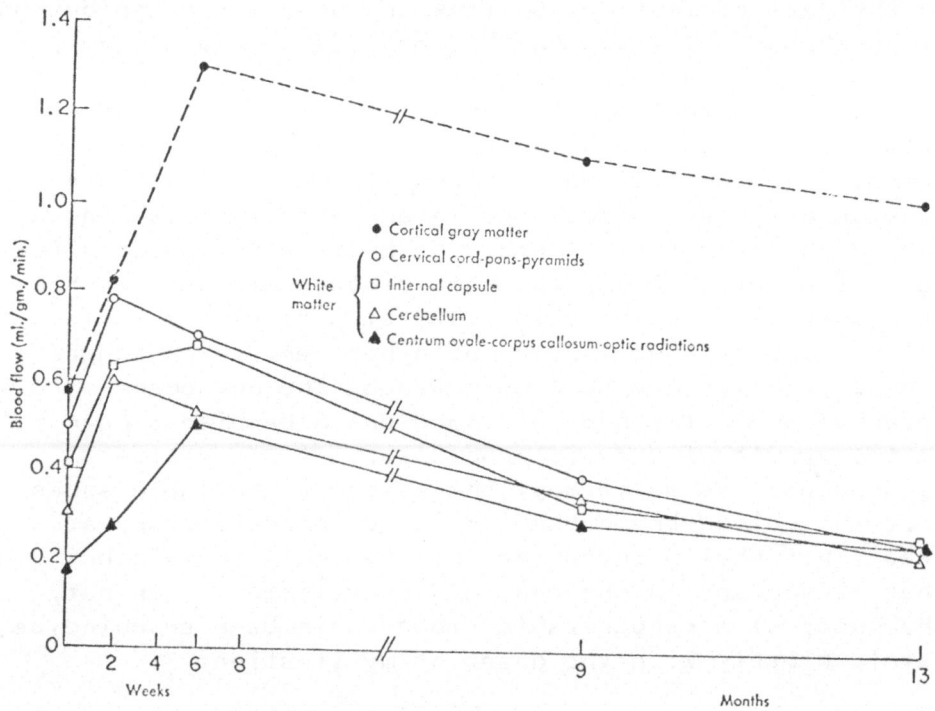

FIG. 4. Blood flow in various structures of the dog
brain from birth to maturity.

terms of the increase in oxygen consumption and it is
necessary to conclude that the immature CNS is intrinsi-
cally more resistant to anoxia than the mature CNS.
Most likely, the requirements of the metabolic processes
related to the fully functional mature CNS are much more
stringent than the requirements of the functionally
silent "immature" tissue.

Among the small molecule metabolites of the CNS,
free amino acids show consistent changes during development. The pattern of change varies from amino acid to
amino acid: some increase markedly in concentration,
such as glutamic acid and, to a lesser extent, aspartic
acid and GABA; others, like glutamine and glutathion,
remain fairly constant; finally, others like tryptophan
and tyrosine decrease markedly.

   The activity of different enzymes shows marked
changes in the course of development. The enzymes most
directly related to function (choline esterase, choline
acetylase, monoaminoxidase, 5-tryptophan decarboxylase,
etc.) increase in most cases at about the same time, although at different rates. The enzymes concerned with
general metabolic activity show a less consistent pattern:
ATP-ases and cytochrome oxidase increase markedly at the
onset of spontaneous electrical activity, which coincides
with the onset of a marked increase in oxygen consumption.
On the other hand, dehydrogenases present varied patterns:
malic dehydrogenase increases markedly, lactic dehydrogenase much less.

   All these morphological, physiological and chemical
changes are related temporally to each other in a pattern
that is consistent from species to species in the various
mammals that have been studied in detail. In all these
species, the changes build up to a "critical period" when
the hitherto functionally silent CNS shows the onset of
function characterized by muscular responses to cortical
stimulation and the onset of spontaneous electrical
activity. The timetable of changes for each species
varies before and after this critical period, according
to longevity and to the degree of maturity of each
particular species at birth. Thus, in guinea pig the
critical period is the 41st to 45th day of pregnancy;
the animal being born at 65 days of pregnancy. On the
other hand, in the rat, the critical period does not begin until 7 to 10 days after birth.

   At this critical period, which constitutes a true
turning point in the development of CNS, the neurons have
already undergone sizeable arborization: myelination
starts, spontaneous electrical activity is established,
vascularization is in fast expansion and overall oxygen
consumption is reaching its steepest increase. It is
worth reemphasizing here that the relationship of this
critical period to the time of birth varies from species
to species. Hence, in comparing the developing CNS of
different species, chronological age may not be a valid
criterion for degree of maturation and it may become
necessary to use other parameters of growth such as

myelination, the histological picture, or the EEG.

It must be pointed out that in the human species, the CNS plays a central role in the overall economy of the body during the period of growth. This is because at that time brain represents a greater part of the body than it does in the adult. Hence its energetic and material requirements absorb a substantial part of what is available to the whole body. To assure optimal conditions for its development, the CNS appears to have first call upon the resources available to the body as a whole. For instance, in starvation during the period of brain development, the weight of this organ falls behind the normal growth curve, but this decrease in growth is proportionally much less marked than the decrease in weight of the body as a whole. In a similar fashion, under conditions of ascorbic acid deprivation, brain and suprarenal cortex are the two tissues that show the smallest reduction in ascorbic acid content, and the first tissues to recover their normal ascorbic acid content upon restoration of an adequate supply of ascorbic acid.

The "priority" enjoyed by the CNS in the satisfaction of its needs during development does not make it immune from deprivation. Indeed, one of the interesting new developments in the field is the study of the effects of total or partial starvation on the development of CNS. This work has already established that CNS can be damaged by starvation during the period of its development, and that this damage may become irreversible if the starvation is sufficiently protracted. These observations on experimental animals stress the possibility that starvation in humans during the first years of life may damage the CNS irreparably and this, in turn, would result in a permanent deficit in performace during adult life. Since a substantial part of mankind lives under conditions of partial or severe food deprivation, it would appear that the study of development of the CNS may suddenly add to its obvious humanistic importance, the significance of research into the practical and tragic problem of the persistence of poverty and human misery in a society with the technical means to wipe out both scourges.

REFERENCES

1. Schadé, J.P., Van Backer, H., and Colon, E., in
    Growth and Maturation (D.P. Purpura and J.P. Schadé,
    Eds.) Elsevier (1964)

2. Brizee, K.R., Vogt, J., and Kharetchko, X., in
    Growth and Maturation (D.P. Purpura and J.P. Schadé,
    Eds.) Elsevier (1964)

3. Craigee's Neuroanatomy of the Rat (revised and ex-
    panded by W. Zeman and J.R.M. Innes) Academic Press
    (1963)

4. Folch, J., Casala, J., Pope, A., Meath, J.A.,
    LeBaron, F.N., and Lees, M., in The Biology of Myelin
    (S. Korey, Ed.) P.B. Howber, Inc. (1959)

5. Kennedy, C., Grave, G.D., Jehle, J.W., and Sokoloff,
    L. Neurology 20:613 (1970)

REFERENCES

SECTION I

STRUCTURAL CONSTITUENTS DURING DEVELOPMENT

ROLE OF NUCLEIC ACIDS IN BRAIN DEVELOPMENT

A. Giuditta

International Institute of Genetics and

Biophysics, Naples, Italy

The aim of this article is to provide a survey of the experimental findings which relate to the role of nucleic acids in brain development, with no pretence at a complete coverage, but also to present some of the evidence obtained with comparable systems wherever this may by useful to illustrate a concept or a possible mechanism. The metabolic behaviour of nucleic acids is at the core of the theory which attempts to explain development in terms of gene action and of its control (repression and derepression) in strictly ordered patterns in space and time. To provide even a cursory appreciation of the complexities which this mechanism is called upon to explain, we will start with a brief description of the main categories of biological events which occur during the growth of the nervous system. This will be also useful in discussing the limits of this theory, at least in its current formulation, and the possible other factors which may contribute to the molding of growing nerve tissues.

FEATURES OF BRAIN DEVELOPMENT

The main events which take place during brain growth may be divided into the following classes: a) cell proliferation, b) cell differentiation, c) formation of specific connections among cells, d) cell migration, e)

cell death. Needless to say, each of these categories is
strictly associated to the other ones in a perfectly ordered spatial and temporal sequence. Differentiation,
for instance, appears to be deeply connected to proliferation since it starts only in cells which have ceased
to divide (1, 2) following a scheme recognized also in
other differentiating tissues. The original matrix cells
are responsible for the proliferative activities which
yield more matrix cells but also, at defined periods,
cells which no longer divide and start differentiating
into the various types of neurons (neuroblasts) and glia
cells (spongioblasts). A key feature of this process is
the formation of the cell lines in sequential waves which
may overlap each other to some extent but which follow
general rules as to the time of appearance of the different cellular types: first, highly differentiated neurons
with long axonal connections, e.g. Purkinje cells; last,
neurons with short axons, e.g. granule cells, and also
glia cells (1). From the morphological point of view,
neuronal differentiation involves striking changes in
the nucleus and in the cytoplasm (3) and the formation
of the most complex and ordered patterns of dendritic
and axonal outgrowth. It is on the specific shapes and
sizes of neurons that the circuitry of the nervous system is based. Differentiation invests of course also the
spongioblasts which give rise to astrocytes and oligodendrocytes and to their main subdivisions (4, 5). The
different phases of differentiation tend to be concentrated in defined periods of development, during which
biochemical analyses show the rapid accumulation of several constituents (RNA, proteins, lipids, enzymes). This
is the case, for instance, of the formation of dendrites
and axons and of myelination (6). Differentiation however
does not proceed in a set of independent cellular entities which do not communicate with each other. On the
contrary, different neurons and also nerve and glia cells
establish specific contacts with each other, again in a
precise temporal sequence, in what is perhaps the most
intriguing aspect of neural ontogeny. Specification may
be in regard only of cellular types, as shown by the association between axons and oligodendroglia (7), but may
also reach extremely high levels of resolution as in the
topographic point to point relations which form between

the retina and the optic tectum of vertebrates (8). The
numerous examples of mutual interactions among developing neural centres and between centres and peripheral
areas (9, 10) may also be considered to fall into the
same category of events, at least formally. Cell migration occurs normally during early and late stages of
development of the nervous system, contributing substantially to its morphogenesis. It may concern single cells,
but more generally involves whole cellular populations
which swarm from one to another region of the growing
nerve tissue, following defined pathways and obeying unknown stimuli (11). Together with cell death, also a phenomenon of normal occurrence (12), cell migration represents one of the most elusive aspects of the development
of the nervous system in terms of underlaying mechanisms
(11). Of all these features which to a different extent
occur also during growth and differentiation of other
organs and tissues, some can be considered of rather special significance in the development of the nervous system. Among them is the uniquely high degree of intercellular connections, most of which are of a highly specific
nature, and the striking influence exerted by the external ambient.

BIOCHEMICAL OBSERVATIONS

1. DNA

There is ample evidence that some neuronal types,
e.g. Purkinje neurons, contain twice the amount of DNA
present in diploid nuclei (13). DNA determinations and
nuclear counts carried out in the growing brain of
Octopus vulgaris have established that DNA content per
nucleus progressively increases with age over and above
the initial diploid value, in the lobes that contain
large motoneurons (14). At least in this case DNA content cannot be taken as a reliable measure of cell number.
In mammals however polyploidy appears to be limited to a
restricted number of neurons and cannot alter estimates
of cell number based on DNA determinations. Since polyploidy appears to be the constant feature of some neuronal types, developing during defined periods of maturation (15), it should be considered the result of normal
differentiation along with other cellular properties.

The rate of DNA increase in the developing vertebrate brain is markedly dependent on the species, on the
region examined and on the time of devélopment (16, 17).
This indicates that proliferative processes occur unevenly and sequentially in different brain regions and at
different times in different species. In many of them the
total number of cerebral cells continues to increase well
after birth, although at a reducing pace.

The histone/DNA ratio of a chromatin fraction isolated from developing chick brain does not vary significantly in the period between 36 h of embryonic life and
1 year-old adults. Similarly, the ratio of total protein
per DNA in the same chromatin fraction shows only a limited decrease during the first days of embryonic growth.
On the other hand, the RNA/DNA ratio decreases considerably
from the 36 h embryo (0.35) to the 8 day-old brain (0.2),
remaining constant thereafter (18). In mouse brain, however, the amount of protein associated with DNA in a nucleoprotein fraction shows considerable age-dependent
fluctuations, decreasing during the first 6 months and
starting to increase again after 15 months. Somewhat parallel changes occur in the melting temperature of the
nucleoprotein which is in part dependent on the association
of protein to DNA (19). The relevance of these findings to
possible mechanisms of gene control remains to be established

DNA polymerase activity similar in many respects to
that described in other animal systems has been detected
in several regions of developing chick brain. The intensity of the enzymatic reaction appears to be generally
proportional to the rate of DNA increase (20).

Several autoradiographic studies have been carried
out in recent years in growing brain exposed to ^{3}H-thymi-
dine (1). The uptake of this precursor just before mitosis and its indefinite permanence in labelled nuclei have
been elegantly exploited in the identification of proli-
ferating cells and of their migratory and differentiative
pathways. It is with these techniques that the principle
of sequential neuronal genesis has been established. Thy-
midine incorporation into rat brain DNA, measured after
a pulse of 12 h, is highest at 6 days of postnatal life

and declines rapidly till 15 days and more slowly thereafter. The rate at birth appears to be considerably lower
than at 6 days, perhaps because of a birth stress (21)
although no information is given on the rate of thymi-
dine incorporation during intrauterine life.

2. RNA

A biphasic accumulation of RNA has been detected in
developing neural tissues analyzed with ultraviolet microscopy. In spinal ganglia and ventral horn cells of the
chick the UV absorption progressively increases from the
2nd day of embryonic life but shows a marked and abrupt
drop at the 4-5th day, which is followed by a new steady
increase till the 8th day (22). In the cerebral cortex of
the guinea-pig a somewhat similar pattern occurs with a
first absorption maximum at about 35 days of foetal life
and a later overlapping phase which corresponds to the
appearance of the first Nissl bodies (23). No Nissl staining material can be detected in the nerve cell bodies
during the first wave, during which however also the primitive neuronal fibers appear as UV-opaque structures,
suggesting the presence of axonal RNA. Ribosomes in
young developing axons have been observed also in electron
microscopic studies (3). Several lines of evidence have
established the relation between the appearance of the
Nissl bodies and the maturation of the nucleolus (24).We
now know that this intranuclear organelle is concerned
with the process of synthesis of ribosomal RNA and with
the assembly of subribosomal particles which are later
transferred to the cytoplasm to form ribosomes and ribosomal aggregates.

Synthesis of RNA in matrix cells and in neuroblasts
of the developing cortex has been studied in chick embryos
with autoradiographic techniques (25). An active incorporation of uridine in cells of the germinal layer (matrix
cells) is present from the 6th to the 10th day of incubation, decreasing afterwards to very low values (15th day).
Most of the labelled RNA remains in the nucleus. In the
neuroblasts of the mantle layer to which matrix cells migrate from the 6th day, uridine incorporation is initially

low (6-8th day) but increases as the neuroblast differen-
tiates and acquires its normal complement of ribosomes.
At the same time there is an enhanced passage of labelled
RNA into the cytoplasm. On the other hand, labelling of
spongioblasts remains low and mostly nuclear. Considering
the neuroblast derivation from matrix cells, it seems that
there are two main periods of RNA synthesis, one at the
stage of matrix cell when most of the newly-made RNA remains nuclear, the other one when the neuroblast starts
differentiating into a nerve cell. During this second phase
there is a flow of nuclear RNA into the cytoplasm. A study of the nuclear RNA accumulating in matrix cells might be
rewarding, in view of its possible role in the control of
nuclear functions.

     Several age-dependent changes have been detected in
the microsomal fraction isolated from postnatal rat brain
(26). The RNA content of cortical microsomes reaches a
maximum at 25 g body weight and then declines. The ratio
of labelled RNA found in nuclear and microsomal fractions
after a pulse of 1 h with radioactive orotic acid decreases from a maximum of 20 at 7 g body weight to 5.0-5.5 at
40 g body weight and beyond. Under the same labelling
conditions deoxycholate (DOC) treatment releases a high-
ly-labelled RNA fraction from young but not from adult
brain microsomes. A further difference is revealed after
incorporation of radioactive valine into cerebral microsomes in vitro. The percentage of labelled protein released by DOC into an easily sedimentable particulate
fraction is approximately 50% of the total at 7 g body
weight but becomes less than 10% at and beyond 25 g body
weight. The percentage of radioactive protein released in
the soluble phase remains however essentially the same
throughout the same period. These findings have been interpreted as suggestive of the presence of two different
classes of microsomal structures one of which, present only
in young brains, is concerned with the synthesis of "structural" proteins, possibly in relation to the process of
myelin formation.

     The biosynthesis of RNA in newborn rat brain has
been recently investigated with slices incubated in presence of $8-^{14}C$-adenine (31). RNA has been fractionated

chemically into a high molecular weight fraction (h-RNA) and a fraction obtained from the phenol interphase layer (n-RNA). These fractions have been respectively equated to cytoplasmic and nuclear RNA on the basis of analogies with other systems and of the similar labelling behavior of h-RNA and polysomal RNA. While this identification should be accepted with some caution, the results confirm previous findings obtained in vivo (32). In the slices of newborn brain labelling of n-RNA proceeds linearly since the beginning of the incubation but h-RNA shows an initial lag period of approximately 30 min, presumably due to the time required for r-RNA synthesis and transfer to the cytoplasm. This contrasts with the in vivo labelling kinetics of microsomal RNA from adult rat brain which do not show any lag (32). Addition of actinomycin D to the brain slices after the start of adenine labelling induces a partial loss of n-RNA most of which, however, remains stable for at least 3 h. The labile n-RNA may have a half life of less than 30 min. Nuclear RNA with similar properties is present in several types of animal cells including developing amphibian embryos (33)and adult rabbit brain (34).

The possible regulatory role of polyamines in nucleic acid metabolism of developing brain has been emphasized by experiments carried out recently with chick embryos (35). Injection of spermine or spermidine into the air space of 10 days-old embryos 2 h before a pulse with ^{3}H-formate induces an increase in labelling which amounts to 35-60% for DNA and nuclear RNA and to 100-150% for microsomal RNA. This last effect is reflected in an enhanced labelling of the light polysomal peaks and in the appearance of labelled heavier polysomes. During normal development similar changes occur only after 16 days of embryonic growth. As brain develops, the rate of RNA synthesis decreases progressively, in part at least because of a dilution of the biosynthetic system by other accumulating structures. The decrease is substantial and has been noted with cellular suspensions (36) and with tissue slices (37). In one of these investigations the reduction in biosynthetic rate has been related to the corresponding rise in activity of uridine nucleosidase (38). RNA polymerase activity has been studied in brain, also in relation to the age of the animal. The results however have been contradictory (39,40).

3. Protein

Little information can be expected to come from gross
analyses of protein accumulation in growing brain structures, except for the overal characterization of this process in its temporal and spatial parameters. A deeper insight into the mechanisms which regulate their appearance
in the course of development may be obtained by analysis
of single proteins. Several of these investigations have
been carried out on the maturing brain, particularly in
periods which correspond to the laying down of highly differentiated structures such as dendrites, synapses or
myelin (6). Quite often the pattern consists in an initial
slow rate of increase followed by a rapid process of accumulation. This last phase occurs at well defined times
depending on the brain region and on the protein examined,
and appears to reflect the concomitant derepression of the
same set of genes in each of the individual genomes present in the structure.

By far the best known example of enzyme induction
during brain development regards the appearance of glutamine synthetase activity (GS) in the embryonic retina of
the chick (41). GS remains at low levels till approximatly
the 15-17th day of embryonic life and then increases rapidly to level off near hatching, at 21 days. Morphological
and functional differentiation of the retina occurs during
the same period. Analysis of this system is not complicated
by concomitant processes of cell multiplication which
stops earlier, at about the 10th day, and has been greatly
facilitated by the finding that GS induction can be repro-
duced in vitro with 16-17 day-old retina and even with 10
day-old retina, if the culture is supplemented with adult
serum. Under these conditions GS appearance can be blocked
with inhibitors of protein synthesis such as puromycin and
cycloheximide, irrespective of the age of the retina or
of the culture, and of the amount of enzyme already expressed. On the other hand, actinomycin D blocks the enzyme increase if added at the time of explant, but becomes
progressively less effective if added 1-2 days later.
Evidently the induction of GS requires protein and RNA
synthesis, but the GS forming system becomes independent
of transcription, as the age of the culture increases.

At this time control appears to involve only the translational phase of gene expression. A similar situation has
been described during the early stages of embryonic development which may proceed normally even in presence of
actinomycin D (42). The insensitivity of the GS forming
system to actinomycin D develops only in presence of adult
serum and if the protein synthesis of retinal explants is
not inhibited. The mechanism of these effects is still
unclear (43). Thyroid and adrenal hormones are capable of
inducing the formation of GS in vivo if injected before
the normal time of enzyme rise. The effect has been mi-
micked by hydrocortisone also under in vitro conditions
(44). The coordinated synthesis of GS by retinal cells
appears therefore to be under hormonal control, a conclusion which seems valid also for other biochemical and
morphological processes of brain maturation (45).

Mechanisms

There is no reason to believe that the role of nucleic acids in brain development differs significantly
from that postulated in other developing organs and tissues.
In all cases efforts have been largely directed at the e-
lucidation of the mechanisms which operate in the control
of gene expression in the well ordered spatial pattern and
temporal sequence which is displayed by normal developing
organisms. To this aim, several possibilities have been
proposed which in turn have emphasized the merits of histones, non-histone nuclear proteins or RNA in the processes of repression and derepression (46-48) or have postulated the occurrence of chemical modifications of the genome
(49). While we cannot expand here on any of these theories
which still wait the consent of experimental proof, we
wish to draw attention to some conclusions, by no means all,
which have come from work carried out with developing systems in general and in some cases also with developing
nervous tissue.

The nuclear activity of the adult, highly differentiated cell is strikingly modified by exposure to a foreign
cytoplasm. This has been shown in elegant experiments of
nuclear transplantation from brain and other adult organs
into enucleated eggs of amphibia (50). Brain nuclei which

are inactive in replicating DNA start an active synthesis
shortly after having been injected into the egg cytoplasm.
Cytoplasmic factors control also the extent and pattern of
nuclear RNA synthesis and are responsible for the early
differentiation of different regions of the embryo (50,
51). The main conclusion to be drawn from these findings
is that nuclear activities are under the control of cytoplasmic factors. The identification of one of these factors
as a protein species specifically inhibiting ribosomal
RNA transcription in amphibian eggs (52) may open the way
to the study of its regulatory mechanism. In a similar
trend recent experimental evidence has suggested the occurrence in the same biological system of protein components which may direct the action of RNA polymerase (53),
with a mechanism similar to that of the bacterial sigma
factors (54).

      The existence of cytoplasmic determinants raises se-
veral questions which relate to the prospective role of
nucleic acids in development. If these factors, as it seems
likely, are synthesized on nuclear templates, why are they
stored in the cytoplasm? If they are produced by cells of
previous generations, for how long can they be stored and
how are they triggered into action? Perhaps the simplest
explanation may be that each factor is produced by the
nucleus and segregated in a region of the cytoplasm which
will go to only one of the daughter cells and, upon its
action on the genome, will induce it to take one line of
differentiation, eventually implying the synthesis of another cytoplasmic factor which will take differentiation
a step further at a following mitotic cycle. A mechanism
of this sort would account for the progressive functional
modification of the genome, without implying any loss of
its original capacities. The sequential specification of
neuroblasts and spongioblasts and of the individual neuronal and glial cell types may be explained by a similar
mechanism.

      On a more formal plane, it may be rewarding to con-
sider the process of formation of the bewildering complexity of interconnected structures in the nervous system
as a sequence of similar steps, each of which determines
the outcome of the next one. This requires however that
the steps are strictly time-coordinated in all the cells

of similar or different specification which participate
in the moulding of a nerve region. There are in principle
two modalities with which to achieve such a coordination,
one involving independent sequences of determinations
within each cell or cell line, the other one calling for
mutual interactions among cells, also at the level of
gene expression. The first modality implies an extremely
tight control along the time axis to insure that complementary events take place in different cells at the right
time. This seems a highly unlikely and contorted device,
unnecessary to postulate in view of the many examples of
mutual interactions which are already known in developing
biological systems and in particular in the nervous system (10, 12). The second modality appears therefore a
more likely one, based as it is on a principle of mutual
determination and control which might easily account for
the self-regulating abilities of developing brain. If
carried to the extreme consequences, this outlook should
change our perspective views, since it would consider the
mechanism of gene expression as only one of the factors
relevant to the understanding of the formation of the
nervous system. Axonal growth, for instance, must depend
on the synthesis of specific proteins which are genetically
determined, but the direction of growth is likely not to be
determined by the genome of the same cell. This last process may be traced ultimately to the expression of other
genomes, through the production of proteins or of protein
products, but for the understanding of how connections are
made it will remain essential to unravel the mechanisms
which direct axonal growth at the phenotypic level. Similar
reasoning should be applied to dendritic growth, formation
of synapses, myelination and in general to all the instances in which proper development depends on the cooperative
efforts of two cells or more, including such morphogenetic
phenomena as cell migration and death. An interesting example of this sort of phenomenon may be considered the
arrest of mitotic activity and the induction of neurites
in cultures of neuroblasts switched to a medium lacking
foetal serum (55). This induction is completely blocked by
colchicine or vinblastine, but proceeds undisturbed under
conditions of almost total inhibition of protein synthesis.
It must be concluded, at least in this case, that the immediate trigger to differentiation operates at a level

beyond translation, even if the inference of a store of
required proteins implies the previous activation of the
whole mechanism of gene expression.

Of a completely different nature appears on the other
hand the mechanism of action of the best known of the
factors which promote growth and differentiation of nerve
cells, the nerve growth factor (NGF), which in vitro eli-
cits also a dramatic outgrowth of neurites from sympathetic and sensory ganglia (56). The biological effects of
NGF in vivo consist in stimulating the proliferation,
differentation, and hypertrophy of their target prospective neurons and the growth of their neurites. Injection
of the specific antibody to newborn mice induces a very
rapid regression of the affected sympathetic nerve cells
(cessation of mitotic activity, degeneration) with a resultant state of immuno-sympathectomy. By electron microscopy
the first changes elicited by the antibody have been traced to the fragmentation of nucleoli, chromatin condensation and folding of the nuclear membrane (57). This result
is in accord with a postulated primary action of NGF on
DNA-primed RNA synthesis which has found experimental
support also in studies carried out with explanted ganglia (58).

Space limitations do not allow treatment of other important aspects of brain development, in particular of
those related to the action of the organism's environment
and to the mutual influence of nerve centers. For these
aspects the interested reader is referred to some reports
of the literature which may introduce him to the field,
even if not all of them touch specifically on nucleic
acids (59-63).

REFERENCES

(1) Altman, J. in Handbook of Neurochem., vol. 2, p.137
    (Ed. A. Lajtha) Plenum Press (1969).

(2) Eceles, J. C. Proc. Natl. Acad. Sci. US 66:294 (1970).

(3) Caley, D. W. and Maxwell, D.S. J. Comp. Neurol.
    133:17 (1968).

(4) Wechsler, A. and Keller, K. Progr.Brain.Res.26:93 (1967

(5) Mori, S., and Leblond, C.P. J. Comp.Neurol.139:1 (1970).

(6) McIlwain, H. in Biochemistry and the Central Nervous System, 3rd Ed., Churchill (1966).

(7) Schonbach, J. et al. J. Comp. Neurol. 134:21 (1968).

(8) Sperry, R.W. Proc.Natl.Acad.Sci.US 50:703 (1963).

(9) Hess, A. Biol. Rev. 32:231 (1957).

(10) Hughes, A.F.W. Aspects of neural ontogeny, Logos Press (1968).

(11) Trinkaus, J.P. Cells into Organs, Prentice Hall (1969).

(12) Saunders, J.W. Science 154:604 (1966).

(13) Lentz, R.D., and Lapham,L.W. J.Neurochem.16:379 (1969).

(14) Giuditta, A. et al. Brain Res., in press.

(15) Sandritter, W. et al. Z. Zellforsch. 80:145 (1967).

(16) Mandel, P. et al. in Comparative Neurochemistry (Ed. D.Richter) p.149, Pergamon Press (1964).

(17) Rappoport, D.A. et al. in Handbook of Neurochemistry vol.1, p.101 (Ed. A.Lajtha) Plenum Press (1969)

(18) Dingman, C.W., and Sporn, M.B. J. Biol. Chem 239:3483 (1964).

(19) Kurtz, D.I., and Sinex, F.M. Biochim. Biophys. Acta 145:840 (1967).

(20) Margolis, F.L. J. Neurochem. 16:447 (1969).

(21) Mori, K. J. Neurochem. 17:835 (1970).

(22) Hughes, A. J. embryol. exper. Morphol. 3:305 (1955).

(23) Hughes, A., and Flexner, L.B. J. Anat. 90:386 (1956).

(24) Hyden, H. Acta Physiol. Scand. 6,Suppl.17:1 (1943).

(25) Sensenbrenner, M., and Mandel, P. Z.Zellforsch. 82:65 (1967).

(26) Adams, D.H., and Fox, M.E. Brain Res. 12:157 (1969).

(27) Dellweg, H. et al. J. Neurochem.15:1109 (1968).

(28) Yamagami, S., and Mori, K. J.Neurochem. 17:721 (1970).

(29) Bernsohn, J., and Norgello, H. Proc. Soc. Exp. Biol.
     Med. 122:22 (1956).

(30) Ringborg, U. Brain Res. 2:296 (1966).

(31) Sharma, S.K., and Singh, U.N. J.Neurochem.17:305 (1970).

(32) Adams, D.H. Biochem. J. 98:536 (1956).

(33) Brown, D.D., and Gurdon, J.B. Proc. Natl. Acad. Sci.US
     51:569 (1964).

(34) Vesco, C., and Giuditta, A. Biochim. Biophys. Acta
     142:335 (1967).

(35) Caldarera, C.M. et al. J. Neurochem. 16:309 (1969).

(36) Johnson, T.C. J. Neurochem. 14:1075 (1967).

(37) Orrego, F. J. Neurochem. 14:851 (1967).

(38) Guroff, G. et al. J. Neurochem. 15:489 (1968).

(39) Barondes, S.H. J. Neurochem. 11:663 (1964).

(40) Bondy, S.C., and Waelsch, H. Life Sci. 3:633 (1964).

(41) Moscona, A.A. , and Kirk, D.L. Science 148:519 (1965).

(42) Brachet, J. in Comprehensive Biochemistry vol.28, p.23
     Eds.M.Florkin and E.H.Stotz) Elsevier Publ.Co.(1967).

(43) Kirk, D.L. Proc. Natl. Acad.Sci.54:1345 (1965).

(44) Piddington, R. Devel. Biol. 16:168 (1967).

(45) Krawiec, L. et al. Brain Res. 15:209 (1969).

(46) Smith, E.L. et al. Physiol.Rev. 50:159 (1970).

(47) Ursprung, H., and Huang, R.C. Progr. Biophys. Mol.
     Biol. 17:151 (1967).

(48) Britten, R.J., and Davidson, E.H. Science 165:349 (1969).

(49) Scarano, E. Ann. Embryol. Morphog.,Suppl. 1:51 (1969).

(50) Gurdon,J.B., and Woodland,H.R. Biol.Rev. 43:233 (1968).

(51) Davidson, E.H. et al. Proc. Natl. Acad. Sci.US 54:696
     (1965).

(52) Crippa, M. Nature in press.

(53) Tocchini-Valentini, G.P., and Crippa, M. Cold Spring Harbor Symp. Quant. Biol. 35 (1970) in press.

(54) Burgess, R.R. et al. Nature 221:43 (1969).

(55) Seeds, N.W. et al. Proc. Natl. Acad. Sci. US 65:160 (1970).

(56) Levi-Montalcini, R., and Angeletti, P. Phsyiol. Rev. 48:534 (1968).

(57) Levi-Montalcini, R. et al. Brain Res. 12:54 (1969).

(58) Toschi, G. et al. J. Neurochem. 13:539 (1966).

(59) Riesen, A.H. in Prog. in Physiological Psychology p.117 Academic Press (1966).

(60) Gyllenstein, L. et al. J.Comp.Neurol.126:463 (1966).

(61) White, R.H. J. Exp. Zool. 166:405 (1967).

(62) Hubel, D.H. Physiologist 10:17 (1967).

(63) Margolis, F.L., and Bondy, S.C. Exp.Neurol.27:353 (1970).

CHOLESTEROL BIOSYNTHESIS IN LIVER TISSUE[*]

Mary E. Dempsey

Department of Biochemistry, University of Minnesota

Medical School, Minneapolis, Minnesota 55455

Most of the early stages of cholesterol biosynthesis in liver tissue are now well defined with regard to control sites, intermediates, and specific enzymes (e.g. (1)). In contrast much basic information is still to be obtained for the stages of cholesterol biosynthesis after squalene cyclization. The purpose of this report is to review recent findings in our laboratory which elucidate mechanisms of liver enzymes catalyzing conversion of squalene and sterol intermediates to cholesterol. In particular, we identified a liver protein which functions uniquely as carrier for water insoluble precursors during cholesterol biosynthesis. This protein is designated squalene and sterol carrier protein or SCP. In this report we describe the properties of SCP; its role as activator of reactions catalyzed by microsomal enzymes; its specificity; and the inhibition of its binding and activation functions by inhibitors of cholesterol biosynthesis.

PROPERTIES OF THE SQUALENE AND STEROL CARRIER PROTEIN (SCP)

In previous studies we showed that conversion of $\Delta^{5,7}$-cholestadienol to cholesterol by purified microsomal $\Delta^{5,7}$-sterol Δ^{7}-reductase of rat liver is stimulated 4-fold or greater, in the presence of excess NADPH, by an activator isolated from the high speed supernatant fraction of homogenates (2-5). This activator was demonstrated to be a heat stable protein (e.g. by heat treatments, gel filtration, salt and perchloric acid precipitation, inactivation by trypsin) and partially purified (Figure 1). The mole-

[*]Supported by U.S.P.H.S. Research Grants HE-8634 and HE-6314

Activator Preparation	Activation of Δ^7-Reductase	Purification	Yield
	-fold	-fold	%
A	5.2	1.0	100
SCP	36.8	300	37

Figure 1.   Purification of the activator of $\Delta^{5,7}$-sterol Δ^7-reductase.

      The upper half of the 105,000 x g supernatant fraction of liver homogenates is designated fraction A.   The purified activator is designated SCP or squalene and sterol carrier protein. SCP was purified by heat treatment, $(NH_4)_2 SO_4$ fractionation, and gel filtration.   For details of purification and assay for Δ^7-reductase activity, see (4-5).

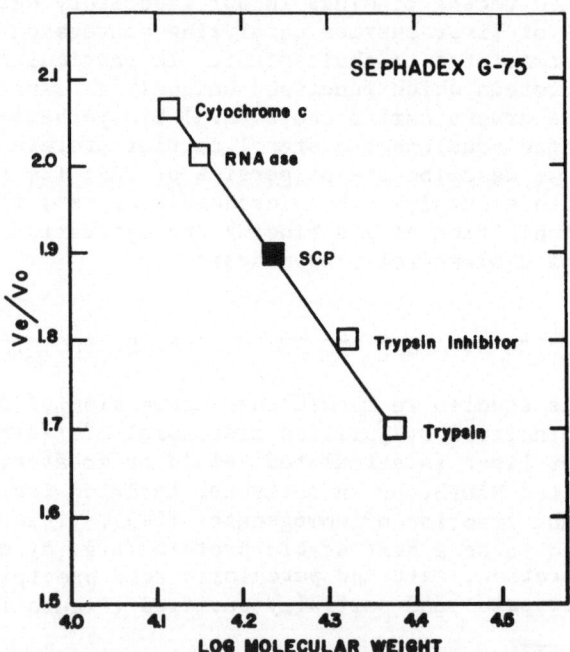

Figure 2.   Migration relative to known proteins of the squalene and sterol carrier protein during gel filtration.

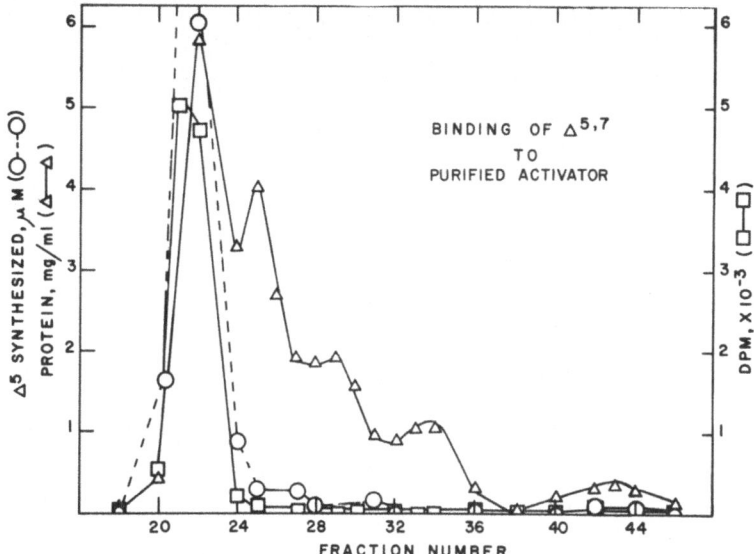

Figure 3. Migration of labeled $\Delta^{5,7}$-cholestadienol with its activator protein (SCP) during gel filtration on Sephadex G-75.

Protein fractions were assayed for the presence of labeled
$\Delta^{5,7}$-cholestadienol and the activator of $\Delta^{5,7}$-sterol Δ^{7}-reductase
as described elsewhere (4-5).

cular weight of the purified activator is approximately 16,000
(Figure 2). The requirement of the activator protein for full microsomal Δ^{7}-reductase activity indicates that this enzyme should be
classified as a member of the growing list of enzymes requiring
more than one protein for their maximum activity (6).

In the course of our work with the activator of $\Delta^{5,7}$-sterol Δ^{7}-
reductase we discovered a unique property, i.e. the activator protein is capable of binding $\Delta^{5,7}$-cholestadienol. In the process of
binding substrate the low molecular weight species of the activator
aggregates to a high molecular weight species (molecular weight
greater than 60,000) (4). If the unpurified high speed supernatant
fraction of homogenates is first incubated with labeled $\Delta^{5,7}$-chole-
stadienol and then purified, only the high molecular weight aggregate of the activator is detected (Figure 3). In addition, substrate tightly bound to the activator (e.g. Figure 3) is converted
to cholesterol by microsomal Δ^{7}-reductase considerably faster than
unbound substrate.

Cholesterol Precursor	% Bound
Acetate	4
Mevalonate	4
Squalene	100
Lanosterol	75
Δ^7-4α-CH_3	83
$\Delta^{7,24}$	97
$\Delta^{5,7,24}$	93
$\Delta^{5,24}$	94
Δ^7	97
$\Delta^{5,7}$	90
Δ^5	90
Δ^7-3-ketone	76

Figure 4. Binding of cholesterol precursors to the squalene and sterol carrier protein (SCP).

Binding assays were performed as described elsewhere (5,7). The synthetic Δ^{24}-sterols were gifts of Professor Ivan D. Frantz, Jr.

SPECIFICITY OF THE SQUALENE AND STEROL CARRIER PROTEIN (SCP)

A variety of known synthetic and biosynthetic cholesterol precursors were tested for their ability to bind to the protein activator of $\Delta^{5,7}$-sterol Δ^7-reductase (Figure 4) (7). We found that all the water insoluble precursors of cholesterol tested were bound to the activator and in bound form readily converted to cholesterol or other intermediates by liver microsomal enzymes. In contrast, cholesterol metabolities (e.g. steroid hormones, bile acids, cholestane derivatives not cholesterol intermediates) were not bound (Figure 5) (7). The only exceptions were the diol and triol derivatives of cholestane and the drug AY-9944. We showed previously that the cholestane derivatives are inhibitors of cholesterol biosynthesis in vitro and our data indicated that these compounds interfere with the binding of cholesterol precursors to the activator-carrier protein (SCP) (8,9). AY-9944, a Δ^7-reductase inhibitor,

Compound	% Bound
Pregnenolone	27
Progesterone	16
Testosterone	4
Estradiol	3
Cholate	7
Taurocholate	7
Δ^4-Cholesten-3-one	<1
Cholestanol	30
Cholestane	<1
3-Ketocholestane	<1
$3\beta,5\alpha$-Cholestanediol	50
$3\beta,5\alpha,6\beta$-Cholestanetriol	94
Phenethylbiguanide	<1
AY-9944	Bound
Tetrahymanol	<1
$\Delta^{5,7,22}$-Cholestatrienol	90

Figure 5. Binding of cholesterol metabolites and other compounds
to the squalene and sterol carrier protein (SCP).

Binding assays were performed as described elsewhere (5,7).
Steroid hormones were the gift of Professor Frank Ungar; bile
acids, Dr. James Carey; cholestane-diol and -triol, Dr. Donald
Witiak; phenethylbiquanide, H. S. Sadow; AY-9944, Dr. D. Dvornik;
tetrahymanol and $\Delta^{5,7,22}$-cholestatrienol, Drs. Robert Conner
and M. J. Koroly.

has been studied extensively by the Ayerst groups (e.g. 10). Labeled AY-9944 was not available for our studies. AY-9944 binding was
dectected by showing inhibition of microsomal Δ^7-reductase by the
activator purified following exposure to the drug under the usual
conditions of the binding test. Phenethylbiquanide (Figure 5) is
also a Δ^7-reductase inhibitor (11). Its lack of binding to the
activator is probably due to its high water solubility. Tetrahymanol and $\Delta^{5,7,22}$-cholestatrienol (Figure 5) are synthesized by
the protoza Tetrahymena pyriformis (12,13). In studies just completed we demonstrated that one of these compounds $\Delta^{5,7,22}$-chole-
statrienol is bound to the activator-carrier protein (SCP) and reduced to $\Delta^{5,22}$-cholestadienol by microsomal Δ^7-reductase (14). The
liver enzymes are not capable of converting the later sterol to
cholesterol (i.e. there is no Δ^{22}-reductase in liver).

Related studies suggested that serum proteins and other liver
proteins could substitute partially for the carrier but not the

Preparation	Activation of Δ^7-Reductase
	-fold
A	5.0
Albumin (BSA)	1.1
Serum (Rat)	0.7
Serum (Human)	0.2
HDL (Rat)*	1.4
HDL_3 (Human)*	1.5

Figure 6.   Specificity of activation of microsomal Δ^7-reductase.

A is unpurified activator-carrier (SCP), see Figure 1.   The apo-high density lipoproteins (HDL) were gifts of Dr. A. M. Scanu.

activation functions of SCP (e.g. Figure 6 for Δ^7-reductase).   However, when apo-high density lipoprotein, in particular, was first preincubated with substrate to induce binding and then tested for activation of Δ^7-reductase it functioned nearly as well as SCP itself (Figure 7).   These findings suggest a possible structural relationship between high density lipoprotein and SCP.

Preparation	Preincubation with $\Delta^{5,7}$*	Relative Activation
		-fold
A	-	1.0
	+	1.9
SCP	+	1.3
HDL_3	-	0.5
	+	1.2

*Preincubation at 37° for 15' under N_2 with $\Delta^{5,7}$-cholestadienol, followed by addition of microsomal enzyme, GSH, and NADPH and further incubation for 45'.

Figure 7.   Effects of preincubation of activators of Δ^7-reductase.

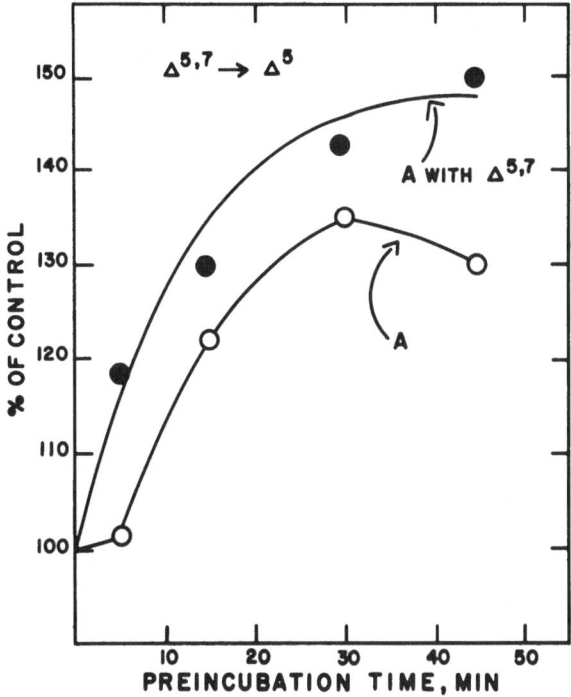

Figure 8. Effects of preincubation at 37° in the absence of
microsomal enzyme on the activation of Δ^7-reductase by its unpurified activator protein (A).

INHIBITION OF SCP ACTIVITY BY INHIBITORS OF CHOLESTEROL SYNTHESIS

     An interesting phenomenon currently under study is illustrat-
ed by the results of Figure 8 (7). Warming of unpurified SCP
(i. e. the high speed supernatant, Fraction A) to 37° in the absence of microsomal enzyme results in a marked increase in the
ability of the protein to activate Δ^7-reductase. The initial presence of substrate ($\Delta^{5,7}$-cholestadienol) causes an additional increase in activation ability. In contrast, AY-9944 either present
initially during warming of the activator in absence of microsomal
enzyme or added after the warming period abolishes the increase
in activation ability (Figure 9) (7). These results indicate that
one mode of inhibition of Δ^7-reductase by AY-9944 is to interfere
with formation of the active conformation of SCP and perhaps also
the binding of substrate to SCP. These results are in agreement
with related observations made with a variety of other inhibitors
of specific enzymic steps in cholesterol biosynthesis (8-9,11).

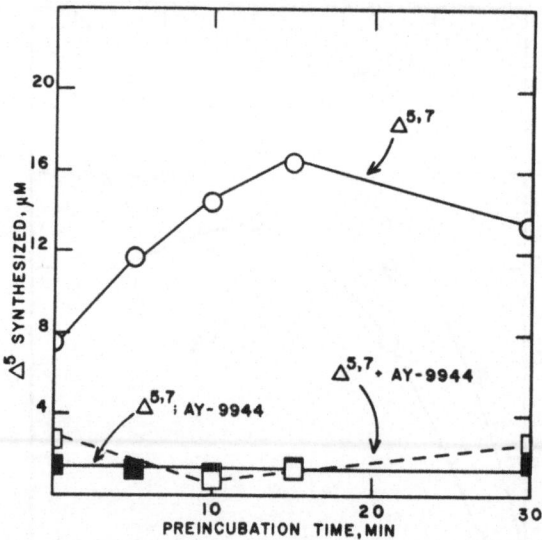

Figure 9. Effect of AY-9944 on the increased activation ability occurring during preincubation of the Δ^7-reductase activator protein at 37°.

SUMMARY

The studies outlined here and related findings suggest a common role for the squalene and sterol carrier protein (SCP) as

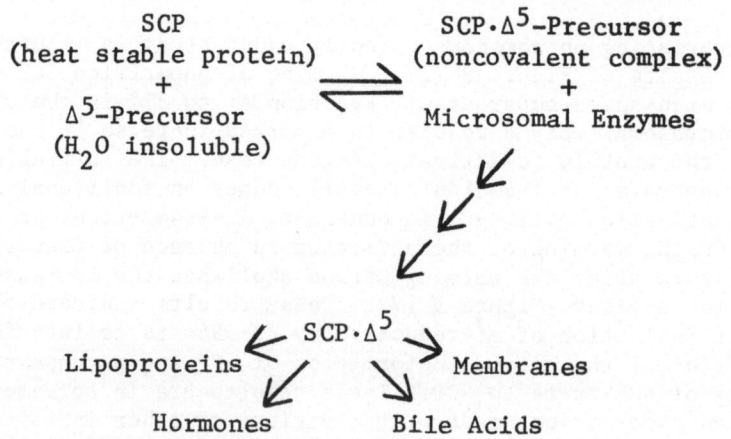

Figure 10. Role of the squalene and sterol carrier protein in cholesterol (Δ^5) synthesis.

vehicle for water insoluble precursors of cholesterol. They also offer new insights into molecular events of cholesterol synthesis and lipid-protein interactions (Figure 10).

REFERENCES

1. Popjak, G. in Methods in Enzymology, Vol. 15 (Ed. R. B. Clayton) (1969).

2. Dempsey, M. E., Seaton, J. D., Schroepfer, G. J., Jr., and Trockman, R. W. J. Biol. Chem., 239:1381 (1964).

3. Dempsey, M. E. J. Biol. Chem., 240:4176 (1965).

4. Dempsey, M. E. in Methods in Enzymology, Vol. 15 (Ed. R. B. Clayton) (1969).

5. Ritter, M. C., and Dempsey, M. E. Biochem. Biophys. Res. Commun., 38:921 (1970).

6. Ebner, K. E. Accounts of Chemical Research 3:34 (1970).

7. Ritter, M. C., and Dempsey, M. E. in preparation.

8. Dempsey, M. E., Ritter, M. C., Witiak, D. T., and Parker, R. A. in Proc. 2nd Inter. Symp. on Atherosclerosis (Ed. R. Jones) (1970).

9. Witiak, D. T., Parker, R. A., Brann, D. R., Dempsey, M. E., Ritter, M. C., Connor, W. E., and Brahmankar, D. M. J. Med. Chem. in press (1970).

10. Dvornik, D., Kraml, M., and Bagli, J. F. Biochemistry 5:1060 (1966).

11. Dempsey, M. E. in Adv. Exptl. Med. Biol., Vol. 4 (Eds. W. L. Holmes,L. A. Carlson, R. Paoletti) (1969).

12. Conner, R. L., Landrey, J. R., Burns, C. H., and Mallory, F. B. J. Protozoology 15:600 (1968).

13. Conner, R. L., Mallory, F. B., Landrey, J. R., and Iyengar, C. W. L. J. Biol. Chem., 244:2325 (1969).

14. Koroly, M. J. and Dempsey, M. E. in preparation.

BIOSYNTHESIS OF STEROLS IN DEVELOPING BRAIN

E. Grossi Paoletti

Institute of Pharmacology, University of Milan

20129 Milan, Italy

During development brain tissue actively synthesizes cholesterol to meet the requirements necessary for formation of its structures (1, 2). The biosynthetic activity in brain varies from species to species, according to the sequence of the developmental stages and is higher during the period of maximal brain growth (3). In adults, however, cholesterol biosynthesis is hardly detectable in brain (2). Very little is known about the mechanisms and control of cholesterol synthesis in brain, although synthesis in liver has been studied extensively (4).

The synthesis of cholesterol in brain is at least as efficient as in the liver during the suckling period in the rat, although relatively very much less cholesterol is synthesized by adult rat brain (1, 2). In fact, as sterol uptake from blood is rather slow (5) brain local synthesis must of necessity be a very important factor during maturation. Cholesterol synthesis in the brain has been studied "in vivo" (6, 7) and "in vitro" using both slices and homogenates (2, 3, 8). A few years ago our group showed, using crude homogenates from 10 day old rats, that mevalonate could be converted into cholesterol (8), and, more recently, Kelley et al. (9) have shown that adult brain homogenates efficiently convert mevalonic acid into squalene but much less sterols are formed.

In the experiments reported here, brain homogenates from immature rats were studied for their capacity to synthesize unsaponifiable and sterols. Brains from 10 day old rats were homogenized with 0.1 M potassium phosphate buffer, pH 7.4 (1:2.5 tissue : medium) and centrifuged for 10 min. at 1,000 x g to eliminate cell debris and nuclei (crude homogenate). The supernatant was further centrifuged at 12,000 x g for 30 min. The resultant pellet comprises the mitochondrial fraction and includes mitochondria, nerve endings and myelin (10). The supernatant (S_{12}) was centrifuged at 105,000 x g for 90 min. to separate the microsomes (in the pellet) and the soluble fraction (S_{105}). In most experiments the S_{12} fraction, containing the complete biosynthetic activity from mevalonate to cholesterol, was used as an enzyme source.

Incubations were performed under the conditions described in the tables and the homogenates were saponi- fied afterwards in N KOH in ethanol 50% at 70°C for 1 hr. The unsaponifiable residue was extracted with petroleum ether (3 extractions) and the extracts were combined, washed and dried. Sterols were purified by thin layer chromatography (TLC), using silica gel G as absorbent. The recovered sterols were acetylated with pyridine: acetic anhydride (10:1) and the sterols with one and two double bonds were separated on silica gel G:$AgNO_3$ TLC (11). In some experiments, cholesterol was purified through the formation of dibromide for control of TLC purification. Aliquots of unsaponifiable fraction and sterols were counted for radioactivity by means of a Packard liquid scintillation counter. The figures re- ported for incorporation of mevalonic acid do not take into account the fact that only one of the two isomers of the supplied ^{14}C-DL-mevalonate (MVA) is utilized; for this reason percentage values should be doubled.

The cofactors required for maximal MVA incorporation were established (Table I). It was observed that the S_{12} fraction was practically incapable of utilizing mevalonate in the absence of cofactors although the same

fraction from liver catalized sterol synthesis well
without any addition (personal observation). Furthermore,
a NADPH regenerating system with added $MgCl_2$, ATP and
GSH permits the best precursor incorporation into both
unsaponifiable material and sterols and for this reason
was used in further experiments.

Various substrates may be chosen for studying
cholesterol synthesis in brain (2, 6). Mevalonate, which
is the best precursor for use in liver synthesis is poorly
utilized by the brain when administered "in vivo" (12).
It may be used by brain homogenates, provided that the
potassium salt (MVA-K) (8) is used instead of the acid-
lactone form (MVA-L). The S_{12} fraction (Table 2)
utilizes MVA-K three to four times more efficiently than
MVA-L, both for the formation of unsaponifiable and
sterols. This difference, which has been discussed previously (8) may be observed in brain homogenates only
when the labeled mevalonate is supplied as acid-lactone

TABLE I.

MVA-L INCORPORATION IN BRAIN HOMOGENATES:
COFACTOR REQUIREMENTS

Each incubation system contained I ml 12,000 x g supernatant of I0 day old rat brain homogenate (1:2.5 tissue:
medium, 0.I M buffer phosphate pH 7.4) added with 0.008M
$MgCl_2$ and system A or B incubated for 2 hr. at 37°C.

Complete System A : I µmole ATP, I0 µmoles GSH, 5 µmoles
NADH, I µmole NADPH, 200 mµmoles DL-Mevalonate (I µC).
Complete System B : as for system A, except NADH and
NADPH are replaced by I µmole NAD, I µmole NADP and
3 µmoles G-6-P.

	UNSAPONIFIABLE d.p.m.	STEROLS d.p.m.
No cofactors	1,500	--
Complete System A	194,900	16,210
Complete System B	303,300	54,770

TABLE 2.

INCORPORATION OF MEVALONIC ACID-LACTONE VERSUS POTASSIUM MEVALONATE

1 ml of rat brain homogenate (S_{12}) was incubated as in Table I with complete system B and 200 mμmoles of DL-mevalonate. Each value represents duplicated experiments.

	UNSAPONIFIABLE MVA mumoles incorporated	% INCORPORATION
Experiment 1:		
MVA-L	13.87	6.91
MVA-K	46.28	23.09
Experiment 2:		
MVA-L	17.57	8.89
MVA-K	50.96	25.74

(Amersham, U.K.) and not as dibenzylethylendiamine salt (New England Nuclear, Cambridge, U.S.A.).

Another difference between liver and brain homogenates is that it is generally suggested that nicotinamide be added as cofactor for sterol synthesis. However, we observed, in agreement with the observation of Kelley et al. (9), that nicotinamide has a strong inhibitory action on the activity of brain homogenates (Table 3) when used at concentrations as low as 0.013M instead of 0.03M, the concentration used to stimulate liver synthesis (13). Nicotinamide was therefore omitted from the following experiments.

Both microsomes and soluble enzymes are necessary for the complete synthesis of cholesterol from mevalonate in the liver (14). To check if the same enzyme distribution is present in the brain various fractions of 10 day old rat brain homogenates were tested and it was established that both the microsomal and soluble fraction are required for cholesterol formation from mevalonate in brain (Table 4), while the crude mitochondrial and

TABLE 3.

MVA-K INCORPORATION INTO THE UNSAPONIFIABLE OF RAT BRAIN HOMOGENATES

Conditions of incubation are as for Table I with system B, except for the use of mevalonate as potassium salt (200 mμmoles); nicotinamide 13 μmoles. Values represent mean of duplicated experiments.

	UNSAPONIFIABLE d.p.m.	% INCORPORATION
With nicotinamide	40.300	1.90
Without nicotinamide	505.300	23.82

TABLE 4.

BRAIN HOMOGENATES - SUBCELLULAR FRACTIONS: INCORPORATION OF MVA-K INTO UNSAPONIFIABLE

Incubation system and conditions as in Table I with system B. S_{12} = Supernatant at 12,000 x g; S_{105} = Supernatant at 105,000 x g. Each value is mean of duplicated experiments.

	% SUBSTRATE INCORPORATION	mμmoles/mg protein/2 hr.
S_{12}	22.76	5.40
Mitochondrial fraction	0.12	0.03
Microsomal fraction	0.03	0.06
S_{105}	6.82	1.71
S_{105} + Microsomes	20.85	4.46

microsomal fractions alone are inactive. The soluble
fraction alone incorporates mevalonate efficiently into
unsaponifiable but not into sterols. In fact, it is
known that the liver's transformation of mevalonate is
stopped at the level of farnesyl-pyrophosphate, which
is converted to squalene by microsomal enzymes (14).
Further studies are necessary to evaluate this point.
However, the present experiments indicate a similar subcellular distribution of the enzymes involved in
cholesterol formation in both brain and liver tissues.

The optimal substrate concentration for mevalonate
has been studied and the results (Fig. 1) indicate that
the unsaponifiable formation is dose dependent; also,
it is much more efficient than the formation of sterols.
The effect of the incubation time on MVA conversion is
also reported (Fig. 2). The incorporation of MVA into
unsaponifiable fraction is maximal 1 hr. after the

MVA INCORPORATION IN IMMATURE RAT BRAIN HOMOGENATES
EFFECT OF MVA CONCENTRATION

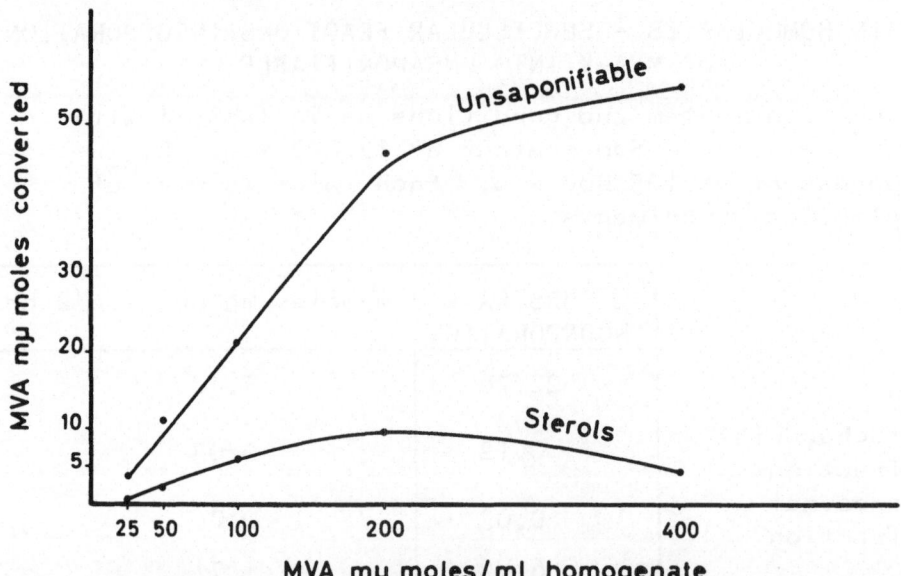

FIG. 1. Conditions as specified in Table 1. MVA-K was
used as substrate. Each point is the mean of two determinations.

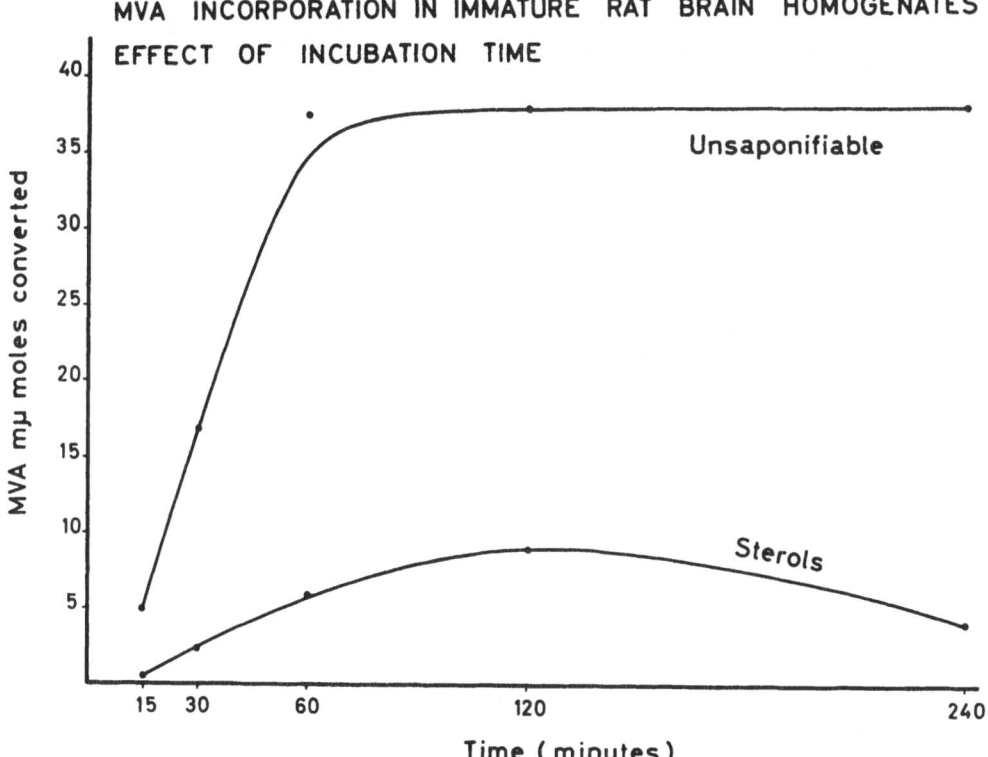

FIG. 2.   Conditions as in Table I.   200 mµmoles MVA-K
were used.   Each point is the mean of two determinations.

beginning of incubation, while the maximal conversion
into sterols is reached after 2 hours.   It is noteworthy
that, while in liver homogenates under the same experimental conditions, at least 90% of the formed unsaponifiable fraction is cholesterol, in brain homogenates no
more than 20% of the total unsaponifiable fraction is
represented by sterols.   Furthermore, a direct comparison
between adult rat (300 g body weight) brain homogenates
and 10 day old rat brain homogenates indicates (Table 5)
that the incorporation of MVA in adult brain homogenates
is only slightly lower than in 10 day old rat brain.
However, the sterol formation is strikingly different,
being more than 7 times greater in the immature brain.
This finding is in agreement with Kelley et al. (9), who
demonstrated that cholesterol synthesis in cell free

TABLE 5.

MVA-K INCORPORATION IN BRAIN HOMOGENATES

Incubation system and conditions as in Table I.   Each
value is mean of duplicated experiments.

	MVA mμmoles incorporated/ml homog./2 hrs.	
	UNSAPONIFIABLE	STEROLS
IMMATURE	49.61	12.45
ADULT	36.19	1.60
RATIO	1.37	7.78

extracts of adult rat brain is largely blocked at the
level of squalene cyclization to lanosterol.

The nature of the sterols formed in "in vitro"
biosynthesis was also investigated. Previous research
from our group showed that in both developing and adult
brain two series of intermediates of cholesterol biosynthesis are present in minute amounts (15, 16). "In
vivo" administration of MVA allowed us to evaluate the
possible biosynthetic role of the sterols with 30, 29,
28 and 27 carbon atoms of the $\Delta^{8,24}$ series (15). This
finding agrees with the detection of large pools of
desmosterol (C27, $\Delta^{5,24}$) in immature brain (17, 18, 19)
and indicates that in immature brain Δ^{24}-reductase
activity is limited. The analysis of the time course
of sterol formation by brain homogenates (Fig. 3) indicates that also in this case a precursor/product
relationship between sterols with two and one double bonds
is present. Bromination of the one double bond sterol
fraction indicates that cholesterol is practically the
only labeled sterol, while desmosterol is the major component of the sterols with two double bonds. However,
some labeled sterols with 30, 29, 28 and 27 carbon atoms
can be detected by using column chromatographic separa-
tions (15).

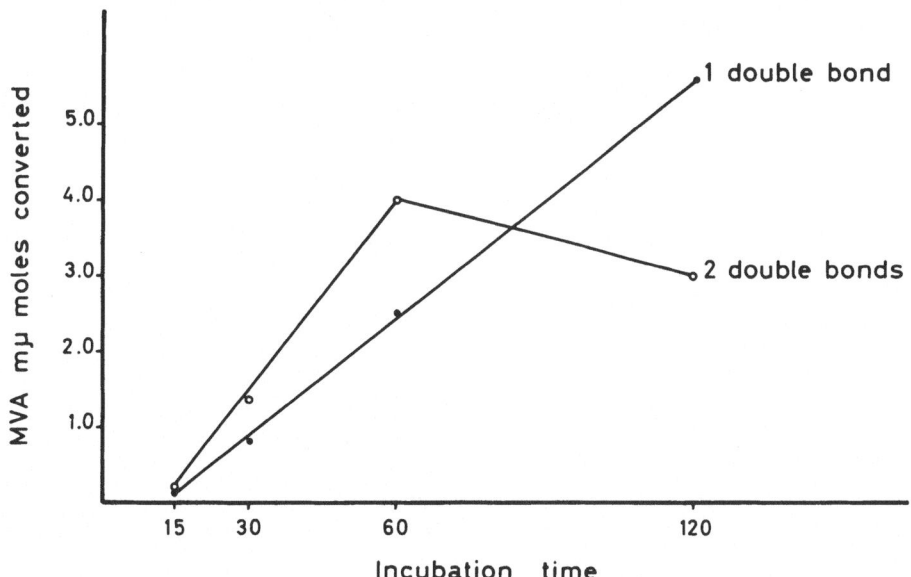

FIG. 3. Conditions as specified in Table I. MVA-K was used as substrate. Each point is the mean of two determinations.

To summarize, the present report shows that sub-cellular fractions from rat immature brain are able to synthesize cholesterol efficiently from mevalonate. However, the formation of the total unsaponifiable material is much more efficient than the sterol formation.

When comparing adult with developing brain, the amount of precursor incorporated into sterols is much slower in the former case while the unsaponifiable formation does not differ considerably.

Given an efficient system for synthesizing and detecting intermediary sterols with their subsequent transformation into cholesterol, it should be possible to evaluate the enzymatic activity of the last steps of cholesterol biosynthesis in immature brain, with the aim of explaining the accumulation of desmosterol during

brain development and the mode of cholesterol formation
at the sterol level.

ACKNOWLEDGEMENTS

This research has been partially supported by a
grant from the Association for the Aid of Crippled
Children, New York, U.S.A.

REFERENCES

1. Srere, P.A., Chaikoff, J.L., Treitmann, S., and
    Burstein, L.S. J. Biol. Chem. 182:629 (1950)

2. Grossi, E., Paoletti, P., and Paoletti, R. Arch.
    Int. Physiol. Bioch. 66:564 (1958)

3. Korey, S.R., Stein, A. in Regional Neurochemistry
    (Eds. S.S. Kety and J. Elkes), C.C. Thomas, Spring-
    field (1961)

4. Olson, J.A. Erg. Physiol. Biol. Chemie und Exp.
    Pharmakol. 56:163 (1965)

5. Clarenburg, R., Chaikoff, I.L., and Morris, M.D.
    J. Neurochem. 10:135 (1963)

6. Kabara, J.J., and Okita, G.T. J. Neurochem. 7:298
    (1961)

7. Davison, A.N. in Adv. Lipid Research vol. 3 (Eds.
    D. Kritchevsky and R. Paoletti) Academic Press (1965)

8. Fumagalli, R., Grossi, E., Poggi, M., Paoletti, P.,
    and Garattini, S. Arch. Biochem. Biophys. 99:529
    (1962)

9. Kelley, M.E., Aexel, R.T., Herndon, B.L., and
    Nicholas, H.J. J. Lipid Res. 10:166 (1969)

10. Aeberhard, E., and Menkes, J.H. J. Biol. Chem.
    243:3884 (1968)

11. Galli, G., and Grossi Paoletti, E. Lipids 2:72
    (1967)

12. Garattini, S., Paoletti, P., and Paoletti, R. Arch. Biochem. Biophys. 84:253 (1959)

13. Bucher, N.L.R., and McGarrahan, K. J. Biol. Chem. 222:1 (1956)

14. Popják, G. in Methods in Enzymology, vol. XV (Ed. R.B. Clayton) Academic Press (1969)

15. Weiss, J.F., Galli, G., and Grossi Paoletti, E. J. Neurochem. 15:563 (1968)

16. Galli, G., Weiss, J.F., and Grossi Paoletti, E. Science 162:1495 (1968)

17. Fish, W.A., Boyd, J.E., and Stokes, W.M. J. Biol. Chem. 237:334 (1962)

18. Kritchevsky, D., and Holmes, W.L. Biochem. biophys. Res. Comm. 7:128 (1962)

19. Paoletti, R., Fumagalli, R., Grossi, E., and Paoletti, P. J. Am. Oil Chem. Soc. 42:400 (1965)

BRAIN AND MYELIN STEROLS STUDIED USING SPECIFIC
INHIBITORS OF STEROL SYNTHESIS

R. Fumagalli

Institute of Pharmacology, University of Milan
20129 Milan, Italy

Sterols are major components of brain cellular and subcellular membranes and they are represented almost exclusively in mature mammalian brain by non-esterified cholesterol. However, during brain maturation and myelin deposition, stages of development characterized by a high rate of cholesterol biosynthesis, considerable amounts of desmosterol, a sterol precursor of cholesterol, are present in the brain of several mammalian species, including man (1).

In recent years pharmacological means have been used to study the sterol pattern of various animal tissues. Triparanol[*] and 20,25-diazacholesterol are the best known selective inhibitors of the Δ^{24}-reductase, an enzyme regulating the last steps of cholesterol biosynthesis and inducing desmosterol accumulation in brain and peripheral tissues (2, 3, 4). A specific inhibitor of 7-dehydrocholesterol-Δ^{7}-reductase, another enzyme involved in cholesterol biosynthesis, is also widely used (5). This compound, named AY-9944**, inhibits the

[*]   1-p(beta-diethylaminoethoxy)-phenyl-1-(p-tolyl)-2-(p-chlorophenyl)ethanol

[**]   trans-1,4-bis(2-dichlorobenzylaminomethyl)cyclo-hexane dihydrochloride.

enzyme at low concentrations (6) and accumulates 7-
dehydrocholesterol ($\Delta^{5,7}$) in various rat rissues (5).
These inhibitors of sterol synthesis, all acting after
squalene cyclization, were not used as hypocholesteremic
agents because the reduced formation of cholesterol (Δ^5)
is paralleled by a concomitant accumulation of the sterol
precursors of cholesterol; however, their use has
provided useful information to the understanding of the
sequence of reactions leading to cholesterol formation
in various mammalian tissues, including brain.

BRAIN AND LIVER STEROLS IN NEW BORN RATS
AFTER AY-9944 TREATMENT

AY-9944 was administered to pregnant rats in an
attempt to compare the changes of the sterol profile in
liver and brain of newborn animals (7). The sterol
pattern in these organs is quite different (Fig. I).
As expected, 7- dehydrocholesterol is largely accumulated in the liver, along with another still unidentified
C 27 sterol which is absent in brain. 7-Dehydrocholes-
terol is also accumulated in brain, where desmosterol
($\Delta^{5,24}$) concentration is reduced and, finally, two
other sterols, absent from the liver of the same animals,
have been detected. One of these has been identified
as 7-dehydrodesmosterol ($\Delta^{5,7,24}$), a sterol previously
considered to be a possible cholesterol precursor (8),
and later isolated in guinea-pig intestine (9) and pig
liver (10).

In liver, our findings are in agreement with the
reported possibility that the reduction of Δ^{24} double
bond may take place before the formation of 7-dehydro-
cholesterol (11) and this sterol is then accumulated
without the concomitant appearance of 7-dehydrodesmos-
terol. On the other hand, in the brain of growing
animals the reduction of Δ^{24} double bond appears to be
a limiting step in cholesterol formation so that some
desmosterol is normally present, suggesting that, in
brain at this age cholesterol is formed mainly from
precursors with a double bond in the side chain. The

LIVER OF NEWBORN RAT

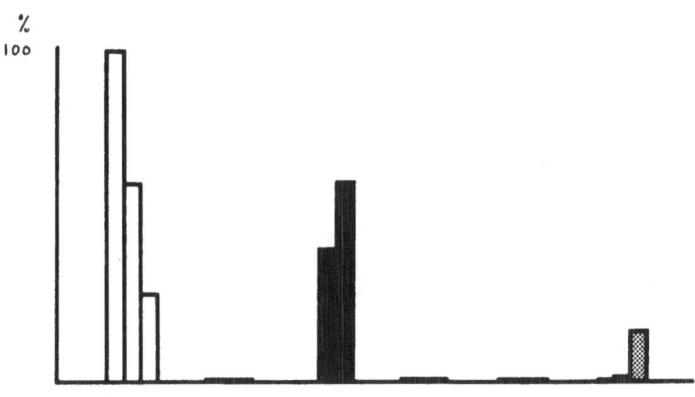

BRAIN OF NEWBORN RAT

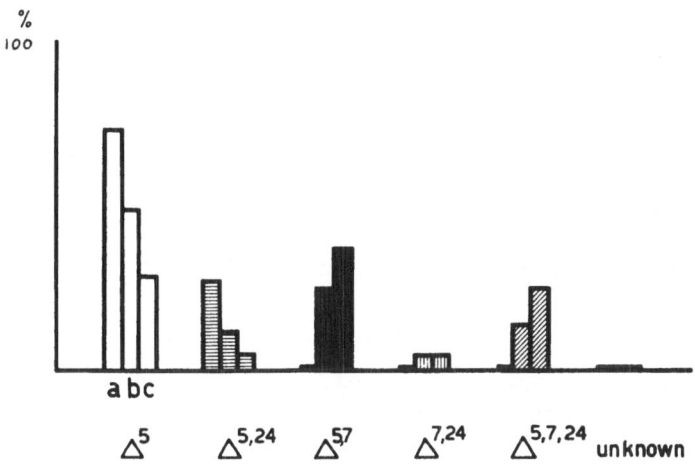

a   Controls

b   AY-9944   3 day treatment

c   AY-9944   6 day treatment
          5 μ moles/kg/day

FIG. I. Sterol composition (as % of total sterols) of newborn rat brain and liver after AY-9944 treatment of the pregnant rat mother.

accumulation of 7-dehydrodesmosterol and the decreased
concentration of desmosterol observed in presence of
AY-9944 suggest that this drug also inhibits the 7-
dehydrodesmosterol-Δ^7-reductase and that the accumulation
of 7-dehydrocholesterol follows, at least partially, the
reduction of Δ^{24} at the level of 7-dehydrodesmosterol
(Fig. 2).

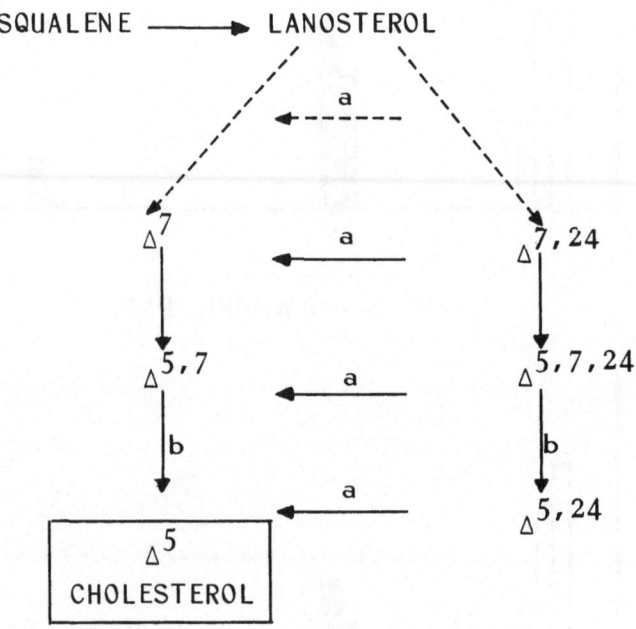

FIG. 2.   Last steps of cholesterol biosynthesis.
The dotted lines refer to intermediate sterols with 30,
29, 28 and 27 carbon atoms.
a = the site of action of Δ^{24}-reductase;   b = the site
of action of Δ^7-reductase; $\Delta^{7,7}$ = lathosterol; $\Delta^{7,24}$ =
7,24-cholestadiene-3β-ol; $\Delta^{5,7}$ = 7-dehydrocholesterol;
$\Delta^{5,7,24}$ = 7-dehydrodesmosterol; $\Delta^{5,24}$ = desmosterol.

STEROLS IN MYELIN OF GROWING RAT BRAIN

The distribution of sterols in subcellular fractions,
especially myelin, in normal conditions and after administration of the above mentioned inhibitors of sterol
biosynthesis, has been investigated.

Desmosterol, a normal component of developing brain, is normally present in the myelin fraction during the myelination period and its concentration decreases during brain maturation. Although desmosterol levels of the soluble, mitochondrial and nuclear fractions are approximately the same as in the whole brain homogenates, myelin of both brain and spinal cord usually contains much less. Brain desmosterol concentrations decrease considerably during the period between 12 and 20 days of age and the decrease is parallel in homogenate and myelin. However, desmosterol appears to be retained longer in the myelin fraction (12). 20,25-Diazacholesterol and AY-9944 administered separately to rats (2-8 mg/animal by stomach intubation between the 15th and 20th day of life) accumulate desmosterol and 7-dehydrocholesterol respectively in brain and spinal cord myelin (13).

Table I shows the effect of AY-9944 treatment on sterol content of rats at different time intervals after final drug administration.

The results obtained after 20,25-diazacholesterol treatment are similar to those reported for AY-9944, with accumulation of desmosterol instead of 7-dehydrocholesterol. The amounts of cholesterol precursors accumulated are not dose-dependent (13). When the total dose varied from 2 to 8 mg/animal the proportion of desmosterol or 7-dehydrocholesterol did not greatly change. This proportion is then the maximum amount replaceable at this age and it may be deduced that the blockade of the corresponding enzymes is maximal at such doses of the drugs.

At the end of treatment, more desmosterol is accumulated than 7-dehydrocholesterol. The levels of both precursors subsequently increase up to 50-60% of total sterols at 30 days of age; 7-dehydrocholesterol is almost completely replaced at 60 days while some desmosterol is still retained, especially in myelin. Taking into account also the increase of myelin in 30 and 60 day old animals, this rate of replacement is greater than expected for myelin cholesterol.

TABLE I.

TOTAL STEROLS (mg/100 mg DRY WEIGHT), DESMOSTEROL (D) AND 7-DEHYDROCHOLESTEROL (7-D) PERCENTAGE IN CNS OF RATS AFTER TREATMENT WITH AY-9944

CONTROLS	23 DAYS		30 DAYS		40 DAYS		60 DAYS	
	Total Sterols	D%	Total Sterols	D%	Total Sterols	D%	Total Sterols	D%
Brain homogenate	7.70	0.8	8.24	0.6	7.87	0.6	8.28	0.4
Brain myelin	15.24	2.5	15.62	0.7	13.17	0.7	13.93	0.7
Spinal cord homogenate	9.67	0.7	9.00	n.d.	9.66	n.d.	9.43	n.d.
Spinal cord myelin	15.94	1.4	15.09	0.5	13.97	n.d.	14.00	n.d.

TREATED*	Total Sterols	7-D%	Total Sterols	7-C%	Total Sterols	7-D%	Total Sterols	7-D%
Brain homogenate	5.83	25.2	4.59	48.6	3.90	34.4	8.06	2.7
Brain myelin	13.23	31.6	14.11	54.4	12.58	45.5	14.29	n.d.
Spinal cord homogenate	6.57	29.2	4.95	61.6	3.00	56.8	9.24	n.d.
Spinal myelin	12.61	32.7	13.45	61.3	12.20	59.3	14.00	n.d.

* 4 doses (2mg/rat/per os) on the 15th, 17th, 19th and 21st days of life.

There are several explanations for the high rate
of replacement of these cholesterol precursors: these
sterols may be reduced in situ in the myelin membrane
or exchanged with cholesterol. It is possible that a
higher rate of sterol metabolism prevails at this age
than has been found for later ages or again these sterols
could be metabolized more rapidly than cholesterol.

Lipids exist in myelin as cholesterol complexes
(14, 15) and, therefore, desmosterol and 7-dehydrocholes-
terol are evidently able to substitute cholesterol in
the complex.

In contrast to newborn animals, the administration
of AY-9944 does not induce accumulation of 7-dehydro-
desmosterol in the brain of 20 to 30 day old rats. This
could mean that at this age the reduction of Δ^{24}-double
bond in the brain is no longer a limiting step and that
such a reduction takes place before the formation of
7-dehydrodesmosterol.   At this age the accumulation
of 7-dehydrodesmosterol can be induced in rats by a
double pharmacological treatment using both diaza-
cholesterol and AY-9944 (16). The former drug allows
cholesterol to be formed mainly from precursors with
unsaturated lateral chain, the latter prevents the
conversion of 7-dehydrodesmosterol to desmosterol (Fig.
2).

STEROLS IN CSF OF PATIENTS BEARING BRAIN TUMORS

The use of drugs in studying sterol metabolism in
the CNS has also been applied to humans. Desmosterol
has been found to be present in low concentrations in
human brain tumors, such as glioblastomas (17). After
a short triparanol treatment, desmosterol accumulates
in brain tumors but not in the normal surrounding brain.
A consequence of this is the possible appearance of
desmosterol in cerebrospinal fluid (CSF) of brain tumor
bearing patients only. In fact, a correlation exists
between the CSF concentration of desmosterol and the

TABLE 2.

AVERAGE CSF CONCENTRATIONS OF CHOLESTEROL AND DESMOSTEROL
IN PATIENTS TREATED WITH TRIPARANOL

PATIENTS	No. OF CASES	CHOLESTEROL µg/ml ± S.E.	DESMOSTEROL µg/ml ± S.E.
Without brain tumor	52	1.92 ± 0.11	0.009 ± 0.002
With brain tumor	30	5.33 ± 0.82	0.204 ± 0.040

presence of a brain tumor (18) (Table 2). The origin of
CSF desmosterol could be due either to a passage of the
sterol from the blood through an altered blood-brain
barrier or to endogenous formation of desmosterol in the
tumor (19) which is then released into CSF, or to both
these possibilities. The observation that some patients
have desmosterol in their CSF in the absence of circulating desmosterol underlines the importance of the latter
possibility.

ACKNOWLEDGEMENTS

This research has been partially supported by a
grant from the Association for the Aid of Crippled
Children, New York, U.S.A.

REFERENCES

1. Paoletti, R., Fumagalli, R., Grossi, E., and
    Paoletti, P. J. Am. Oil Chem. Soc. 42:400 (1965)

2. Avigan, J., Steinberg, D., Vroman, H.E., Thompson,
    M.J., and Mosettig, E. J. Biol. Chem. 235:3123 (1960)

3. Thompson, M.J., Dupont, J., and Robbins, W.E.
    Steroids $\underline{2}$:99 (1963)

4. Fumagalli, R., and Niemiro, R. Life Sci. $\underline{3}$:555
    (1964)

5. Dvornik, D., Kraml, M., Dubuc, J., Givner, M., and
    Gaudry, R. J. Am. Chem. Soc. $\underline{85}$:3309 (1963)

6. Niemiro, R., and Fumagalli, R. Biochim. Biophys.
    Acta $\underline{98}$:624 (1965)

7. Fumagalli, R., Niemiro, R., and Paoletti, R. J. Am.
    Oil Chem. Soc. $\underline{42}$:1018 (1965)

8. Johnstone, J.D., and Bloch, K. J. Am. Chem. Soc.
    $\underline{79}$:1145 (1957)

9. Frantz, I.D., Jr., Sanghvi, A.T., and Clayton, R.B.
    J. Biol. Chem. $\underline{237}$:3381 (1962)

10. Dvornik, D., Kraml, M., and Bagli, J.F. J. Am.
     Chem. Soc. $\underline{86}$:2739 (1964)

11. Goodman, D.S., Avigan, J., and Steinberg, D. J.
     Biol. Chem. $\underline{238}$:1287 (1963)

12. Smith, M.E., Fumagalli, R., and Paoletti, R. Life
     Sci. $\underline{6}$:1085 (1967)

13. Fumagalli, R., Smith, M.E., Urna, G., and Paoletti,
     R. J. Neurochem. $\underline{16}$:1329 (1969)

14. Finean, J.B. Experientia (Basel) $\underline{9}$:17 (1953)

15. Vandenheuvel, F.A. J. Am. Oil Chem. Soc. $\underline{40}$:455
     (1963)

16. Smith, M.E., Hasinoff, C., and Fumagalli, R. Lipids
     $\underline{5}$:665 (1970)

17. Fumagalli, R., Grossi, E., Paoletti, P., and
     Paoletti, R. J. Neurochem. $\underline{11}$:561 (1964)

18. Paoletti, P., Vandenheuvel, F.A., Fumagalli, R.,
     and Paoletti, R. Neurology $\underline{19}$:190 (1969)

19. Fumagalli, R., Grossi-Paoletti, E., Paoletti, R.,
     and Paoletti, P. Ann. N.Y. Acad. Sci. $\underline{159}$:472
     (1969)

PHOSPHOLIPID METABOLISM IN THE DEVELOPING BRAIN

G. B. Ansell

Department of Pharmacology, The Medical School

Birmingham, B15 2TJ, England

The adult mammalian brain contains more phospholipid per unit wet weight than any other tissue and it is distributed among the sub-cellular components. During the course of development the phospholipids are deposited not only as a result of the increase in brain size but also as a result of the structural changes during maturation, notably the deposition of myelin. The rate of deposition in the brain as a whole is well documented especially for the rat and human and the literature up to 1964 has been summarised by Ansell and Hawthorne (1). A more recent comprehensive study of the rat brain has been made by Wells and Dittmer (2). During development the fatty acid composition of some phospholipids changes (e.g. ref. 3) which may be due to preferential synthesis or loss of certain phospholipids or exchange of fatty acids within the molecules themselves.

Phospholipid Levels in Developing Brain

Other contributions will have dealt with the different periods of brain development. In Fig.1 are indicated the levels of different phospholipids in the rat brain expressed as a percentage of the adult level. Both ethanolamine plasmalogen and triphosphoinositide, though not exclusive to myelin, are very important components of this material whereas phosphatidylethanolamine and phosphatidylcholine are more general components of cell membranes. It is not surprising therefore, that the former show the larger increase during the period when myelin is most actively deposited and are relatively scarce before this process. Cerebrosides are even more closely associated with the myelination process.

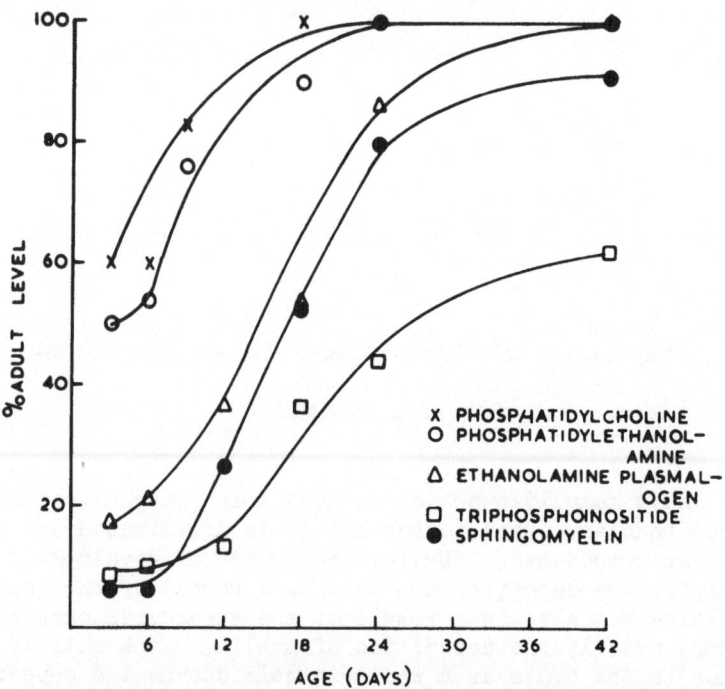

Fig.1   The levels in μmoles/gram fresh weight of some phospholipids in developing rat brain expressed as a percentage of the adult level. (Values from Ref. 2).

It is clear from Fig.1 that there is an obvious increase in the amount of sphingomyelin deposited during myelination in the rat, a rate which decreases subsequently.   If the amounts of sphingomyelin are plotted against the $\log_{10}$ of the age as suggested by Rouser and Yamamoto (4) it can be seen (Fig.2) that there are three very clear stages and active deposition apparently occurs from about 10 days to 70 days of age.   For ethanolamine plasmalogen the active deposition occurs for 100 days, (Fig.3).   Though there are measurable increases in both these lipids after the active period, they are slow.   In this connection it is of interest that the total period of myelination probably extends well into adult life (5).   According to Rouser and Yamamoto (4) the sharp inflections in the semi-logarithmic plots are not seen if grey and white matter are analysed separately.

General Comments on Metabolism

It is of some interest to know the parameters of phospholipid metabolism which change during the period of active myelination and growth.   Early studies _in vivo_ suggested that, during myelination, there was an increased uptake of labelled precursors into phospho-lipids (for references see ref. 1).   However, it was subsequently appreciated that measurements of the specific radioactivity of the

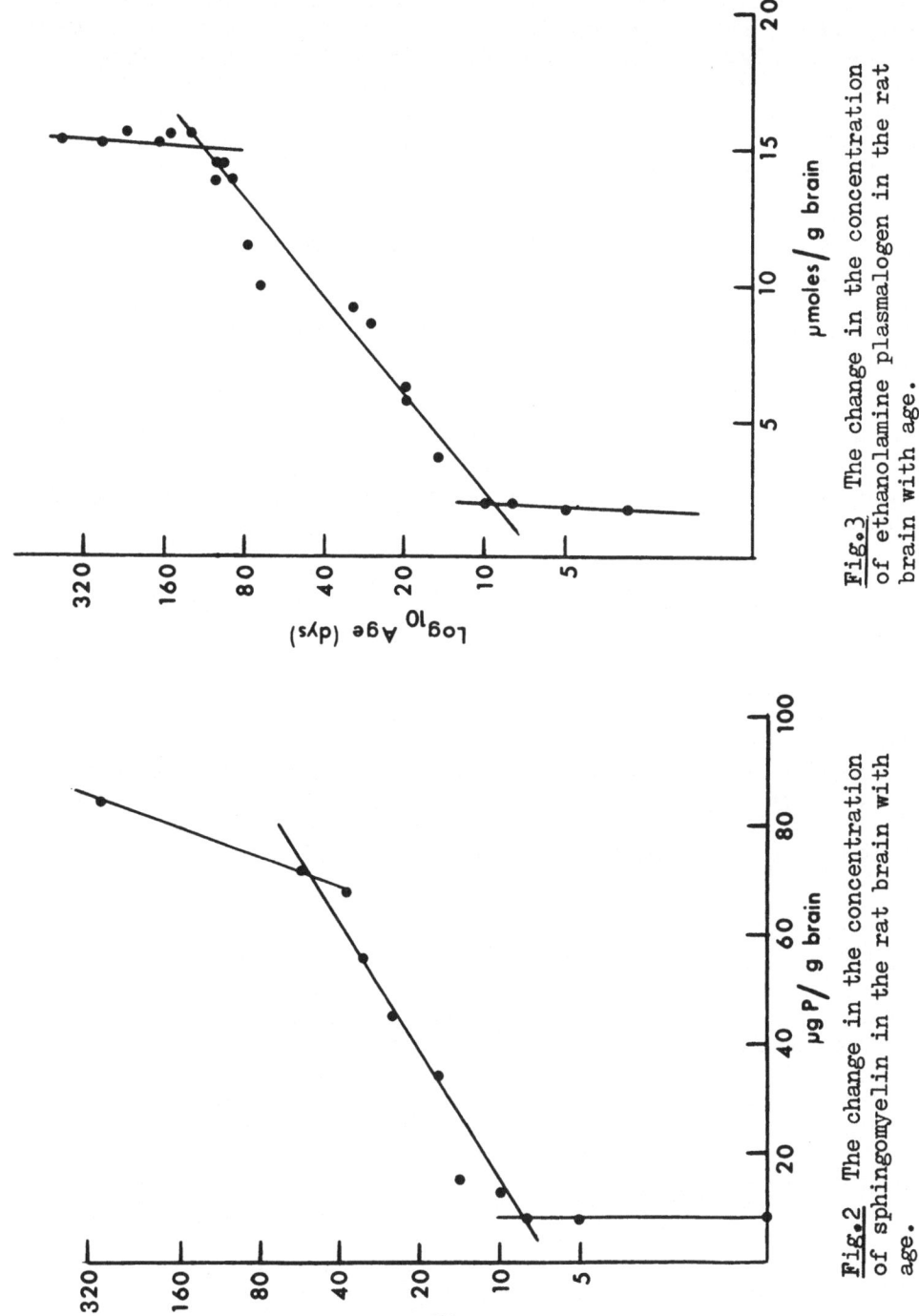

Fig.3 The change in the concentration of ethanolamine plasmalogen in the rat brain with age.

Fig.2 The change in the concentration of sphingomyelin in the rat brain with age.

phospholipid formed were deceptive because material already
synthesised and increasing in amount diluted the newly synthesised
material, giving a lower value for older animals than younger
animals.   Thus the rate of sphingomyelin formation, as measured by
the incorporation of ^{32}P in a 2-week-old rat, was apparently much
greater than in an adult rat but if corrected for the amount present
it might be concluded that the rates were similar (6).

Turning to studies in vitro, it was shown by Pritchard (7) that
certain precursors e.g. labelled acetate were incorporated very much
more rapidly into the phospholipids of slices of immature brains
than adult; the incorporation of others e.g. ethanolamine was
unaffected by age.   Maximum incorporation occurred before the onset
of myelination.   In the case of the typical neuronal glycolipids,
the gangliosides, incorporation of the radioactivity of labelled
glucose fell from birth to a very low level at 25 days; this was in
sharp contrast to incorporation into a typical myelin glycolipid,
cerebroside (8) though in this instance the maximum incorporation of
labelled precursor occurred before the peak of deposition measured
chemically (9).

There is presumably an induction period in which phospholipids are synthesised and not broken down followed by a period in
which synthesis is balanced by catabolic processes.   The extensive
work of Davison and his colleagues (10) has indeed shown that, when
certain labelled precursors are injected into young myelinating
animals, the material so deposited remains in the myelin sheaths for
very long periods.   It is now clear, however, that the phospholipid
composition of myelin changes during development (11, 12) and that
certain myelin lipids turn over faster than others (13).   Can we
detect any changes in the metabolic systems responsible for the
synthesis or catabolism of phospholipids during this myelination
period?

There have been few attempts to measure levels of phospholipid precursors in the brain of developing animals.   If one
accepts that the cytidine pathway is important in the formation of
ethanolamine phospholipids in the brain then it is of interest that
the concentration of phosphorylethanolamine in the brain of the newborn rat is five times that of the 21-day-old animal (14).   Nothing
appears to have been reported on the changes in levels of other
precursors.   Phosphatidylglycerophosphate, a precursor of phosphat-
idylglycerol, rises in the pre-myelination period in the rat, stays
constant during myelination and then falls rapidly on maturation;
it is the only phospholipid to do so (though the relative amounts of
some phospholipids e.g. phosphatidylcholine do fall).   Wells and
Dittmer (2) comment that this "is consistent with it having an
intermediary role in a pathway with a rate-limiting reaction
subsequent to its formation".   Of more general significance may be

the finding of Laatsch (15) that the activity of glycerophosphate dehydrogenase (L-glycerol-3-phosphate: NAD oxidoreductase, EC 1.1.1.8) in rat fore-brain rises rapidly from birth to reach a peak at 25 days of age, whereafter it stays constant. Since brain tissue cannot phosphorylate glycerol (16) and therefore presumably contains no glycerol kinase (ATP:glycerolphosphotransferase, EC 2.7.1.30) and since at neutral pH the dehydrogenase reaction is in favour of glycerophosphate production, the increase in the dehydrogenase could be a response to an increased requirement for glycerophosphate.

The changes in metabolism of two myelin phospholipids, ethanolamine plasmalogen and triphosphoinositide have been studied in some detail.

Phosphoinositide Metabolism

The fact that triphosphoinositide is largely found in the myelin sheaths (17) and that the specific radioactivity of this lipid after subarachnoid injection of $[^{32}P]$ orthophosphate into young rats was highest in the myelin fraction (18) prompted Hawthorne and his colleagues (19) to investigate the enzymes associated with the synthesis of this lipid in the maturing rat brain. Essentially they found that, whereas the increase in the level of the enzyme CDP-diglyceride inositol phosphatidate transferase which forms phosphatidylinositol was not associated with myelination and the optimal activity of the enzyme phosphatidylinositol kinase preceded myelination, there was a definite association of the increase in diphosphoinositide kinase with the deposition of myelin. Of further interest was the observation that the level of triphosphoinositide monoesterase, an enzyme responsible for the breakdown of triphosphoinositide, also increased with the increase in triphosphoinositide (see below).

Plasmalogen Metabolism

The possible function of alkenyl acyl phospholipids (plasmalogens) in excitable tissues is an intriguing one but we are as yet uncertain of the definitive pathway for their synthesis. Undoubtedly ethanolamine plasmalogen (1-alkenyl-2-acyl-sn-glycero-3-phosphorylethanolamine) is the most characteristic phospholipid of the myelin and, as we have seen, its rate of deposition in the myelinating brain is considerable. Evidence from experiments with $[^{32}P]$ orthophosphate and $[^{14}C]$ ethanolamine suggest that the whole molecule can be assembled from smaller ones. Suggestions that it is formed from the 1,2-diacyl analogue or the 1-alkyl-2-acyl analogue by conversion at the 1-position have not been confirmed (see, for example, the recent elaborate study by Debuch and her colleagues (20) who injected, intracerebrally, labelled phosphatidylethanolamine and the 1-alkyl-2-

acyl analogue prepared from rat brain).   Evidence from experiments
in vivo by Ansell and Spanner (21) strongly suggests that labelled
ethanolamine is incorporated into ethanolamine plasmalogen by the
cytidine pathway.   This is supported by findings in vitro (22) that
an ethanolamine phosphotransferase (CDP-ethanolamine-1,2-diglyceride
ethanolamine phosphotransferase, EC.2.7.8.2.) in brain can transfer
phosphorylethanolamine to 1-alkenyl-2-acyl glycerol to yield
ethanolamine plasmalogen.   Since Bickerstaffe and Mead (23) have
suggested that the alkenyl ether linkage is formed at the glyceride
stage and Joffe (24) has suggested independent synthetic pathways
for the three ethanolamine phospholipids in brain, the cytidine
pathway could be significantly involved in plasmalogen synthesis
de novo.

    There are certain aspects of the synthesis of ethanolamine
plasmalogen and phosphatidylethanolamine which are puzzling.
Although the transfer of phosphorylethanolamine to ethanolamine
lipid from CDP-ethanolamine proceeds rapidly in the presence of
homogenates or microsomal preparations, the incorporation of free
phosphorylethanolamine is feeble even when the cytidylyltransferase
(CTP:ethanolamine-phosphate cytidylyltransferase, EC.2.7.7.14), an
enzyme of the cell sap, is present.   The transfer of free ethanolamine via the cytidine pathway is even more difficult to demonstrate
in vitro.   Indeed the kinase in the cell sap responsible for the
phosphorylation and which may be similar to choline kinase (ATP:
choline phosphotransferase, EC.2.7.1.32) has a poor affinity for
ethanolamine (S.G. Spanner, unpublished observation) which is
surprising.

    Another anomaly is the fact that L-serine is the precursor of
ethanolamine in the mammalian organism (25) and can yield the
ethanolamine for brain ethanolamine phospholipid (26) although
decarboxylation of free or lipid-bound serine has never been
demonstrated in brain.   The formation of ethanolamine from serine
has recently been explained by Stoffel and Henning(27) who showed
that, when $[1-^3H]$ erythro-DL-sphingosine($[1-^3H]$ 4t sphingenine) was
injected intraperitoneally into myelinating rats, the radioactivity
in the brain was almost all in the lipid-bound ethanolamine or
choline.   No free radioactive sphingosine could be detected in the
brain.   Presumably the formation of the ethanolamine and choline
took place in the liver.   This would explain how ethanolamine
derives from C_2 and C_3 of serine since the latter provides C_1 and
C_2 of sphingosine.   Subsequent studies by Stoffel, Sticht and
Le Kim (28) have shown that liver tissue can degrade dihydrosphingosine phosphate to palmitaldehyde and phosphorylethanolamine.
There is no evidence for the transfer of phosphorylethanolamine to
the brain from the liver but if such a reaction takes place in the
brain as suggested by very recent experiments by Stoffel, Le Kim
and Heyn (29), it could provide phosphorylethanolamine for the

cytidine pathway, though the rationale for the observed rapid phosphorylation of the base in vivo then becomes obscure. A more important aspect of the recent experiments by Stoffel et al (29) is their finding that $[3-^3H]$-DL-sphinganine (dihydrosphingosine), when injected intracerebrally into myelinating rats, can donate the alkenyl ether chain of plasmalogen. A series of intermediate steps has been suggested, which include the formation and subsequent dehydrogenation of the alkyl ether analogue.

Although the activity of the phosphotransferase is maximal in the brain of the 16-day-old rat when plasmalogen deposition is greatest, the activity in the adult animal is also high (22). Furthermore, it has been shown that the half life of the ethanol- amine plasmalogen of the myelin of the adult rat brain, as measured by the incorporation of intracerebrally injected $[^{14}C]$ ethanolamine is as low as 15 days (30). This suggests that even one of the characteristic myelin phospholipids is metabolically active in adult life (compare the results of Smith and Eng (13)) and that ideas on the metabolic inertness of myelin might have to be modified.

Another interesting feature of the metabolism of ethanolamine plasmalogen in the developing brain is the observation that a plas- malogenase which cleaves the alkenyl ether linkage is almost absent from the new-born rat brain but increases during the period of maximum deposition of its substrate (31). Significantly this enzyme is found in white matter rather than grey matter (32). Mention has already been made of the increase in level of tri- phosphoinositide monoesterase during myelination; Bowen and Radin (33) have shown that cerebrosidase (cerebroside galactosidase) which hydrolyses a third myelin lipid, cerebroside, also increases maximally during myelination in the rat. The significance of the increase of these three enzymes during myelination remains to be established.

The metabolism of the small amount of choline plasmalogen in the brain has been little studied but it may have a significant role in development. McMurray (34) showed that the labelling of choline plasmalogen from CDP-$[^{14}C]$ choline in dispersions of myelinating brain was much greater than that of phosphatidylcholine. Marshall (unpublished experiments) has confirmed this and has also demon- strated that, whereas the labelling of choline plasmalogen by $[^{32}P]$ orthophosphate in vivo was significantly greater than that of phosphatidylcholine in young animals, the rates were similar in adult animals.

Localisation of Phospholipid Synthesis in The Developing Brain

No experiments to decide whether systems capable of synthesising

phospholipids are localised in glial cells or neurones appear to
have been performed, though it would be interesting to know whether
lipids are transferred from glial cells to neurones or <u>vice versa</u>.
Freysz, Bieth and Mandel (35) found that the half lives of all the
neuronal phospholipids of adult rats were always shorter than those
of glial lipids; whether this would be true for the oligodendroglial
cells of young animals which produce myelin is as yet unsolved.

The formation of triphosphoinositide by the stepwise phosphorylation of phosphatidylinositol is unusual in that the phosphatidylinositol kinase is located in the plasma membrane (36),
but the diphosphoinositide kinase is in the cell sap (37) as is
the triphosphoinositide monoesterase (38).   Therefore, if differs
from other myelin phospholipids as far as is known in that its
final formation is in the cell sap.   It is believed, for example,
that ethanolamine plasmalogen is transferred from the endoplasmic
reticulum to the myelin pool (39).

Conclusion

In this brief account it has not been possible to consider in
detail many other interesting features of the topic.   There are,
for example, changes in the fatty acid composition of the phospholipids during development.   Thus, in the developing rat brain the
proportion of C_{18} and $C_{18:1}$ fatty acids in phosphatidylcholine
increases at the expense of C_{16} and $C_{16:1}$ (3) and this is probably
true for most glycerophospholipids (40).   When white and grey
matter are considered separately, the changes are complicated. The
proportion of $C_{22:6}$ in the ethanolamine lipids of human grey matter
rises from 20% at birth to 35% in the adult (41).   Before myelination C_{18} predominates in the sphingomyelin of human grey and white
matter but, after myelination C_{22-26} fatty acids amount to 70% of
the total (42).   Subtle changes in the metabolic pathways must
account for these changes.   Finally the possibility must be
considered that although nervous tissue can synthesise phospholipids
<u>de novo</u>, certain lipids, including phospholipids are brought to the
brain from outside.   This was first demonstrated for cholesterol in
the chick and rabbit (43,44).   Cuarón and his colleagues (45)
injected [^{32}P] orthophosphate into pregnant rabbits and found that
the labelling of the foetal brain phospholipids lagged significantly
behind their net deposition; this could imply that the phospholipids
were labelled outside the brain before being transported there.
Recently Hoelzl and Franck gave evidence for the direct incorporation
of doubly labelled phosphatidylcholine into the brain pool after
intraperitoneal injection (46) and indirect evidence has been found
by Ansell and Spanner (47).   It follows that there are many
features of brain phospholipid metabolism as yet little understood.

Acknowledgements: The work carried out by Drs. Spanner,
R.F. Metcalfe, E.F. Marshall and myself has been partly supported
by the Medical Research Council and the Multiple Sclerosis Society
of Great Britain.   The interest of Professor P.B. Bradley is
appreciated.

References

1. Ansell, G.B., and Hawthorne, J.N. Phospholipids: Chemistry,
    Metabolism and Function, Elsevier (1964)

2. Wells, M.A. and Dittmer, J.C. Biochemistry $\underline{6}$:3169 (1967)

3. Marshall, E., Fumagalli, R., Niemiro, R. and Paoletti, R. J.
    Neurochem. $\underline{13}$:857(1966)

4. Rouser, G. and Yamamoto, A. Lipids $\underline{3}$:284 (1968)

5. Norton, W.T., Poduslo, S.E. and Suzuki, K. in Biochemical
    Factors Concerned in the Functional Activity of the Nervous
    System (Ed. D. Richter) Pergamon Press (1969)

6. Ansell, G.B. and Spanner, S. Biochem. J. $\underline{79}$:176 (1961)

7. Pritchard, E.T. Canad. J. Biochem. Physiol. $\underline{40}$:353 (1962)

8. Maker, H.S. and Hauser, G. J. Neurochem. $\underline{14}$:457 (1967)

9. Davison, A.N. and Dobbing, J. in Applied Neurochemistry p.253
    (Eds. A.N. Davison and J. Dobbing) Blackwell (1969)

10. Davison, A.N. in Applied Neurochemistry, p.178 (Eds. A.N.
    Davison and J. Dobbing) Blackwell (1969)

11. Horrocks, L.A. J. Neurochem. $\underline{15}$:483 (1968)

12. Eng, L.F. and Noble, E.P. Lipids $\underline{3}$:157 (1968)

13. Smith, M.E. and Eng, L.F. J. Amer. Oil Chem. Soc. $\underline{42}$:1013
    (1965)

14. Agrawal, H.C., Davis, J.M. and Himwich, W.A. J. Neurochem.
    $\underline{13}$:607 (1968)

15. Laatsch, R.H. J. Neurochem. $\underline{9}$:487 (1962)

16. Wieland, O. and Suyter, M. Biochem. Z. $\underline{329}$:320 (1957)

17. Sheltawy, A. and Dawson, R.M.C. Biochem. J. 100:12 (1966)

18. Kai, M. and Hawthorne, J.N. Biochem. J. 98:62 (1966)

19. Salway, J.G., Harwood, J.L., Kai, M., White, G.L. and
     Hawthorne, J.N. J. Neurochem. 15:221 (1968)

20. Debuch, H., Friedemann, H. and Müller, J. Hoppe Seyler's
     Z. physiol. Chem. 351:613 (1970)

21. Ansell, G.B. and Spanner, S. J. Neurochem. 14:873 (1967)

22. Ansell, G.B. and Metcalfe, R.F. J. Neurochem. in the press.

23. Bickerstaffe, R. and Mead, J.F. Lipids 3:317 (1968)

24. Joffe, S. J.Neurochem. 16:715 (1969)

25. Arnstein, H.R.V. Biochem. J. 48:27 (1951)

26. Wilson, J.D., Gibsoh, K.D. and Undenfriend, S. J.biol. Chem.
     235:3539 (1960).

27. Stoffel, W. and Henning, R. Hoppe Seyler's Z. physiol. Chem.
     349:1400 (1968)

28. Stoffel, W., Sticht, G. and Le Kim, D. Hoppe Seyler's Z.
     physiol. Chem. 349:1745 (1968)

29. Stoffel, W., Le Kim, D. and Heyn, G. Hoppe Seyler's Z.
     physiol. Chem. 351:875 (1970)

30. Ansell, G.B. and Spanner, S. J. Neurochem. 15:1371 (1968)

31. Ansell, G.B. and Spanner, S. in Variation in Chemical
     Composition of the Nervous System as Determined by Develop-
     mental and Genetic Factors, p.7 (Ed. G.B. Ansell) Pergamon
     (1966)

32. Ansell, G.B. and Spanner, S. Biochem. J. 108:207 (1968)

33. Bowen, D.M. and Radin, N.S. J. Neurochem. 16:501 (1969)

34. McMurray, W.C. J. Neurochem. 11:315 (1964)

35. Freysz, L., Bieth, R. and Mandel, P. J. Neurochem. 16:1417
     (1969)

36. Kai, M., White,G.L. and Hawthorne, J.N. Biochem.J. 101:328
     (1966)

37. Kai, M., Salway, J.G. and Hawthorne, J.N. Biochem.J. 106:791 (1968)

38. Salway, J.G., Kai, M. and Hawthorne, J.N. J. Neurochem. 14:1013 (1967)

39. Horrocks, L.A. J. Neurochem. 16:13 (1969)

40. Kishimoto, Y., Davies, W.E. and Radin, N.S. J. Lipid Res. 6:532 (1965)

41. Svennerholm, L. J.Lipid Res. 9:570 (1968)

42. Stålberg-Stenhagen, S. and Svennerholm, L. J.Lipid Res. 6:146 (1965)

43. Davison, A.N., Dobbing, J., Morgan, R.S. and Payling Wright, G. J. Neurochem. 3:89 (1958)

44. Davison, A.N., Dobbing, J., Morgan, R.S. and Payling Wright, G. Lancet 1:658 (1959)

45. Cuarón, A., Gamble, J.,Myant, N.B. and Osorio, C. J. Physiol. 168:613 (1963)

46. Hoelzl, J. and Franck, H.P. in Proceedings of the Second International Meeting of the International Society of Neurochemistry p.219 (Eds. R. Paoletti, R. Fumagalli and G. Galli) Tamburini Editore (1969)

47. Ansell, G.B. and Spanner, S. in Proceedings of the International Symposium on the Effect of Drugs on Cholinergic Mechanisms in the CNS (Ed. E. Heilbronn) in the press (1970)

BRAIN GANGLIOSIDES IN DEVELOPMENT

G. Tettamanti

Department of Biological Chemistry

Medical School, Milan , Italy

INTRODUCTION

Gangliosides are glycosphingolipids containing sialic acid (primarily N-acetyl-neuraminic acid). They are present in brain in different forms, which differ in the pattern of fatty acids and sphingosines, and in the number and sequence of units forming the carbohydrate portion of the molecule. The main series of known gangliosides are distinguished on the basis of the common neutral glycosphingolipid moiety (*): a) G 3 series, galactosyl-β(1-4)glucosyl-β(1-1)ceramide; b) G 2 series, N-acetyl-galactosaminyl-β(1-4)galactosyl-β(1-4)glucosyl-β(1-1)ceramide; c) G 1 series, galactosyl-β(1-3)N-acetyl-galactosaminyl-β(1-4)galactosyl-β(1-4)glucosyl-β(1-1)ceramide. The G 3 and G 2 series contain gangliosides having one (GM3 and GM2) or two (GD3 and GD2) residues of sialic acid; the G 1 series contains gangliosides having one (GM1), two (GD1a and GD1b), three (GT1a and GT1b), and four (GQ1) residues of sialic acid.

Brain gangliosides are located in neurons (3). The subcellular fractions of brain having the highest gangli-

(*) The ganglioside nomenclature according to Svennerholm (1,2) has been adopted.

oside content are: a) the subfraction containing nerve
endings (4,5,6) and nerve ending membranes (7); b) the
microsomal fraction (4,8,9); c) the myelin subfraction
(5,10).

For details concerning the nomenclature, chemistry
and biology of gangliosides see the reviews by Svennerholm (1,2), Ledeen (11), Wiegandt (12,13), Mc Cluer (14)
and Gielen (15).

CHANGES OF THE TOTAL GANGLIOSIDE CONTENT IN BRAIN DURING
DEVELOPMENT

The early appearance of gangliosides in the various
stages of embryological growth has been examined in Ra-
na pipiens by Yiamouyiannis and Dain (16). It was established that gangliosides appear first at the late gastrula, a stage well before any morphological appearance
of the nervous system. Later on the ganglioside level
rises till hatching, and furthermore in the tadpole period of life, in parallel with the morphogenesis of the
nervous system.

The changes in the content of brain gangliosides occurring with age in chick embryo and in human and rat
foetuses are also known. In all cases a sharp increase
of ganglioside content (expressed as lipid bound sialic
acid) has been observed : 20-fold from the 5th day to
hatching in chick (17); 2-fold from the end of the 3rd
foetal month to the time of delivery in human (Table I)
(18); 3-fold from the 8th day before birth to birth in
rat (19). The increase of lipid bound sialic acid can be
assumed to represent an increase of the moles of gangliosides since, at least in chick embryo, where both parameters were determined, the rise with age of ganglioside
sialic acid and of ganglioside sphingosine are parallel
(17).

The increase of gangliosides in brain during the embryological and foetal stages of development is related
to two main phenomena: a) the increase in the number of

TABLE I

Content of lipid bound sialic acid (g/100 g dry weight)
in human foetuses, premature, newborn and adult subjects

foetuses			premature			
cm 8	cm 15	cm 23	24 weeks	35 weeks	newborn	adult
0.21	0.23	0.31	0.33	0.38	0.39	0.80

All data (a part from 'adult') from L. Svennerholm (18)

neurons, occurring in the same period of life; b) the
accumulation of the product inside the individual neurons
as demonstrated by Garrigan and Chargaff (17).

After birth a further increase of the ganglioside
content occurs in brain. In human brain a 2-fold increase from birth to adulthood has been reported by Suzuki
(20). In rat brain (figure 1)the maximum level is reached at the 18-20th day, the increase having a slow rate
for the first 6-8 days, followed by a more rapid rate
till the 12-14th day (19,20). According to most authors
(19,20,21,22,23) the adult level is 3 times higher that
at birth (a much higher rise -50-fold- has been reported by Rosenberg and Stern,24). The actual levels of gangliosides at the different ages, reported by the various
authors, are somewhat different, the discrepancies being
probably due to the different procedures used for the
extraction and quantitation of gangliosides.

The problem of the authenticity of the increase of
gangliosides (expressed as lipid bound sialic acid) has
been carefully examined by Spence and Wolfe (21), who
concluded that the increase of rat brain gangliosides
from birth to the 20th day of life corresponds to a true
accumulation (figure 2). This assumption has been confirmed by Kishimoto et al. (25) who quantitated gangliosides
by determining ganglioside stearate: the developmental

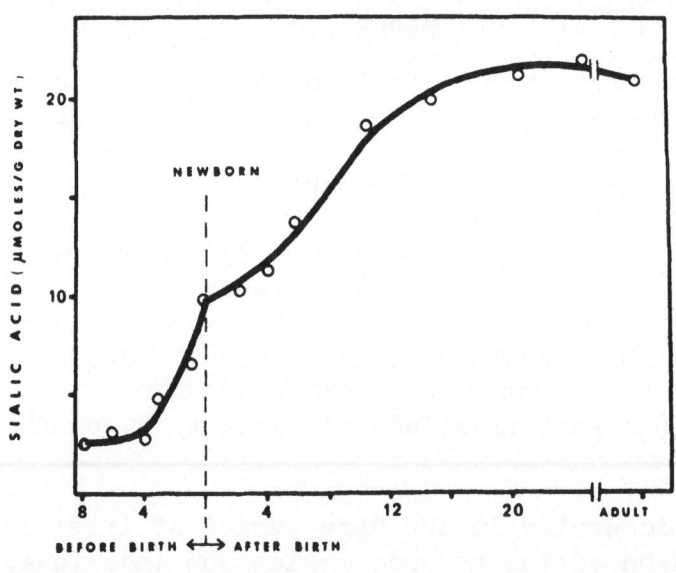

Figure 1. Changes in total gangliosides (expressed as
μmoles lipid bound sialic acid) of rat brain during development. From Tettamanti and Bonali (19).

profiles of ganglioside sialic acid and ganglioside stearate are quite parallel.

        Since gangliosides are assumed to be located entire-
ly in neurons and since the formation of new neurons sl-
ows down greatly at an early postnatal age, it may be
concluded that during the postnatal life neurons are able
to accumulate additional gangliosides for an appreciable
time. Attempts to establish a correlation between this
process of accumulation and the morphological changes of
the neurons during postnatal development have been made
by Rubiolo de Maccioni and Caputto (26). On the basis of
the experimental data provided by these authors the accumulation of gangliosides in rat brain (from birth to the
20-25th day) is contributed by the following processes:
a) the formation of myelin (which contains gangliosides);
b) the formation of new synaptic complexes (which are very rich in gangliosides); c) the formation of ganglioside
rich membranous structures, preparatively included in the
microsomal fraction (possibly dendrites, in rapid growth
and branching, 27).

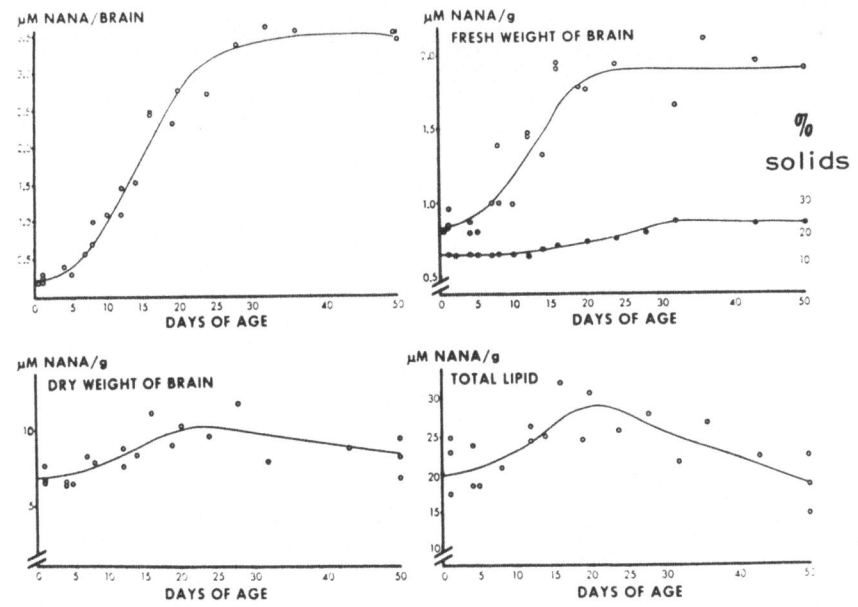

Figure 2. The accumulation of gangliosides, as μmoles lipid bound sialic acid (NANA) in rat brain during post-natal development. From Spence and Wolfe (21).

Some attention is also to be paid to the developmental changes of the lipid components of gangliosides. Rosenberg and Stern (24) showed that, in rat brain, ganglioside sphingosine changes from almost exclusively C 18 at birth to nearly equal quantities of C 18 and C 20 with organ maturation. The same authors provided evidence that C 18:0 fatty acid (the most abundant fatty acid in gangliosides) reaches the maximum level at the end of the 1st month of age, followed by a gradual diminution (about 30 %) till the adulthood. Conversely C 18:1 undergoes a 2-fold increase from birth to the 5th month (this level being later maintained), and C 16:0 rises rapidly up (100 % increase) from birth till 1 ½ month, while its level in the adult is approximately the same as at birth.

CHANGES IN THE PATTERN OF BRAIN GANGLIOSIDES DURING DEVELOPMENT

The pattern of gangliosides changes during development.

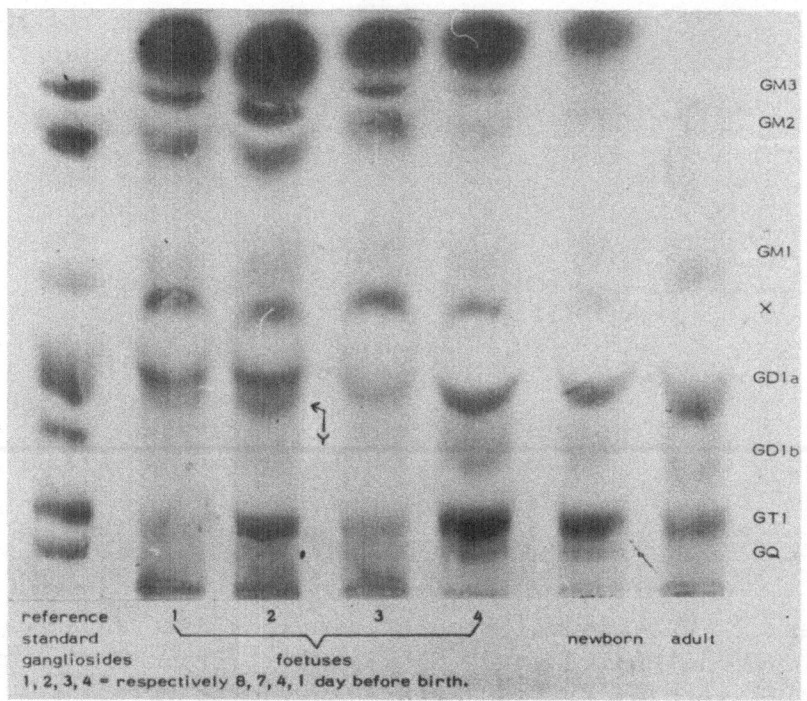

Figure 3. Brain ganglioside patterns of rat foetuses,new-
born and adult subjects. Precoated silica gel thin layer
plate (Merck); solvent: Chloroform/methanol/water = 60/
35/8 (v/v/v); 14 hour run at 18°C; location of the spots
with an anisaldehyde spray reagent. From Tettamanti and
Bonali (19).

     Recent investigations, carried out on rat foetuses,
(19) showed that at an early stage of brain morphogenesis
(8 days before birth) the following gangliosides are present in considerable amounts (figure 3): GM3, GM2, GM1,
GD1a, and two uncommon gangliosides (X and Y in figure 3)
not well characterized yet. GD1b is practically absent,
GT1b in very small amounts. During the further foetal development the ganglioside pattern changes: GD1b and GT1a
increase markedly; GQ1 appears and undergoes a considerable increase; GM3 and GM2 gradually diminuish reaching,
at birth, a very low level; the two uncharacterized gangliosides gradually decrease till an almost zero level.
The ganglioside pattern of the newborn rat is thus cha-

racterized by the presence of four major gangliosides
(GM1, GD1a, GD1b and GT1b),followed by GQ1 and traces of
GM3 and GM2.

After birth, in the period of the accumulation of
gangliosides, the most peculiar event is the drastic increase of GD1a (19,20) (figure 4): therefore its percentage on the total ganglioside content goes sharply up.
After the 20th day of age till adulthood GD1a decreases,
GM1 remains unchanged, GD1b and GT1b undergo a slow but
constant increase. The maximum level of GQ1 is reached
at the 15th day, followed by a decrease till the very
low level of the adult animal (19). In newborn human brain Svennerholm (28) found a ganglioside pattern characterized by a very large GD1a fraction, followed by GM1,
with small amounts of GD1b and GT1b. After birth, according to Suzuki (20), the % content of GD1a gradually di-
minuishes (with higher rate between 10 and 30 years),the
level of GM1 remains unaffected, while GD1b and GT1b increase markedly, just in parallel with the diminution of
GD1a. In foetal human brain the pattern is reported to be
closer to that of the adult (18). Berra and Galeone-Bertona (29) found similar developmental changes in the ganglioside pattern of pig brain.

Therefore the main features of the developmental changes of the ganglioside pattern in mammalian brain can tentatively be outlined as follows: a) GM3, GM2, GM1 and
GD1a appear at a very early stage. Among them GD1a undergoes a very rapid increase (the maximum level being reached around birth in human and at the 18-20th day after
birth in rat), followed by a drastic decrease. GM1 inver-
sely follows the behaviour of GD1a. GM3 and GM2 rapidly
decrease and reach very early (before birth) the final,
low, level; b) GD1b and GT1b, which probably occur at a
later stage, undergo a gradual increase till the adulthood; c) GQ1 has the latest appearance.

The changes with age of the ganglioside pattern at
the level of brain subcellular fractions have not been
sufficiently studied yet. Suzuki et al. (10) revealed the
occurrence of changes in the pattern of gangliosides con-

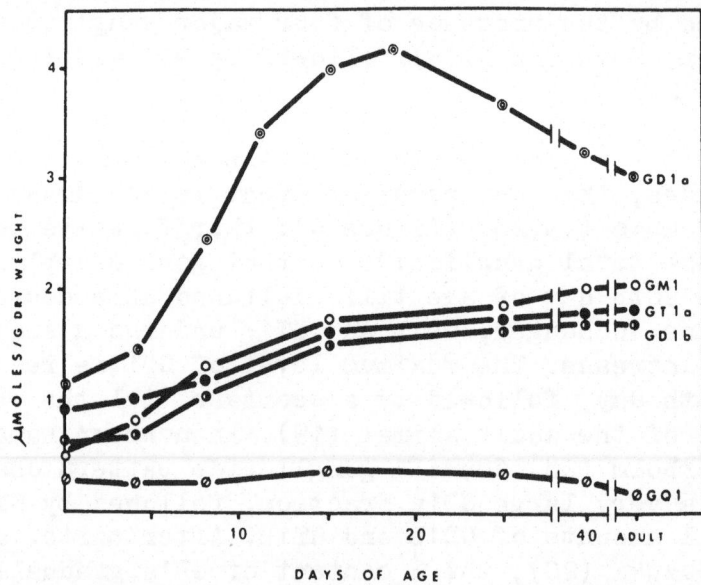

Figure 4. Developmental patterns of rat brain gangliosides. Each ganglioside is expressed as μmoles / g dry
weight. From Tettamanti and Bonali (19).

tained in purified rat brain myelin in relation to age.
The main feature is the rise with age of GM1, the content
of which shifts from the 56% of the total gangliosides at
the 15th day to the 90-93.5 % in the adult animal. Correspondingly the levels of GD1a, GD1b, and GT1b decrease.

METABOLISM OF GANGLIOSIDES DURING DEVELOPMENT

The absolute content of gangliosides and their pattern are necessarily the expression of precise metabolic
equilibria.Therefore it is conceivable that the developmental changes in the total content and in the pattern of
gangliosides reflects developmental changes in the enzyme activities involved in ganglioside synthesis and de-
gradation.

The biosynthesis of the different gangliosides takes
place schematically along three routes. The first route,
elucidated by the Roseman's (30,31,32,33,34), and Dains's
(35,36) groups, takes origin from lactosylceramide to

which a residue of sialic acid, N-acetyl-galactosamine, galactose and sialic acid are added by a stepwise process, the following gangliosides being sequentially formed: GM3, GM2,GM1,GD1a. The addition of each individual glycosyl residue is catalized by a specific glycosyltransferase. The second route, suggested but only partly confirmed by Kaufman et al. (34) and Arce et al. (37), originates from GM3 to which an additional residue of sialic acid is attached: GD3 is formed, in which the two molecules of sialic acid are linked by $\alpha(2-8)$ glycosydic linkage. N-acetyl-galactosamine, galactose and sialic acid are then added by the intervention of the correspondent glycosyltransferase, GD2,GD1b and GT1b being formed. In the third route, demonstrated by Handa and Burton (38) and Yip and Dain (39) the entire neutral carbohydrate chain of the G 1 series is formed, then a residue of sialic acid is attached, GM1 being produced.

Gangliosides are degraded by a complex and not completely known mechanism, the main role of which is played by neuraminidase. Brain neuraminidase, as showed by Leibovitz and Gatt (40), and Tettamanti and Zambotti (41), splitts gradually off sialic acid from oligosialogangliosides, monosialoganglioside GM1 being obtained as a provisional end product. After the enzymatic removal of galactose (42) and N-acetyl-galactosamine (43) GM3 is formed, from which neuraminidase is able to remove the last residue of sialic acid (41). Lactosylceramide is the final product. For details see the Svennerholm's review(44).

CHANGES OF THE TURNOVER RATE OF GANGLIOSIDE DURING DEVELOPMENT

Moser and Karnovsky (45) provided evidence that, after injection of D-glucose-6-C 14, the specific radioactivity of mouse brain gangliosides reaches the maximum level around the 10th day of postnatal age and then decreases till the adult level (roughly the 60%). A similar behaviour in the incorporation of radioactivity into gangliosides has been shown in rat brain either in the case in which the radioactivity was counted in ganglioside stearate (after administration of acetate-1-C 14) (46) or

in the case in which the total radioactivity of gangliosides was measured (D-glucose-UC 14 was used) (47).

Attempts have been made by Suzuki(48)to correlate the
developmental changes in the ganglioside pattern with underlying metabolic reactions. The experiments,carried out
lying metabolic reactions. The experiments, carried out
on rat brain, showed that the relative rates of synthesis
of the major gangliosides (GM1,GD1a,GD1b,GT1b) and the
distribution among them of the radioactivity, once incorporated at 2 days of age, change with age in parallel with
the developmental changes of the analytical pattern. The
same parallelism was shown to be valid also in the case
of tetrasialoganglioside GQ1 (49). These results demonstra
ted the limitation of the in vivo approach to the problem,
in other words, the inability of the in vivo experiments
to individuate among developmental changes of the biosynthetic, the interconversion, and the preferential degra-
dation rates.

DEVELOPMENTAL PATTERNS OF THE ENZYMES INVOLVED IN GAN-
GLIOSIDE METABOLISM

Yiamouyiannis and Dain (16) showed, in the course of
their studies on frog embryos, that from late gastrula
throught hatching the only detectable gangliosides are
GM1 and GD1a. During the tadpole period of life a further
couple of gangliosides appears, GD1b and GT1b. In the a-
dult animal, as reported by Tettamanti et al. (50), tetrasialoganglioside GQ1 not only is present but takes up
the 54 % of the total ganglioside content. Reminding the
three biosynthetic pathways outlined before, the following
interpretation of the above data seems possible : at the
earliest developmental stage (embryo) only the enzymes
involved in the first (and /or third) biosynthetic route
are maturating, the expression of this being the accumulation of the end products, GM1 and GD1a. At a later developmental stage (tadpole) the enzymes of the second route appear, thus GD1b and GT1b also start accumulating.At
the level of adult animal a new line of enzymes becomes
operating , bringing to GQ1. The qualitative changes in
the ganglioside pattern occurring in developing rat foetuses may also be consistent with the hypothesis that the

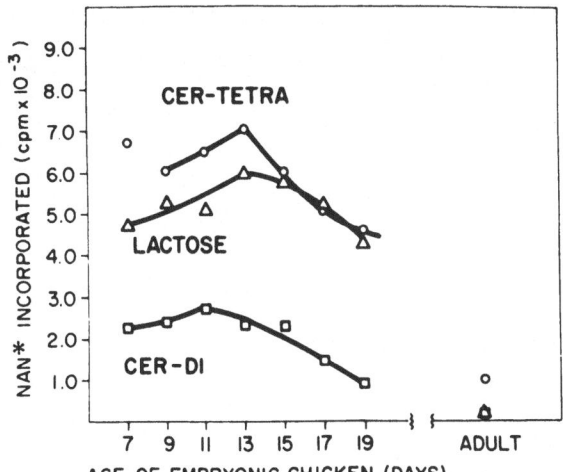

Figure 5. Sialyltransferases activities at various stages
of embryonic development and in the adult animal (chicken
brain).The values are plotted as c.p.m./mg protein/2 hours
NAN = N-acetylneuraminic acid; CER-TETRA,CER-DI,LACTOSE
= respectively tetraglycosyl-ceramide,diglycosyl-cerami-
de and lactose as acceptors. From Kaufman (32).

maturation times of the enzymes involved in the biosynthesis of the variuos ganglioside lines are different.

     Kaufman et al. (32) observed (figure 5) that three
different sialyltransferases of chick embryo brain reach
mawimum level of activity around the 11-13th day of embryonic life, while the levels of the same enzymes in the a-
dult animal are much lower (from 1/5 to 1/10). Two galac-
tosyltransferases from rat brain were found by Hildebrand
et al. (51) to perform a similar behaviour: high levels
of activity at birth, followed by a decrease till an almost zero level in the adult. A different developmental
profile for UDP-galactose: GM2 galactosyltransferase has
been reported by Yip and Dain (36). The enzymatic activity rises sharply up from the embryonic life (5 days before birth) to birth, reaches the maximum level between the
13th and the 17th day of postnatal life, then rapidly decreases to the adult level, which approximates that of the
1-day animal. This developmental pattern of the enzyme,
which correlates with the concentration of gangliosides
at the correspondent ages, may reflect the potentiality
of the tissue to synthesize gangliosides.

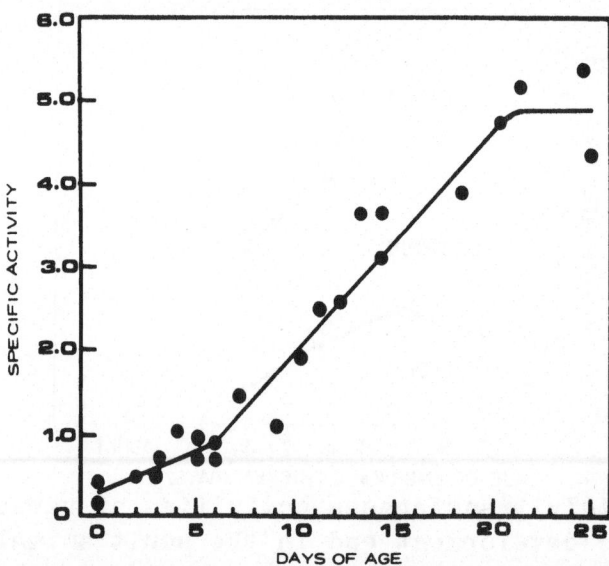

Figure 6. Changes of soluble neuraminidase from rat brain
during development. Specific activities are expressed as
μmoles sialic acid released from neuraminlactose/ mg protein / 3 hours. From Carubelli (53).

        Glycohydrolases involved in ganglioside catabolism
appear to have much higher levels in the adult animal than
glycosyltransferases. Lactosylceramide galactosidase, according to Radin et al. (52) undergoes a rapid increase in
activity from birth to the 20-22th day of age, then decreases slowly to the adult level which remains, however,con-
siderably high. The developmental profile of rat brain soluble neuraminidase, reported by Carubelli (53) and confirmed by Quarles and Brady (54), is very similar (figure
6) . The sharp and relevant increase (8-fold) of neuraminidase activity after birth, till reaching the maximum
plateau around the 20th day of life, may be correleted to
the dramatic inversion of the GD1a / GM1 ratio occurring
in rat brain after the 20th day, that is, after the establishment of a high and constant neuraminidase activity.
These data, if confirmed on particle-bound neuraminidase
(which accounts for more than 90 % of total neuraminidase
activity, 41), could explain, at a molecular level, the
correlation between the developmental pattern of gangliosides, al least in a limited period of life,and a precise
enzyme entity.

The future experiments will surely be able to explain in terms of individual enzyme involvement all the problems connected with the changes with age of the gangliosi- de pattern.

REFERENCES

(1) Svennerholm,L., J. Lipid Res. 5:145 (1964).
(2) Svennerholm,L., in Handbook of Neurochemistry,vol.3 (Ed. A. Lajtha) p. 425 (1970).
(3) Derry D.M., Wolfe, L.S. Science 158:1450 (1967).
(4) Wolfe,L.S. Biochem.J. 79:348 (1961).
(5) Eichberg,J.,Whittaker, V.P.,and Dawson,R.M.C. Biochem- J. 92:91 (1964).
(6) Seminario L.M.,Hren,N.,and Gomez,C.J. J. Neurochem. 11:197,(1964).
(7) Lapetina, E.G.,Soto,E.F.,and de Robertis,E. Biochim. Biophys. Acta 135:33 (1967).
(8) Wherret,J.R.,and Mc Ilwain,H. Biochem.J. 84:232 (1962)
(9) Wiegandt,H. J. Neurochem. 14: 671 (1967).
(10) Suzuki,K., Poduslo,S.E.,and Norton, W.T. Biochim. Biophys. Acta 144:375 (1967).
(11) Ledeen,R. J. Am. Oil Chem. Soc. 43:57 (1966).
(12) Wiegandt,H. Ergeb.Physiol.biol.Chem.exptl.Pharmakol. 57:190 (1966).
(13) Mc Cluer, R. in Proceedings of the 4th International Conference on Cystic fibrosis of the pancreas, Part II (Eds. E. Rossi and E. Stoll) p.121 (1968).
(14) Wiegandt, H. Angewandte Chemie 7:87 (1968).
(15) Gielen, W. Z. Clin. Chemie, 5:97 (1967).
(16) Yiamuoyiannis, J.A.,and Dain,J.A. J. Neurochem. 15: 673 (1968).
(17) Garrigan,O.W.,and Chargaff,E. Biochim.Biophys. Acta 70:452 (1963).
(18) Svennerholm,L. J. Neurochem. 11:839 (1964).
(19) Tettamanti ,G.,and Bonali, F. unpublished results.
(20) Suzuki,K. J. Neurochem. 12:969 (1965).
(21) Spence,M.W.,and Wolfe,L.S. Canad.J.Biochem.45:671 (1967).
(22) Pritchard,E.T.,and Cantin,P.L. Nature 193:580 (1962).
(23) James,F.,and Fotherby,K. J. Neurochem. 10:587 (1963).
(24) Rosenberg,A.,and Stern,N. J.Lipid Res. 7:122 (1966).

(25) Kishimoto,Y.,Davies,W.E.,and Radin,N.S. J. Lipid
     Res. 6:532 (1965).
(26) Rubiolo de Maccioni,A.H.,and Caputto,R. J. Neurochem.
     15:1257 (1968).
(27) Hicks,S.P.,Cavanaugh,M.C.,and O'Brien,E.D. Amer.J.
     Path. 40:615 (1962).
(28) Svennerholm,L. J. Neurochem. 10:613 (1963).
(29) Berra,B.,and Galeone-Bertona,L. Biochim.Biol.Sper.
     5:466 (1966).
(30) Basu,S.,Kaufman,B.,and Roseman,S. J. Biol. Chem.
     240:PC 4115 (1965).
(31) Basu,S. Ph.D. Thesis,The University of Michigan
     (1966).
(32) Kaufman,B.,Basu,S.,and Roseman,S. in Inborn Disord-
     ers of Sphingolipid Metabolism (Eds. S.M. Aronson
     and B.W. Volk)p.193 (1967).
(33) Roseman,S. in Proceedings of the 4th International
     Conference on Cystic Fibrosis of the Pancreas,Part
     II (Eds. E. Rossi and E. Stoll)p. 244 (1968).
(34) Kaufman,B.,Basu,S.,and Roseman,S. J. Biol. Chem.
     243:5804 (1968).
(35) Yiamuoyiannis,J.A.,and Dain,J.A. Lipids 3:378 (1968).
(36) Yip,G.B.,and Dain,J.A. Biochim.Biophys.Acta 206:252
     (1970).
(37) Arce,A.,Maccioni,H.F. and Caputto,R.Arch.Biochem.
     Biophys. 116:52 (1966).
(38) Handa,S.,and Burton,R.M. in Proceedings 1st Interna-
     tional Congress of Neurochemistry,Strasbourg (1966).
(39) Yip,M.C.M.,and Dain,J.A. Lipids 4:270 (1969).
(40) Leibovitz,Z.,and Gatt,S. Biochim.Biochim.Acta 152:
     136 (1968).
(41) Tettamanti,G.,and Zambotti,V. Enzymol.35:61 (1968).
(42) Gatt,S. Biochim.Biophys.Acta 137:192 (1967)
(43) Frohwein,Y.Z.,and Gatt,S. Biochemistry 6:2783 (1967).
(44) Svennerholm,L. in Comprehensive Biochemistry,vol.18
     (Eds. M. Florkin and E. Stotz)p.201 (1970).
(45) Moser,H.W.,Karnovsky,M. J. Biol.Chem. 234:1990 (1959).
(46) Kishimoto,Y.,Davies,W.E.,and Radin,N.S. J.Lipid Res.
     6:525 (1965).
(47) Maker,H.S.,and Hauser,G. J. Neurochem. 14:457 (1967).
(48) Suzuki,K. J. Neurochem. 14:917 (1967).

(49) Bonali,F.,Tettamanti,G.,and Cervato,G. Report XIIIth
     International Conference on the Biochemistry of Li-
     pids, Athens (1969).
(50) Tettamanti,G.,Bertona,L.,Gualandi,W.,and Zambotti,
     V. Rendiconti Ist.Lomb.Sc.Lett.(B) 99:173 (1965).
(51) Hildebrand,J., Stoffyn,P., and Hauser,G. J. Neuro-
     chem. 17:403 (1970).
(52) Radin,N.S.,Hof,L.,Bradley,R.M.,and Brady,R.O.
     Brain Research 14:497 (1969).
(53) Carubelli,R., Nature 219:955 (1968).
(54) Quarles,R.H.,and Brady,R.O. J. Neurochem. 17:801
     (1970).

LIPIDS OF THE NERVOUS SYSTEM: CHANGES WITH AGE, SPECIES VARIATIONS, LIPID CLASS RELATIONSHIPS, AND COMPARISON OF BRAIN TO OTHER ORGANS

George Rouser, Akira Yamamoto[*], and Gene Kritchevsky

City of Hope National Medical Center

Duarte, California 91010

ABSTRACT

Changes with age of the lipid composition of human and animal brains were investigated. Individuals of the same species were found to differ and fall into one of four groups differing in lipid content. Species differences among animals were small and fungal mycelia were similar to animals. The total amount of lipid in brain and other organs increases during differentiation (i.e. with advancing age) and then, as shown for humans, declines (after 26 years of age). The changes were defined by two forms of graphic analysis. In one, the total amount of lipid, the amount of each lipid class, or sums of various classes were plotted as a function of age. In the other, the amount of one lipid and the sums of various lipids were plotted against total lipid or sums of various lipid classes. The latter plots were used to define the interrelationships of the lipid classes and to provide information necessary to fit lines through points on lipid vs. age plots. During development, some brain lipid classes make up progressively less of the total, whereas the relative amount of other classes increases. Thus, some lipid classes substitute for others. Two general substitution groups were recognized by graphic analysis. In substitution group I, the changes with age consist of an increase of cerebroside, sulfatide, and sphingomyelin with a decrease of phosphatidyl choline and phosphatidyl ethanolamine. In group II, a decline of phosphatidyl inositol and other acidic lipids is balanced by a rise of phosphatidyl serine. Equations defining the relationship of each lipid class to total lipid of brain and other organs are presented. Equations defining the changes in the total amounts of each lipid class of human brain as a function of age are also presented.

[*] Present address: Osaka University Medical School, Osaka Japan

I. INTRODUCTION

It has become apparent from recent data that the lipid composition of the nervous system changes throughout life (1,2). These
changes are disclosed only by analysis of representative samples of
whole brain rather than samples of grey and white matter used in
most earlier studies. Brain lipid changes are not unique. Changes
throughout life have also been found for lipids of human aorta (3)
and lens (4) as well as cerebral blood flow, cerebral vascular
resistance, and cerebral oxygen consumption (5).

Our objective has been to measure accurately and precisely
all lipid classes of brain and other major organs at different ages
in different species in order to define precisely the lipid class
interrelationships as a function of age in different species. Determination of all components (lipid classes) of the biological
system assures that data analysis will disclose all interrelationships among the lipid classes. Lipid metabolism is a complex series
of biosynthetic and degradative steps with many control mechanisms
not yet elucidated. The levels of enzymes may differ in different
organs and/or species, but, through balancing of enzyme levels, the
same end result (lipid composition) may be achieved. In addition,
since the rate of polar lipid synthesis depends to some extent upon
availability of membrane binding sites, enzyme levels may not match
lipid class levels. Thus, analysis of all lipid classes is a con-
venient means for measuring the overall balance of factors and
facilitates comparison of different organs and species.

In this report, human brain lipid composition changes with
age are considered in detail. A novel method of graphic analysis
is presented as is its use to define precisely lipid class relationships in brain and other organs of man, several animal species, and
some fungi.

II. METHODS OF LIPID ANALYSIS AND GRAPHIC
ANALYSIS OF LIPID COMPOSITION DATA

Lipid analysis was accomplished as described previously (1,6).
Two types of graphic analysis were used: plots of lipid vs. age and
plots of lipid classes and sums of different lipid classes vs. total
lipid, total polar lipid, or total phospholipid. Large graph paper
was used and scales selected for accuracy with four significant
figures. Arithmetic, semilogarithmic, and logarithmic plots were
compared and arithmetic plots were selected as generally most useful.
Lines were fitted to points as described under 'Results'. Comparison
of plots made from values expressed as molar percentage of the total
lipid, mM/100 gm dry weight, and mM/100 gm fresh weight disclosed
that the latter were most informative, although similar results were
obtained on all plots.

III. RESULTS AND DISCUSSION

A. Relationships of the Amount of Each Lipid Class to Total Lipid Content of Brain and Other Organs

Several important features of brain lipid composition are disclosed by plots of each lipid class against total lipid (total phospholipid + total glycolipid + cholesterol), total polar lipid (total lipid - cholesterol) and total phospholipid. First, the plots disclose that humans differ in the amounts of the lipid classes present. Second, they can be used to determine the age at which the phase of decline of brain lipid begins. Third, brain lipids of different species can be compared even though each may be on a different developmental time scale. Fourth, brain can be compared with other organs.

Figure 1 shows the results obtained by plotting values (mM/100 gm fresh weight) for total phospholipid against total lipid (% fresh weight) of representative samples of human whole brain. A fairly broad general range is clearly defined. Total phospholipid clearly rises at a uniform rate and then the rate of increase changes so that phospholipid represents less of the total lipid in the second phase that begins at the time of onset of active myelination. When total glycolipid is plotted against total polar lipid, the two different phases seen on the total phospholipid plot are also observed, but the rates of change are exactly the opposite. Thus, in the first phase, the glycolipid increases less rapidly than total phospholipid and, in the second phase, total glycolipid increases more rapidly than total phospholipid. It is thus clear that glycolipid replaces phospholipid during myelination. When cholesterol values are plotted against total polar lipid, no change of slope is found and hence there is no rate change. Thus, cholesterol does not appear to replace polar lipids in cell membranes.

The question arises as to whether or not the broad range for total phospholipid, glycolipid, and cholesterol indicates reproducible differences among individuals. That different persons differ in a reproducible manner is indicated by the fact that samples differing in total lipid content from different regions of the same brain give values that fall on only one of the four lines shown in Figure 1. Thus, there do appear to be reproducible individual differences. The finding that values for different regions of one brain form a single straight line which is also formed by values for whole brains from different individuals was used to define the proper fitting of lines through points on all plots.

When values for individual lipid classes were plotted against total polar lipid, the same two general relationships noted for total phospholipid and glycolipid were observed. In the second

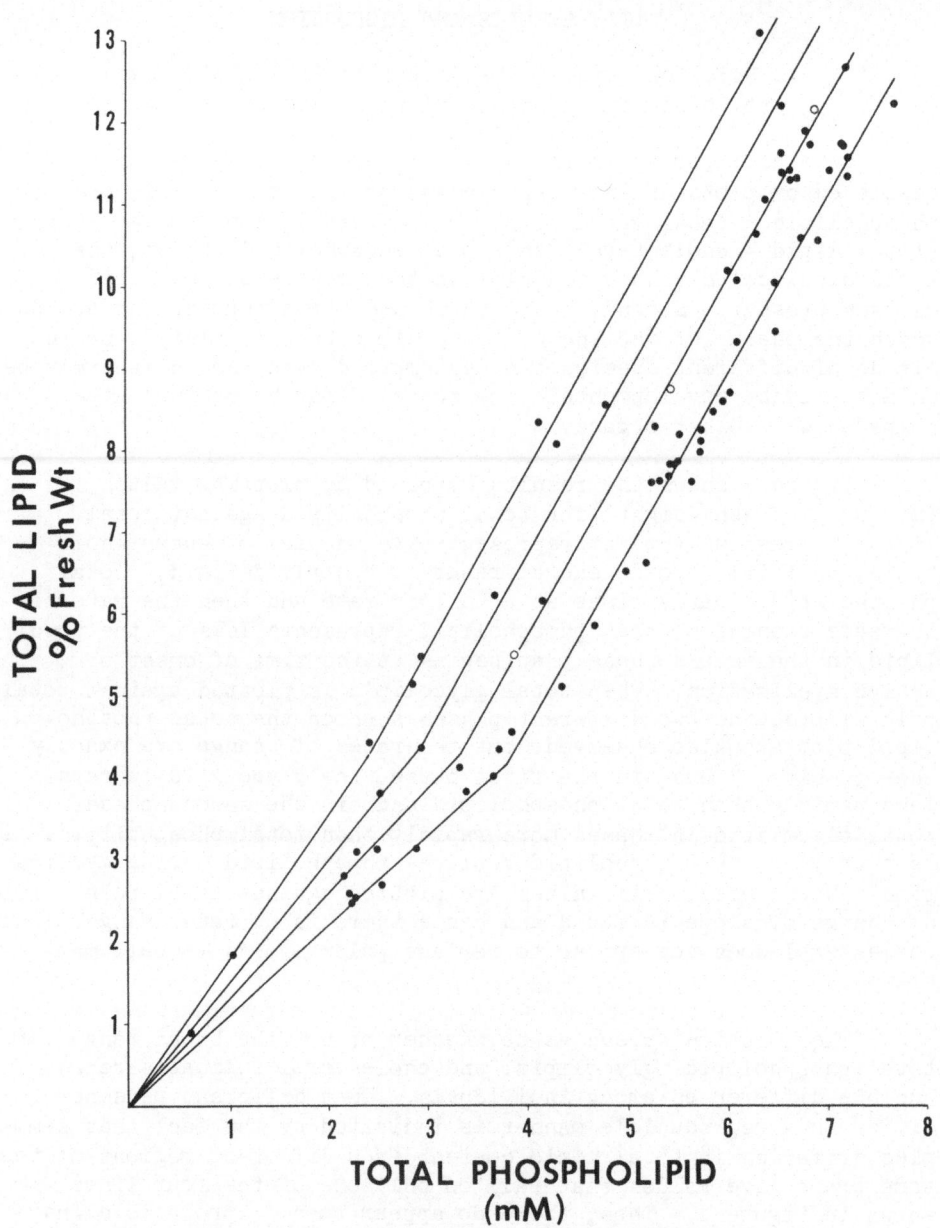

Figure 1

Human brain total phospholipid plotted against total
lipid (% fresh weight). Solid circles are values for
representative samples of whole brain and open circles are values for small pieces taken from different
parts of the same brain. See text for comments.

phase of development (after onset of active myelination), the rate
of increase of phosphatidyl choline, phosphatidyl ethanolamine,
phosphatidyl inositol, and minor acidic phospholipids relative to
total polar lipid was less than in the first phase and the rate for
cerebroside, sulfatide, sphingomyelin, and phosphatidyl serine was
greater. The two types of relationships are illustrated in Figures
2 and 3. Equations for the relationships (Table I) show that there
are four different groups for each lipid class of human brain.

Brains of some animal species differ from normal human brain
in three ways. First, values for phosphatidyl choline, phosphatidyl
ethanolamine, and sphingomyelin may fall on period 1 lines characteristic for other organs. It is of interest that brain values
from some human pathological states also fall on period 1 lines
characteristic for other organs. Second, lines for some lipid
classes of some species do not change slope (i.e. go from period 1
to 2) like those of human brain. This is the case for mouse brain
cerebroside, sulfatide, and sphingomyelin. Third, period 2 lines
may have a different slope. This is true for mouse brain phosphatidyl choline and phosphatidyl ethanolamine (but not for phosphatidyl
serine, phosphatidyl inositol, and diphosphatidyl glycerol).

Species differences among mammals are very small for most
other organs. In general, species differences among vertebrates
and invertebrates are also relatively small and the differences observed do not appear to be specific for any one species. Differences
that can be misinterpreted as species variations arise when animals
at different stages of development are compared and when the large
individual differences among animals of the same species are not
appreciated. Brains of some invertebrate species do not contain
cerebroside or sulfatide. Since the total lipid levels are much
lower in these brains, the differences may perhaps be traced in the
future, at least in part, to the stage of development. It is clear
that graphic analysis is useful in determining which species are
most like humans. We have found bovine brain lipid class relationships to be like those of human brain.

Despite the fact that species may differ in the extent to
which particular lipids are used in brain, the amounts of the
different substitution groups (defined below) do not differ for
different species. Thus, determination of the relative rates of
increase of the substitution groups with age can be used to define
the relative rates of development of species. This can be of value
in experimental pathology because species regulating lipid composition like humans can be selected and differences in time scales
defined so that animals at the proper stage of development for any
desired age in the human can be chosen. Using this criterion, our
data indicate various strains of laboratory mice to be on a time
scale 9.3 times and a strain of Beagle dogs 3.9 times as rapid
as that of the human, whereas the scale for bovine brain

appears to be the same as that of the human.

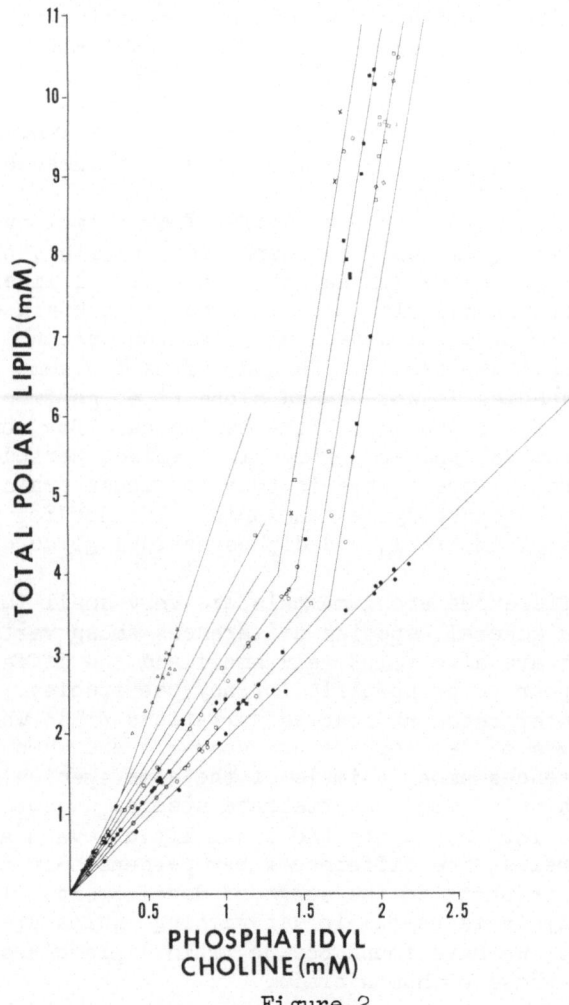

Figure 2

Plot of values for phosphatidyl choline (plus lysophosphatidyl choline) against total polar lipid.
Whole brain values are shown as ■ and □ to differentiate points on different lines; values from
different regions of the same brain as x; values
for Drosophila melanogaster phosphatidyl choline as
Δ (from unpublished data of Lees, Rouser, Kaplan and
McCaman); other organs (heart, lung, kidney, liver,
spleen, skeletal muscle, aorta), cells (blood lymphocytes, erythrocytes, endothelial cells) and fungal
mycelia as ● and o. See text for comments.

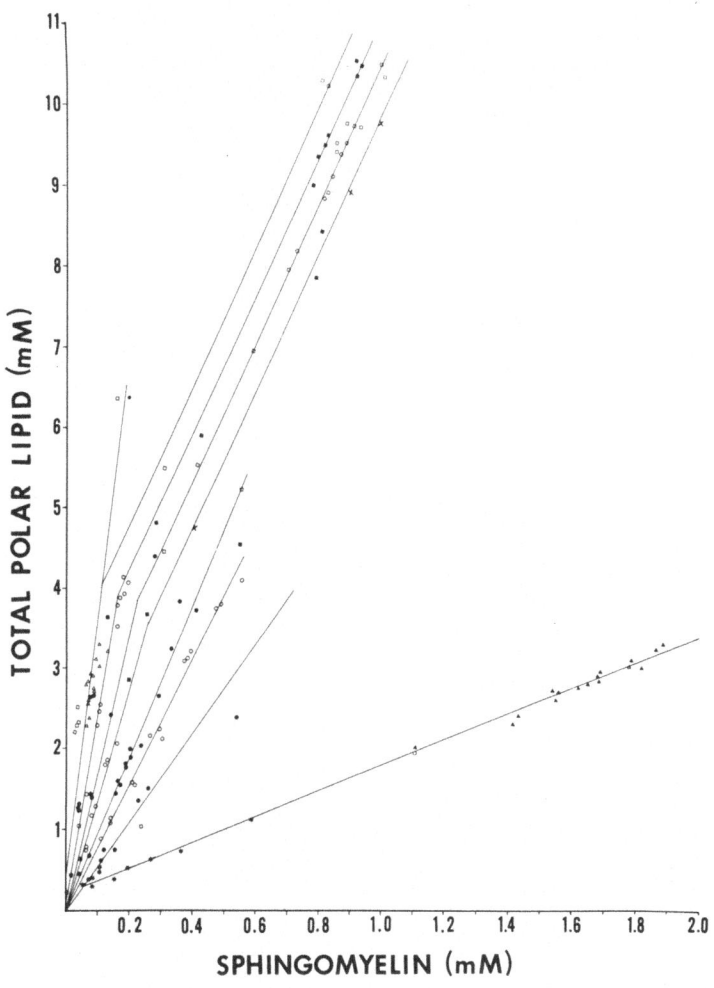

Figure 3

Plot of sphingomyelin values against total polar lipid.
Point designation as for Figure 2 except that Δ are
values for Drosophila ceramide phosphorylethanolamine
which replaces sphingomyelin (unpublished data of Lees,
Rouser, Kaplan, and McCaman), ▲ are Drosophila phosphatidyl ethanolamine values, and * designates human aorta
sphingomyelin values. Note that values for sphingomyelin
of aorta and Niemann-Pick disease spleen (o) and the
values for Drosophila phosphatidyl ethanolamine fall on
the same exceptional line. The fruit fly uses phosphatidyl ethanolamine to replace other phospholipids that
are replaced in human aorta and Niemann-Pick disease
spleen by sphingomyelin. See text for additional comments.

G. ROUSER, A. YAMAMOTO, AND G. KRITCHEVSKY

TABLE I

Equations for Calculation of Lipid Class[a]
Relationships to Total Polar Lipid

Group	Footnote	Period 1	Up To	Period 2
		Phosphatidyl Choline[b]		
1A		0.268(TPL)	--	--
B	c	0.268(TPL)	0.26	0.212(TPL) + 0.010
2		0.308(TPL)	--	--
3	d	0.342(TPL)	4.00	0.0801(TPL) + 1.029
4	d	0.368(TPL)	3.90	0.0817(TPL) + 1.142
5	d	0.412(TPL)	3.80	0.0846(TPL) + 1.246
6	d	0.465(TPL)	3.60	0.0833(TPL) + 1.382
7		0.526(TPL)	--	--
		Phosphatidyl Ethanolamine[e]		
1A		0.168(TPL)	--	--
B	f	0.168(TPL)	0.030	0.070(TPL) + 0.030
2		0.200(TPL)	--	--
3		0.228(TPL)	--	--
4	d	0.246(TPL)	4.00	0.227(TPL) + 0.124
5	d	0.282(TPL)	3.90	0.227(TPL) + 0.252
6	d	0.318(TPL)	3.80	0.227(TPL) + 0.354
7A	d	0.353(TPL)	3.60	0.227(TPL) + 0.473
B	g	0.353(TPL)	0.108	0.625(TPL) − 0.100
		Sphingomyelin		
1	d,h	0.0292(TPL) − 0.012	4.00	0.116(TPL) − 0.336
2	d,h	0.0414(TPL) − 0.004	3.90	0.116(TPL) − 0.280
3	d	0.059(TPL)	3.80	0.116(TPL) − 0.222
4	d	0.072(TPL)	3.60	0.116(TPL) − 0.159
5		0.106(TPL)	--	--
6		0.127(TPL)	--	--
7A		0.180(TPL)	--	--
B	i	0.180(TPL)	0.26	0.625(TPL) − 0.100
		Phosphatidyl Serine		
1	j	0.015(TPL)	--	--
2	j	0.027(TPL)	--	--
3	j	0.040(TPL)	--	--
4	k	0.062(TPL)	4.00	0.129(TPL) − 0.269
5	k	0.077(TPL)	3.90	0.129(TPL) − 0.197
6	k	0.098(TPL)	3.80	0.129(TPL) − 0.137
7	k	0.112(TPL)	3.60	0.129(TPL) − 0.620

TABLE I cont'd

Group	Footnote	Period 1	Up To	Period 2

Phosphatidyl Inositol

Group	Footnote	Period 1	Up To	Period 2
1	l	0.014(TPL) - 0.005	4.00	0.010(TPL) + 0.008
2	l	0.023(TPL) - 0.005	3.90	0.010(TPL) + 0.046
3	l	0.030(TPL)	3.80	0.010(TPL) + 0.084
4	l	0.041(TPL)	3.60	0.010(TPL) + 0.117
5	m	0.058(TPL)	--	--
6	m	0.084(TPL)	--	--
7	n	0.250(TPL)	--	--

Diphosphatidyl Glycerol

Group	Footnote	Period 1	Up To	Period 2
1	o	0.007(TPL)	3.80	0.003(TPL) + 0.016
2	o	0.018(TPL)	3.60	0.003(TPL) + 0.049
3	p	0.035(TPL)	--	--
4	p	0.048(TPL)	--	--
5	p	0.066(TPL)	--	--
6	p	0.093(TPL)	--	--
7	p	0.127(TPL)	--	--

Cerebroside[q]

Group	Period 1	Up To	Period 2
1	0.0253(TPL) - 0.063	4.00	0.355(TPL) - 1.442
2	0.0964(TPL) - 0.175	3.90	0.355(TPL) - 1.259
3	0.127(TPL) - 0.128	3.80	0.355(TPL) - 1.056
4	0.170(TPL) - 0.102	3.60	0.355(TPL) - 0.939

Sulfatide[q]

Group	Period 1	Up To	Period 2
1	0.017(TPL) - 0.042	4.00	0.095(TPL) - 0.429
2	0.025(TPL) - 0.046	3.90	0.095(TPL) - 0.377
3	0.035(TPL) - 0.035	3.80	0.095(TPL) - 0.304
4	0.044(TPL) - 0.026	3.60	0.095(TPL) - 0.235

Substitution Group I[r]	Substitution Group II[r]	Cholesterol Group[s]
Subgroup	Subgroup	
1 0.695(TPL)	1 0.110(TPL)	1 0.570(TPL)
2 0.782(TPL)	2 0.152(TPL)	2 0.635(TPL)
3 0.812(TPL)	3 0.195(TPL)	3 0.692(TPL)
4 0.842(TPL)	4 0.228(TPL)	4 0.750(TPL)
5 0.872(TPL)	5 0.280(TPL)	

Legend for Table I

(a) Values as mM/100 gm fresh weight of tissue. TPL = total polar lipid (excludes cholesterol and other less polar lipids). Drosophila melanogaster data from unpublished results of Lees, Rouser, Kaplan, and McCaman.

(b) Includes lysophosphatidyl choline.

(c) Valid for Niemann-Pick disease spleen and Drosophila melanogaster.

(d) Period 1 valid for many organs including human brain, but period 2 valid for human brain only.

(e) Includes lysophosphatidyl ethanolamine.

(f) For human aorta and Niemann-Pick disease spleen.

(g) Valid for Drosophila only.

(h) Valid for Drosophila melanogaster ceramide phosphorylethanolamine which replaces sphingomyelin completely.

(i) For human aorta only; note that the equation is the same as 7B period 2 for phosphatidyl ethanolamine of Drosophila melanogaster.

(j) Valid for liver, some human aortas, some Drosophila melanogaster mutants, and bovine skeletal muscle.

(k) Period 1 valid for human and other animal brains as well as lung, kidney, spleen, fungi, some aortas, and some Drosophila mutants; period 2 valid for human brain only.

(l) Period 1 valid for human and animal brains as well as lung, Drosophila, aorta, heart, and skeletal muscle of some species; period 2 valid for human brain only.

(m) Valid for liver, kidney, spleen, and skeletal muscle of some species.

(n) For some fungal mycelia only.

(o) Period 1 valid for human and animal brains as well as lung, spleen, aorta, and skeletal muscle of some species; period 2 valid for human brain only.

(p) Valid for liver, kidney, heart, and skeletal muscle of some species.

(q) Period 1 valid for mouse brain at all lipid levels and for human brain up to values shown; period 2 valid for human brain only.

(r) Valid at all lipid levels for all animal species studied thus far, fungal mycelia, and at least some bacteria that have phosphatidyl ethanolamine.

(s) For human brain at all lipid levels.

B. Significance of the Dependence of Lipid Class Levels Upon the Total Lipid Level and Limited Species Variability

Our data for human brain lipid composition changes with age show clearly that the amount of each lipid class is directly related in a constant manner to the total amount of lipid in brain at all ages. Thus, it appears that all lipids of the nervous system, including those of myelin, are in dynamic equilibrium. Data for fatty acid composition changes of whole brain and myelin of rats on essential fatty acid deficient diets (presented elsewhere in this volume) also show a dynamic equilibrium for all lipids of brain and purified myelin. Large changes in fatty acid composition were induced on diets deficient in essential fatty acids and the changes were reversed when the animals were subsequently placed on a normal diet. Recent studies with radioactive phosphorus are in agreement with these findings (10). Data obtained with different approaches indicate that, although myelin is not inert and myelin lipid can exchange with lipids of other membranes in the nervous system, some lipid classes are degraded slowly and parts of molecules are reutilized. Thus, the nervous system does not appear to be as unique as previously supposed with regard to lipid exchange and turnover.

The proof that brain developmental changes are similar for all animal species is of special significance. Lipid vs. lipid plots show vertebrate and invertebrate species to follow similar courses of development. Thus, such plots can be used to find species exactly like humans and substitution group vs. age plots can be used to determine the developmental time scales for animal species as noted above. The great similarity of various species for lipid composition of heart, skeletal muscle, kidney, liver, spleen, and lung was noted previously (7-9).

C. Lipid Class Substitution Groups

As the amount of human brain lipid increases, some lipid classes represent progressively more and some less of the total. Thus, some lipid classes substitute for others in membranes. Substitution rather than simple addition to previously unused sites is shown by the fact that the lipid/protein ratio of myelin does not increase during development (11).

Which lipid classes substitute for others can be determined by plotting sums of lipid classes against total polar lipid. When the values for sums of the lipids that increase at higher lipid levels are added to the values for those that decrease, the proper

sums that demonstrate the substitution groups must fulfill the following criteria: 1) the lines must pass through the origin;
2) straight lines must be obtained that do not change slope at any
lipid level; 3) the organ specificity for particular lines found for
lipid class plots of acidic phospholipids must not be observed;
and 4) the substitution groups for subcellular particulates must
be the same as those of whole organs. Plots of human brain lipid
data showed clearly that sphingomyelin and cerebroside substitute
for phosphatidyl choline, sulfatide substitutes for phosphatidyl
ethanolamine, and phosphatidyl serine substitutes for other acidic
phospholipids and gangliosides. Only the sums of these three
groups gave straight lines without change of slope at any total
lipid level. Our data for human aorta demonstrates a progressive
rise in total lipid with an increase in the relative amount of
sphingomyelin and a decrease of phosphatidyl ethanolamine. Thus,
phosphatidyl ethanolamine was replaced by sphingomyelin. Our data
for Drosophila melanogaster (whole organism) showed that phosphatidyl ethanolamine substitutes for phosphatidyl choline in exactly
the same way (see Figure 3 and equations in Table I) that sphingomyelin replaces phosphatidyl ethanolamine in human aorta. Drosophila, like some other invertebrate species, substitutes ceramide phosphorylethanolamine for sphingomyelin, whereas in other
invertebrates sphingomyelin may be replaced in part or entirely by
ceramide aminoethylphosphonate. We have found uncharacterized
acidic phospholipids in some fungi. These appear to replace some
of the more common acidic phospholipids. Some bacteria substitute
amino acid esters of phosphatidyl glycerol for other group I lipids.
Glycosyldiglycerides are group I lipids of plants and some bacteria,
but in animals ceramide polyhexosides are found instead (glycosyldiglycerides are only trace components of human brain). Phospholipid composition data for bioluminescent bacteria (12) show that
the amounts of substitution groups I and II of several strains that
have phosphatidyl ethanolamine as the only lipid of group I are the
same as those of animal cells. In contrast, some bacteria that do
not contain phosphatidyl ethanolamine appear to have the amounts of
groups I and II reversed.

Substitution group II clearly contains all the acidic
phospholipids and gangliosides and group I the nonacidic lipids
including those with one free amino group and one phosphate group.
Our data indicate that the values found for the subgroups of the
two major groups (Table I) are found for subcellular particulates
from all organs as well as whole cells, organs, and organisms.

D. Feedback Control in the Nervous System

As the total lipid level of a differentiating organ rises,
the relative amount of each lipid class present in membranes changes.
Brain lipid data indicate that, at high lipid levels, the limit of

the ability of cells to synthesize some lipid classes is reached
and other lipid classes are synthesized as replacements. The
failure of cells in the nervous system to synthesize some lipid
classes at lower total lipid levels is explained in a simple manner
by differences in the strength of binding of different lipid classes to lipid-binding protein and product inhibition of biosynthesis.
Most of the lipids appear to be synthesized in the endoplasmic
reticulum and transported to binding sites on other membranes by
attachment to a water soluble transport protein. Competition for
attachment to a transport protein and lipid-binding membrane protein
will take place and the most firmly bound lipid classes will be
found in membranes until their levels become sufficiently high that,
in the dynamic equilibrium that exists, product inhibition of biosynthesis becomes a prominent factor. Product inhibition will be
more apparent at low total lipid levels for the least firmly bound
lipids.

Differences in binding affinity to proteins are easily visualized for phosphatidyl choline and cerebroside of substitution
group I because the phospholipid can bind ionically, whereas cerebroside can bind through hydrogen bonds only. Also, binding
strength for phosphatidyl ethanolamine can be visualized as greater
than that of phosphatidyl choline and cerebroside if the lipids are
linked through divalent ions because only phosphatidyl ethanolamine
forms relatively stable calcium and magnesium salts. We have defined the relative stability of magnesium salts of lipids by the
extent of ion exchange with diethylaminoethyl cellulose. The order
found was phosphatidyl choline = sphingomyelin = sulfatide < phosphatidyl ethanolamine < acidic phospholipids with the phosphatidyl
serine salt being the most poorly dissociated.

E. Relationship of Human Brain
Lipid Composition to Age

In our previous reports (1,2), semilogarithmic and logarithmic
plots of lipids vs. age were presented. Although they have the advantage of expanding the scale in the early stages of development
where changes are most extensive, compression of the scale at later
ages tends to obscure the fact that significant changes take place
during this period. Also, the changes do not appear to be exponential
in nature. Thus, where straight lines are obtained on arithmetic and
logarithmic plots, curves are obtained on semilogarithmic plots.
Arithmetic plots on an expanded age scale should be used when changes
during the first four years of life only are considered. Also, conceptual age is preferable to postpartum age used previously because
the fetal period can be included and compared to postnatal development.

Lipid vs. conceptual age plots showed individuals to vary as
noted also from the plots of lipid classes vs. total polar lipid etc.

This individual variability could not be recognized in our first
analysis with a smaller number of samples. In general, three
different periods could be recognized in age plots. The first
period is characterized by a rapid rate of increase, the second by
a less rapid rate of increase, and the third by a decline in lipid
content. It was not possible to fit lines through points in only
one way by simple inspection of lipid vs. age plots. The criterion
used for lipid vs. lipid plots (noting the line made by values
from samples removed from different parts of one brain) could not
be used for lipid vs. age plots. Other criteria for age plots were
derived, however, from lipid vs. lipid plots. Thus, the age of onset
of the period of decline of lipid was determined unambiguously by
noting the age at which values on lipid vs. lipid plots ceased to
rise along a line and began to fall back along the same line. This
proved to be about 26 years (conceptual age) for all lipid classes.
With this turning point known, lines could then be drawn unambiguously through points from 26 years and above on lipid vs. age plots.
Lines could also be drawn through points in the last part of the
second period of increase of lipid, but recognition of the point at
which the lines stopped required other criteria. The information
necessary for the entire period from conception to 26 years was
derived from lipid vs. total polar lipid plots. One method was the
use of plots of the amounts of substitution groups vs. age. Lines
were drawn through points in such a way that the ratios of the
substitution groups did not change since lipid vs. lipid plots had
shown them to make up a constant percentage of the total lipid at
all ages. Lines were then drawn through points on lipid class vs.
age plots by noting which fits gave values that, upon addition,
gave the correct sum for the amount of each substitution group
read from substitution group vs. age plots.

     Since the relative amounts of the lipid classes change with age,
it is apparent that all lipid classes do not follow exactly the
same course of change with age on lipid vs. age plots. The ages at
which slope changes occur and the equations for the lines are
shown in Table II. The changes with age are rather well defined
for the group with the most lipid. The relatively small number of
samples available fail to define as precisely the early changes
(up to 4 years) for the other groups.

     Clearly, brains of different humans differ significantly in
total lipid and lipid class content and the developmental changes
up to 26 years follow different courses. Myelination has been
shown to continue through the third or fourth decade of life (13).
Since the values for lipid content of different parts of brain
differ and the same lipid class interrelationships are found as for
representative samples of whole brain at different ages where lipid
content is also different, it appears that the developmental changes
arise mostly from the different rates of proliferation of plasma
membrane projections of neurones in the early stages and of glial

TABLE II

Equations for Human Brain Lipid Composition Changes with Age[a]

	Period 1[b]	Up To (yrs)	Period 2[c]	Period 3[d]
Total Lipid[e]				
Group I	2.95(A)	2.10	0.193(A) + 5.83	11.74 − 0.030(A)
II	3.71(A)	1.90	0.193(A) + 6.67	12.52 − 0.030(A)
III	4.59(A)	1.70	0.193(A) + 7.60	13.26 − 0.030(A)
IV	5.81(A)	1.50	0.193(A) + 8.25	13.96 − 0.030(A)
Total Polar Lipid[f],[g]				
Group I	2.78(A)	2.45	0.089(A) + 6.63	9.48 − 0.027(A)
II	3.37(A)	2.25	0.089(A) + 7.53	10.38 − 0.027(A)
III	4.23(A)	2.00	0.089(A) + 8.43	11.23 − 0.027(A)
IV	5.60(A)	1.70	0.089(A) + 9.50	12.27 − 0.027(A)
Total Phospholipid[f]				
Group I	2.57(A)	2.10	0.044(A) + 5.41	6.63 − 0.006(A)
II	3.08(A)	1.90	0.044(A) + 5.82	7.05 − 0.006(A)
III	3.64(A)	1.50	0.044(A) + 6.20	7.52 − 0.006(A)
IV	4.64(A)	1.40	0.044(A) + 6.61	7.92 − 0.006(A)
Phosphatidyl Choline[f],[h]				
Group I	1.20(A)	1.50	0.0032(A) + 1.77	1.88 − 0.0010(A)
II	1.40(A)	1.30	0.0032(A) + 1.87	1.98 − 0.0010(A)
III	1.66(A)	1.20	0.0032(A) + 1.97	2.08 − 0.0010(A)
IV	1.95(A)	1.10	0.0032(A) + 2.07	2.18 − 0.0010(A)
Phosphatidyl Ethanolamine[f],[i]				
Group I	0.90(A)	2.20	0.0045(A) + 1.93	2.06 − 0.0010(A)
II	1.10(A)	2.00	0.0045(A) + 2.13	2.26 − 0.0010(A)
III	1.30(A)	1.90	0.0045(A) + 2.33	2.46 − 0.0010(A)
IV	1.60(A)	1.60	0.0045(A) + 2.53	2.66 − 0.0010(A)

TABLE II cont'd.[a]

	Period 1[b]	Up To (yrs)	Period 2[c]	Period 3[d]
Phosphatidyl Serine[f]				
Group I	0.326(A)	2.50	0.0106(A) + 0.75	1.14 − 0.0046(A)
II	0.410(A)	2.20	0.0106(A) + 0.85	1.25 − 0.0046(A)
III	0.530(A)	1.90	0.0106(A) + 0.98	1.36 − 0.0046(A)
IV	0.685(A)	1.60	0.0106(A) + 1.09	1.47 − 0.0046(A)
Phosphatidyl Inositol[f]				
Group I	0.087(A) − 0.094	2.50	0.0020(A) + 0.099	0.19 − 0.0012(A)
II	0.087(A) − 0.052	2.20	0.0020(A) + 0.125	0.21 − 0.0012(A)
III	0.087(A)	1.90	0.0020(A) + 0.160	0.24 − 0.0012(A)
IV	0.127(A)	1.50	0.0020(A) + 0.192	0.27 − 0.0012(A)
Sphingomyelin[f]				
Group I	0.278(A) − 0.252	1.40	0.010(A) + 0.42	0.77 − 0.0029(A)
II	0.278(A) − 0.128	1.30	0.010(A) + 0.52	0.88 − 0.0029(A)
III	0.265(A)	1.20	0.010(A) + 0.63	0.98 − 0.0029(A)
IV	0.370(A)	1.10	0.010(A) + 0.73	1.08 − 0.0029(A)
Cerebroside[f]				
Group I	0.600(A) − 0.433	2.30	0.039(A) + 0.71	2.12 − 0.013(A)
II	0.600(A) − 0.256	2.65	0.039(A) + 1.17	2.54 − 0.013(A)
III	0.600(A) − 0.137	3.15	0.039(A) + 1.58	2.95 − 0.013(A)
IV	0.600(A)	3.65	0.039(A) + 2.01	3.38 − 0.013(A)
Sulfatide[f]				
Group I	0.174(A) − 0.103	2.80	0.000(A) + 0.35	0.78 − 0.000(A)
II	0.174(A) − 0.071	3.15	0.000(A) + 0.53	0.88 − 0.000(A)
III	0.174(A) − 0.033	3.45	0.000(A) + 0.66	0.99 − 0.000(A)
IV	0.174(A)	3.75	0.000(A) + 0.78	1.10 − 0.000(A)

TABLE II cont'd[a]

	Period 1[b]	Up To (yrs)	Period 2[c]	Period 3[d]
Total Gangliosides[f,j]				
Group I	0.0547(A)	2.30	0.125 - 0.0012(A)	0.102 - 0.00035(A)
II	0.0705(A)	1.90	0.135 - 0.0012(A)	0.112 - 0.00035(A)
III	0.0935(A)	1.50	0.145 - 0.0012(A)	0.124 - 0.00035(A)
IV	0.1390(A)	1.10	0.155 - 0.0012(A)	0.135 - 0.00035(A)
Cholesterol[f]				
Group I	1.72(A)	2.80	0.0280(A) + 4.69	5.62 - 0.0113(A)
II	2.16(A)	2.50	0.0280(A) + 5.36	6.40 - 0.0113(A)
III	2.72(A)	2.20	0.0280(A) + 5.96	7.02 - 0.0113(A)
IV	3.62(A)	1.80	0.0280(A) + 6.53	7.54 - 0.0113(A)

(a) A = conceptual age; (b) from conception to age shown in column 2; (c) begins at age shown in column 2 and continues to 26 years for all groups; (d) from 26 years on for all groups; (e) Sephadex fraction 1 (excludes gangliosides) as percentage of the fresh weight; (f) mM/100 gm fresh weight; (g) total phospholipid plus cerebroside and sulfatide (gangliosides not included); (h) includes lysophosphatidyl choline; (i) includes lysophosphatidyl ethanolamine; (j) from Sephadex fraction 2.

cells that produce myelin in later stages. As the amount of membrane mass that each cell must support increases, the relative
amounts of the lipid classes change with some lipid classes substituting for others. Differences in binding strength to lipid-binding
transport and membrane proteins and product inhibition of biosynthesis can explain the findings as noted in Section III-D.
The data of Davison reported in this volume indicate the presence
in animal brains at all ages of a type of myelin that is devoid of
cerebroside and sulfatide. This is in keeping with our findings
since glial cells that support smaller amounts of plasma membrane
extensions (including myelin) do not need to synthesize sphingolipids to replace glycerolphospholipid. Cells supporting only a
small amount of plasma membrane appear to occur in different parts
of brain at all ages.

ACKNOWLEDGEMENTS

    This work was supported in part by U S. Public Health Service
Grants NS 01847 and NS 06237 from the National Institute of Neurological Diseases and Stroke.

REFERENCES

(1) Rouser, G. and Yamamoto. Lipids $\underline{3}$: 284 (1968).

(2) Rouser, G. and Yamamoto, A. in Handbook of Neurochemistry,
    Vol. I (Ed. A. Lajtha), Plenum Press, 121-169 (1969).

(3) Rouser, G. and Solomon, R.D. Lipids $\underline{4}$: 232 (1969).

(4) Broekhuyse, R.M. Biochim. Biophys. Acta $\underline{187}$: 354 (1969).

(5) Kety, S.S. in Biochemistry of the Developing Nervous System
    (Ed. H. Waelsch), Academic Press, 208-216 (1955).

(6) Rouser, G., Kritchevsky, G. and Yamamoto, A. in Lipid
    Chromatographic Analysis, Vol. I (Ed. A.V. Marinetti),
    Marcel Dekker Inc., 99-162 (1967).

(7) Baxter, C.F., Rouser, G., and Simon, G. Lipids $\underline{4}$: 243 (1969).

(8) Rouser, G., Simon, G., and Kritchevsky, G. Lipids $\underline{4}$: 599 (1969).

(9) Simon, G. and Rouser, G. Lipids $\underline{4}$: 607 (1968).

(10) Smith, M.E. Biochim. Biophys. Acta $\underline{164}$: 285 (1968)

(11) Eng, L.F., Chao, F.C., Gerstl, B., Pratt, D. and Tavaststjerna,
     M.G. Biochemistry $\underline{7}$: 4455 (1968).

(12) Eberhard, A. and Rouser, G. In preparation.

(13) Yakolev, P.I. and Lecours, A.R. in Regional Development of the
     Brain in Early Life (Ed. A. Minkowski), Blackwell Sci. Publ.
     3-70 (1967).

ADDENDUM: Our discovery of seven different groups for each lipid
class and five subgroups for each substitution group is analogous to
the finding of seven different combinations of five different iso-
zymes of acid phosphatase in Tetrahymena pyriformis (S.L. Allen,
Ann. N.Y. Acad. Sci. $\underline{151}$: 190, 1968). Our data fit the Allen tetramer
model very closely. Thus, if cellular membrane lipid-binding protein
is a tetramer and there are only 2 different peptide chains, A^- (acidic)
and B^+ (basic), the values calculated for the percentages of the subgroups of substitution group II are 11.0, 15.3, 19.5, 23.7, and 28.0.
The values are in close agreement with those found which are 11.0,
15.2, 19.5, 22.8, and 28.0. The five subgroups would arise from differences in strength of binding for different lipid classes, acidic
phospholipid binding preferentially to B^+ and binding strength for A^-
being slightly greater for lipids other than acidic phospholipid. The
seven different groups for each lipid class as percentage of the total
lipid would arise from differences in metabolism (G. Rouser, this vol.).

The model predicts a maximum of five tetramers differing in net
charge. Units of this type do not appear to have been isolated. The
model also predicts the occurrence of membranes having only A^- or B^+.
Sarcoplasmic reticulum has been found to have a single acidic peptide
chain (B.P. Yu and E.J. Masoro, Biochemistry $\underline{9}$: 2909, 1970). This leads
to the prediction that substitution group I which preferentially binds
to acidic proteins should be present in the maximum amount of 87.2%.
In agreement with this expectation, the value found was 87.0 (M.Sheetz
and G. Rouser, unpublished). The finding of miniproteins with molecular
weights of about 5000 (Laico et al., Proc. Natl. Acad. Sci. $\underline{67}$: 120,
1970) and sarcoplasmic reticulum protein of about 6500 molecular weight
leads to the expectation that some tetramers would have a molecular
weight in the 20 to 26,000 range. The basic protein of myelin is a
single peptide chain of molecular weight 18,500 (E.H. Eylar, Proc.
Natl. Acad. Sci. $\underline{67}$: 1425, 1970) whereas the Folch-Lees protein weight
is about 35,000 ($\overline{\text{E.}}$ Mehl and A. Halaris, J. Neurochem. $\underline{17}$: 659, 1970).
Variations in the relative amounts of these proteins have been re-
ported. The model predicts ratios of A^-/B^+ of 1/1, 1/3, and 3/1 with
the 1/1 ratio being most common. This is in good agreement with the 1/1
molar ratio for myelin proteins for several species (Mehl and Halaris,
loc. cit.) and the values of 68%, 51-54%, and 23% for the Folch-Lees
protein reported by Eng et al. (Biochemistry $\underline{7}$: 4455, 1968).

We propose that lipid-binding protein of membranes is a tetramer
and that there are only two different peptide chains, one acidic and
one basic, which can combine to give five tetramers of different
charge. Also, it appears that the basic peptide chain preferentially
binds acidic phospholipids whereas the acidic chain has a slight
preference for binding of other types of polar lipids.

SOME ASPECTS OF CARBOHYDRATE METABOLISM IN THE DEVELOPING BRAIN

F. Pocchiari

Laboratori di Chimica Biologica

Istituto Superiore di Sanità, Rome, Italy

The ground that I was called to cover is large and mainly unexplored. If you consider that most data on carbohydrate metabolism in nervous tissue have lately undergone drastic reappraisal in the light of recent experimental advances, you will understand why my exposé will look somewhat fragmentary. My only hope is that the very nature of the subject and its intrinsic complexity will prompt my colleagues to take the few data I shall review as a starting point for fruitful and critical discussion.

The primary activity of adult brain is transmission and storage of specific information at several intracellular and intercellular levels, which requires a rapid topical release of energy.

There is consensus among investigators that glucose is the main source of energy for nervous tissue. The developing brain, on the other hand, needs large amounts of energy for synthetic purposes. In the immature tissue this is supplied mainly by glycolysis and the pentose phosphate pathway, which are more elementary and less differentiated metabolic routes. At the time when the brain is called to give more articulated and specific cortical responses to sensory

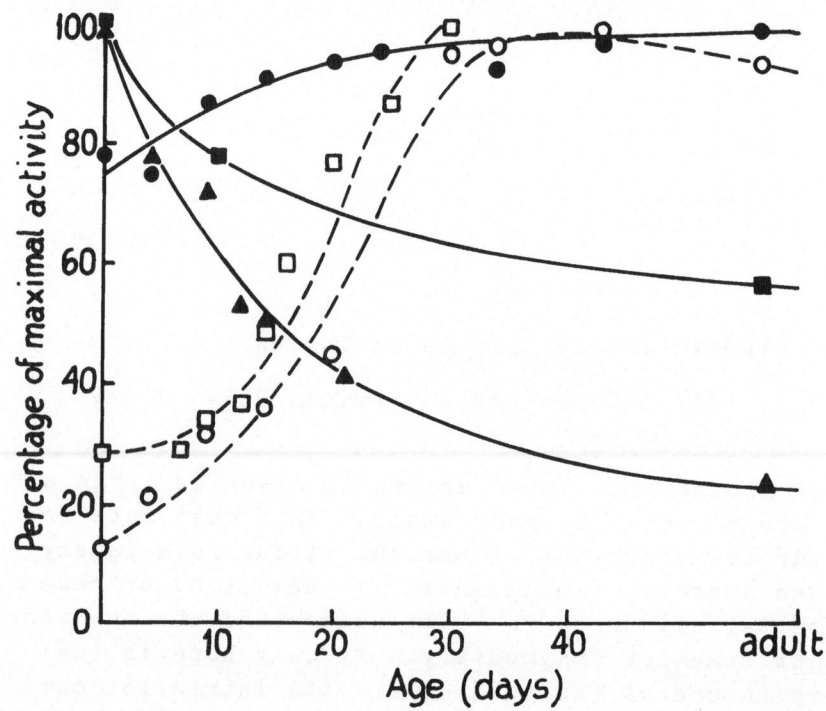

Fig. 1: Changes in representative glycolytic and oxidative enzymes in developing brain, expressed as percentage of maximal activity. ●——● glyceraldehyde phosphate dehydrogenase. O——O glycerol phosphate dehydrogenase. ■——■ glucose-6-phosphate dehydrogenase. ▲——▲ isocitric dehydrogenase. □——□ succinic dehydrogenase.

stimuli, a sharp increase in synthesis of enzymatic protein is observed. To be able to face the enormously increased energy requirements, all the enzymes of the oxidative cycle are synthesized very rapidly.

OXIDATIVE METABOLISM

The consumption of O_2 in adult human brain in normal condition is about 100 μmoles/g of fresh tissue/h, which represents 1/5 of the total amount of

O_2 needed by the whole body. Adult brain is extremely
sensitive to a decrease in O_2 availability. Irreparable
functional damage can be produced by very short periods
of anoxia. Foetal and newborn mammalian brain are much
more resistant, suggesting some degree of independence
from aerobic processes, probably connected with their
lower functional activity as compared to that of the
adult. In the rat the resistance to anoxia falls to
adult values around the 17th day after birth (1). O_2
consumption in the cerebral tissue of newborn rat is
about half that of the adult and remains practically
constant for about ten days. After this period it increases sharply, reaching the adult values in three or
four days.

Another measure of metabolic differences between
the newborn and adult brain (rat and mouse) is the high
ATP/ADP ratio in newborn tissue (2) (3). The balance
between these products reflects the actual equilibrium
between energy-requiring synthetic processes and energy-
yielding catabolic processes. During development there
is a sixfold increase in ATPase activity (4), which
reaches its maximum slope between ten and twenty days
of age, when brain enters its functional maturity.

The rapid variation in resistance to anoxia and in
oxygen requirement indicates that profound changes must
occur in the enzymatic makeup of the newborn nervous
tissue within reasonably short time after birth. If we
take into consideration some enzymes representative of
the oxidative metabolism we see that in rat cerebral
cortex succinic dehydrogenase activity increases considerably between the 10th and the 25th day of life
(Fig. 1). Cytochrome oxidase activity follows the same
pattern, showing a rapid three- to six-fold increase
during the same period (4). Malic dehydrogenase
parallels this behaviour in rat cerebral (5) and cerebellar cortex (6).

On the other hand, total isocitric dehydrogenase
activity, measured in homogenates, showed a marked decrease during the same period (6). Its level in the

newborn is almost four times that of the adult (Fig. 1).
This may be explained by the fact that this activity
reflects mainly that of the cytoplasmatic enzyme, connected with the formation of NADPH, necessary for synthetic processes. A few months ago the hypothesis was
advanced that NADP dependent isocitric dehydrogenase may
have a specific function in the synthesis of the small
glutamate pool (7), which predominates in the immature
brain. Another enzyme which is higher in the newborn
is glucose-6-phosphate dehydrogenase, which is also
involved in the formation of NADPH (5) (Fig. 1).

Increases in oxygen consumption and in oxidative
enzymatic activity of rat brain during development are
paralleled by an increase in mitochondrial protein.
It seems rather unlikely, in the light of the most
recent findings (8) that this increase could be due to
an increment in the number of mitochondria (9). On the
other hand, during development there is an almost ten-
fold increase in wet weight of the whole brain, which
means that an absolute increase of mitochondrial mass
does occur. It is known that mammalian mitochondria
can synthesize protein, to be incorporated in their own
structure (10). Although appearing morphologically
similar to those from adult animals, newborn rat brain
mitochondria show a very low succinic dehydrogenase
activity, about 20% of the adult value (8). As this
enzyme can be considered representative of one of the
major functions of the organelle, that is oxidative
phosphorylation, it must be assumed that substantial
amounts of enzymatic protein are synthesized locally
during development.

PENTOSE PHOSPHATE CYCLE

The pentose phosphate pathway is an alternate way
for glucose oxidation in brain as well as in other
tissues. The activities of some representative enzymes
involved in this cycle, glucose-6-phosphate dehydrogenase
and 6-phosphogluconate dehydrogenase, were found in
brain (11) (12), but the contribution of this pathway to

glucose metabolism is to be considered minimal in the adult animal (13) (14) (15) (16). Until very recently it was calculated to be in the order of 1% or less of the metabolic flux of glucose (16): the latest studies (17) indicate a maximum contribution of around 5%. The technique applied was to measure the distribution of ^{14}C derived from $2-^{14}C$ glucose in the different carbon atoms of the glucose molecules obtained by hydrolysis of the glycogen fraction.

The pentose shunt could be particularly important during brain development, because it provides NADPH, a coenzyme necessary for lipid synthesis, and pentose phosphates, which are precursors of nucleotides. In fact Liuzzi and Angeletti (18) in our laboratories showed that in embryonic chicken brain the amount of $^{14}CO_2$ derived from $1-^{14}C$ glucose was considerably higher than that derived from $6-^{14}C$ glucose. In control adult chicken brain this ratio was found to be equal to unity. A similar observation was made by O'Neill and Duffy (19) working on dog cortex slices. In accord with the isotope findings, glucose-6-phosphate dehydrogenase was found to decrease from the newborn to the adult rat (5) (Fig. 1).

GLYCOLYSIS

The uptake of glucose from the blood by adult human brain is about 18 μmoles/g of fresh tissue/hr. From the initial rates of change of adenosine triphosphate, phosphocreatine, glucose, glycogen and lactate, metabolic rates were calculated for adult and ten days old mice (20). The values for the newborn were found to be half those of the adult. Similar results had been obtained on the cortex of eight days old rats by other researchers (21).

At variance with the oxidative enzymes, which are present in very small amounts, the level of activity of brain enzymes involved in glycolysis in ten days old mice was found to be about 2/3 of the adult values, with the exception of lactic dehydrogenase, which practically

did not vary during development (3). Although there are
very few direct observations on these levels in the newborn animal, one can extrapolate from the behaviour of
glyceraldehyde phosphate dehydrogenase (Fig. 1), which
in the first ten days of life does not vary very much
(22), and assume that the totality of glycolytic enzymes
follow the same pattern. In fact it was recently found
in the rat that hexokinase and lactate dehydrogenase
show a moderate increase during the first week of life
(23). A rise in activity for the whole group occurs
between the 10th and 20th day after birth. One colla-
teral enzyme, glycerol phosphate dehydrogenase, showed
a marked increase from birth to about the 30th day (22),
in concomitance with lipid incorporation into brain
tissue and onset of myelination.

GLYCOGEN METABOLISM

    Adult brain contains relatively small amounts of
glycogen: about 1 mg/g fresh tissue. Studies using
^{14}C labelled precursors showed a rapid turnover, pointing
to some dynamic function for this polysaccharide (24).
Synthesis and breakdown of glycogen occur in brain, as
in other tissues, through different pathways: UDPG
pyrophosphorylase and glycogen synthetase are responsible
for incorporation of glucose molecules into glycogen via
glucose-1-phosphate; phosphorylase for breakdown of the
glycogen chain into glucose-1-phosphate.

    At birth the activity of the enzymes responsible
for glycogen synthesis is roughly 50% of that of the
adult, while phosphorylase and phosphoglucomutase activities are practically negligible. During development
there is a gradual increase of all these enzymes with a
sharp eight-fold increase in activity of the catabolic
enzymes between the 10th and the 20th day after birth
(25) (26).

    Speaking in terms of amount of glycogen, its level
in mammalian cerebral cortex at birth is roughly one
fifth of that of the adult. The sharp metabolic increase

after birth is presumably correlated with the beginning of cortical function.

FATE OF ^{14}C-GLUCOSE

Brain glucose metabolism is characterized by a high rate of incorporation of carbon from the skeleton of glucose into free amino acids. This was observed for the first time in 1955 in our laboratories (27). Our experiments on rat cortex slices were confirmed by other investigators working on perfused brain (28) (29) and in vivo (30) (31) (32). At variance with other organs, brain contains a particular enzymatic setup capable of transforming glucose into pyruvate very rapidly. The carbon atoms of pyruvate via α-oxoglutarate then enter the large pool of glutamate. Balázs (33) recently made an extensive review of these mechanisms, summarizing the factors involved in the transformation of glucose into glutamate: they include the glucose and glycogen pools which in brain are relatively small, the large glutamate pool, the high speed of glycolysis and the limited contribution of other substrates to oxidation.

The amino acids which are found labelled after incubation of rat cortex slices with ^{14}C-glucose are: glutamate, aspartate, alanine, γ-aminobutyrate and glutamine. In slices 20% of the total glucose is metabolized this way, another 20% goes to CO_2 and 60% to lactate (27). In vivo in the adult animal 22 min. after administration more than 70% of the glucose is found in brain free amino acids (31). In the newborn animal the incorporation, in the same experimental conditions, was only 2% (7); it increases slowly up to the 13th day after birth and much more rapidly between the 14th and 22nd day, when it approximates the adult values (34) (35). This sharp increase parallels that observed during the same period for oxidative enzymes and for aminotransferase (36).

If we look at glutamate we see that between the 10th and the 19th day after birth there is an increase of 60%

in its total amount, while its specific activity increases five times. This means that a dilution of glucose carbon must occur at the level of acetyl-CoA, as a
result of oxidation of other substrates (35).

This was also shown by the fact that the incorpor-
poration rates of ^{14}C from acetate, phenylalanine, tyrosine and proline into glutamate, aspartate and glutamine
are the same in 10 days old and in adult mice brain,
whereas glucose is a poorer precursor in the developing
animals (7).

It is an accepted view that not all the glutamate
in the nervous cells is accessible metabolically at the
same time, which means that there are functional compartments for this amino acid (36) (37). Most recent results
indicate that this compartmentation develops along with
the capacity for incorporation of ^{14}C glucose into amino
acids (35).

I hope to have given you in this short survey an
idea of what kind of data have been collected up to now
on carbohydrate metabolism in the growing nervous system.

It is our impression that in developing brain it is
even more difficult than in the adult to devise and carry
out significant experiments, capable of discerning the
subtle changes accompanying development in mammalian
nervous tissue. We must not forget that from the instant
of birth the impact of reality on an organism presents
its nervous system with a dual task. On one side it has
to grow, on the other it has to sustain basic functional
activity.

REFERENCES

(1) Fazekas, J. F., Alexander, F.A.D., and Himwich, H. E. Am. J. Physiol. $\underline{134}$:281 (1941).

(2) Lolley, R. N., Balfour, W. M., and Samson, F. E. J. Neurochem. $\underline{7}$:289 (1961).

(3) Lowry, O. H., and Passonneau, J. V. J. Biol. Chem. $\underline{239}$:31 (1964).

(4) Potter, V. R., Schneider, W. C., and Liebl, G. J. Cancer Res. $\underline{5}$:21 (1945).

(5) Kuhlman, R. E., and Lowry, O. H. J. Neurochem. $\underline{1}$:173 (1956).

(6) Robins, E., and Lowe, I. P. J. Neurochem. $\underline{8}$:81 (1961).

(7) Van den Berg, C. J. J. Neurochem. $\underline{17}$:973 (1970).

(8) Gregson, N. A., and Williams, P. L. J. Neurochem. $\underline{16}$:617 (1969).

(9) Samson, F. E., Balfour, W. M., and Jacobs, R. J. Am. J. Physiol. $\underline{199}$:693 (1960).

(10) Roodyn, D. B. Biochem. J. $\underline{97}$:782 (1965).

(11) Buell, M. V., Lowry, O. H., Roberts, N. R., Chang, M. W., and Kapphahn, J. I. J. Biol. Chem. $\underline{232}$:979 (1958).

(12) Guerra, R. M., Melgar, E., and Villavicencio, M. Biochim. Biophys. Acta. $\underline{148}$:356 (1967).

(13) Bloom, B. Proc. Soc. Exp. Biol. (N.Y.). $\underline{88}$:317 (1955).

(14) Tower, D. B. J. Neurochem. $\underline{3}$:185 (1958).

(15) Hoskin, F.C.G. Arch. Biochem. $\underline{91}$:43 (1960).

(16)   Sacks, W. J. Appl. Physiol. 20:117 (1965).

(17)   Hostetler, K. Y., Landau, B. R., White, R. J.,
       Albin, M. S., and Yashon, D. J. Neurochem.
       17:33 (1970).

(18)   Liuzzi, A., and Angeletti, P. U. Experientia.
       20:512 (1964).

(19)   O'Neill, J. J., and Duffy, T. E. Life Science.
       5:1849 (1966).

(20)   Lowry, O. H., Passonneau, J. V., Hasselberger,
       F. X., and Schultz, D. W. J. Biol. Chem. 239:18
       (1964).

(21)   Greengard, P., and McIlwain, H. in Biochem.
       Developing Nervous System, Proc. 1st Intern.
       Neurochem. Symp. Oxford 1954 (1955).

(22)   Laatsch, R. H. J. Neurochem. 9:487 (1962).

(23)   Biesold, D., and Leonard, B. E. in Ontogenesis of
       the Brain. The biochemical, functional and
       structural development of the nervous system,
       Proc. Intern. Symp. Neuroontogenicum, Praha 1967
       (1968).

(24)   Coxon, R. V. in Handbook of Neurochemistry, Vol.
       3 (Ed. A. Lajtha) (1970).

(25)   McIlwain, H. Biochemistry and the Central
       Nervous System (1966).

(26)   Friede, R. L. Topographic Brain Chemistry (1966).

(27)   Beloff-Chain, A., Catanzaro, R., Chain, E. B.,
       Masi, I., and Pocchiari, F. Proc. Roy. Soc.
       (London) Ser. B 144:22 (1955).

(28)   Geiger, A., Kawakita, Y., and Barkulis, S. S.
       J. Neurochem. 5:323 (1960).

(29)   Barkulis, S. S., Geiger, A., Kawakita, Y., and
       Aguilar, V. J. Neurochem. 5:339 (1960).

(30)   Gaitonde, M. K., Marchi, S. A., and Richter, D.
       Proc. Roy. Soc. (London) Sez. B 160:124 (1964).

(31)   Gaitonde, M. K., Dahl, D. R., and Elliott, K.A.C.
       Biochem. J. 94:345 (1965).

(32)   Minard, F. N., and Mushahwar, I. K. J. Neurochem.
       13:1 (1966).

(33)   Balázs, R. in Handbook of Neurochemistry, Vol. 3
       (Ed. A. Lajtha) (1970).

(34)   Gaitonde, M. K., and Richter, D. J. Neurochem.
       13:1309 (1966).

(35)   Cocks, J. A., Balázs, R., Johnson, A. L., and
       Eayrs, J. T. J. Neurochem. 17:1275 (1970).

(36)   Berl, S., and Clarke, D.D. in Handbook of Neuro-
       chemistry, Vol. 2 (Ed. A. Lajtha) (1969).

(37)   Balázs, R., Machiyama, Y., Hammond, B. J.,
       Julian, T., and Richter, D. Biochem. J. 116:445
       (1970).

THE RESPONSE OF A BRAIN SPECIFIC

PROTEIN AT LEARNING

Holger Hydén and Paul W. Lange

Institute of Neurobiology, Faculty of Medicine

University of Göteborg, Göteborg, Sweden

Abstract

The brain specific acidic protein S100 in the pyramidal nerve cells of the hippocampus was investigated as a possible correlate to learning during transfer of handedness in rats. The amount of S100 increased during training. Intraventricular injection of antiserum against the S100 protein during the course of training prevented the rats from further learning but did not affect motor function in the animals. Antibodies against the S100 protein could be localized after the injection to hippocampal structures by immunofluorescence. By contrast, control animals subjected to the same training and injected with S100 antiserum absorbed with S100 protein or with other antisera against γ-globulins showed no decrease in their ability to learn. The conclusion is that the brain specific protein S100 is linked to the learning process within the training used.

During the last few years we have investigated the acidic brain protein S100 in hippocampal nerve cells during a behavioral test in rats. We wish to report that the amount of nerve cell S100 protein increases in trained animals and that the S100 protein is specifically correlated to learning. This linkage was demonstrated by the use of antiserum against the S100 protein which was injected intraventricularly during the course of the training and localized in the hippocampus by specific fluorescence. The presence of antiserum against the S100 protein in the hippocampus prevents further learning during continued training.

The S100 protein is a defined and brain specific protein and

123

its correlation to learning seems important since brain specific
protein can be supposed to mediate neural functions. This protein
described in 1965 by Moore et al. (1) has a molecular weight of
21,000, a high content of glutamic and aspartic acid and, therefore, moves closest to the anodal front in electrophoresis at pH
$>$ 8. The S100 protein is mainly a glial protein but occurs also
in the nerve cells (2), and constitutes about 0.2 % of the total
brain proteins. The anodal band containing S100 can be separated
in at least three components, two of which precipitate with antiserum against S100 and have a high turnover (3). The S100 protein seems to be composed of three subunits of 7,000 molecular
weight (4). Its appearance in the human frontal cortex parallels
the onset of neurophysiological function (5).

Moore and Perez have described another acidic brain specific protein (14-3-2) localized in nerve cells (6). Still another
acidic protein ("antigen α") unique to the brain has been characterized by Bennett and Edelman (7). In addition, evidence for
the existence of other brain specific soluble proteins has been
presented by Bogoch (8), MacPherson (9), Kosinski (10), Warecka
and Bauer (11).

The training of animals involves a number of variables,
such as motor and sensory activity, motivation, orientation
reflexes, stress, and the learning processes, per se. Active
controls are, therefore, essential to the experimental animals
where these factors have become equated, except learning.

In a well planned behavioral test surgical, mechanical or
electrical measures to the body should be avoided and the stress
factor should be small. For these reasons, we have chosen
reversal of handedness in rats as behavior experiment (12).
The active controls perceive and act similarly to the experimental animals.

Eightyone Sprague-Dawley rats weighing 150-175 g were
used. The experimental set-up has been described previously
in detail (12). It may only be pointed out that the rat retrieved
one food pill at a time by reaching into the glass tube housing the
pills. The rats were induced to use the non-preferred paw by
arranging a wall parallel and close to the glass tube on the opposite side of the preferred paw. The controls used the preferred
paw and received the same amount of reward as the experimental
animals. The rats were trained during two sessions of 25 min
per day. The performance was linear up to the 8th day defined
as number of reaches per day. (Fig. 1) The rats used in our
experiments all showed performance curves similar to that in
Fig. 1. Once learned, this new behavior will remain for a long
time (13).

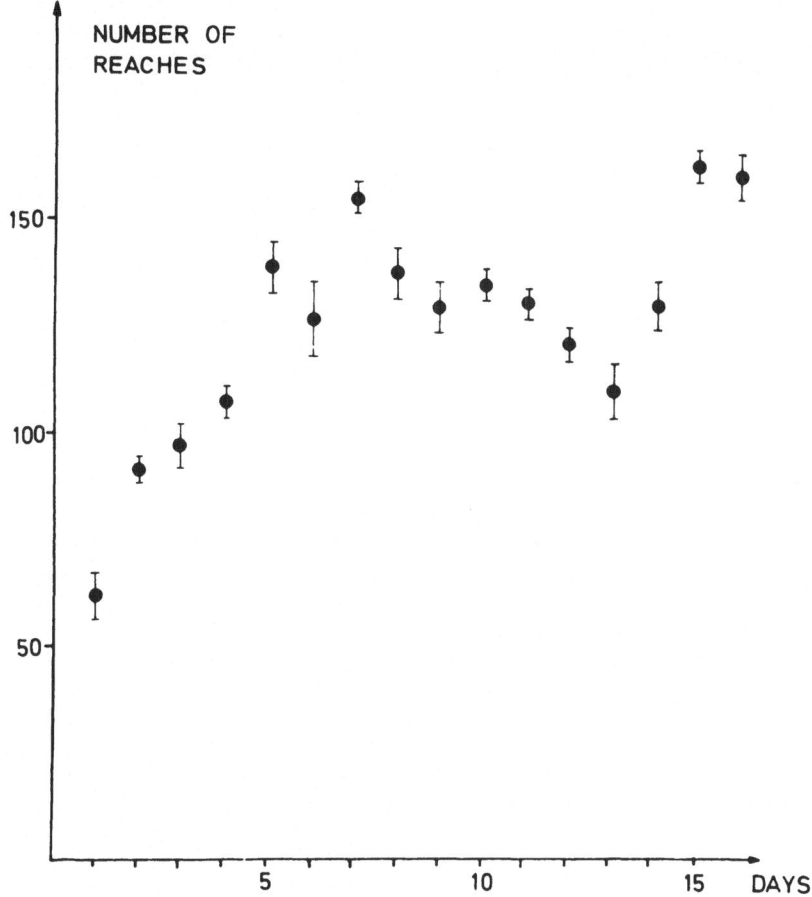

Fig. 1. Performance curve of a group of twelve rats given as the average of the number of reaches as function of number of training sessions (2 x 25 min per day).

The three persons handling the training of the rats and registering the performance, did not know with what serum the rat was injected. Neither did the three persons carrying out the injections of sera and chemical analysis know about the performance of the rats.

In the present experiments, fresh pyramidal nerve cells of the CA3 region of the hippocampus were used. The method for dissection and electrophoretic procedure has been described elsewhere (14, 15).

Increase of the S100 protein. In the trained rat material
the presence of a double anodal protein was seen. The electro-
phoretical pattern from the samples of the control rats, however,
showed only one single anodal protein band (16). Densitometric
recordings were made of 75 electrophoretic patterns from 23 rats.
Table 1 shows the presence of two frontal protein bands in 30 recordings out of 55 (16 rats) from trained animals. One front band
only was observed in all 20 recordings from controls (7 rats).

Protein extracted from pyramidal nerve cells of the CA3
hippocampal region from 5 trained rats was now separated, and
the gel cylinders were placed for 15 minutes in saturated ammonium sulphate solution and briefly rinsed. The protein was fixed in
sulfosalicylic acid and stained with brilliant blue. This treatment
with ammonium sulphate caused the protein band closest to the
anodal front to disappear which is a characteristic solubility
property of the S100 protein. The band immediately behind that
diminished in size but did not disappear completely. Hippocampal
nerve cell protein from another group of 5 trained rats was electro-
phoretically separated on 400 μ diameter gels and precipitated
with 80 % alcohol for 3 minutes. The gel cylinders were placed
in fluorescein-conjugated antiserum against S100 (dilution 1:4)
for 24 hours and then examined in a fluorescence microscope,
and photographed. Both front protein bands showed specific
fluorescence and had thus reacted positively with the anti-S100
antiserum. Thus, the second protein band contains S100 protein
and the original S100 band contains proteins other than the S100
component.

Table 1

Frequence of single and double front anodal protein
fractions in the electrophoretic pattern of 75 polyacrylamide gels from 23 rats (7 controls, 4 resumed training
on 14th day, 12 resumed training on 14th day and 30th day)

Controls		Resumed training on 14th day		Resumed training on 30th day	
One fraction	Two fractions	One fraction	Two fractions	One fraction	Two fractions
20	0	5	10	20	20

The question then arose if the amount of S100 protein had increased in the nerve cells of the trained animals. Measurements of the absorbance of the single protein band from controls and the two bands of the trained animals were performed and compared with an integrating micro-photometer. The same amount of protein from trained and control animals was used for the electrophoresis and the procedure was identical in all experiments. It was found that the amount of protein contained in the two anodal bands of the trained rats was 10 % greater than the amount of protein contained in the one band of the controls.

The effect of antiserum against S100 protein. The next question was whether the increase of the S100 protein during reversal of handedness specifically relates the S100 protein to learning processes occurring in the hippocampal nerve cells. As we pointed out above, training involves several factors not related to learning per se. In the reversal of handedness experiments, such unspecific factors have been eliminated or reduced to a minimum. The motor and sensory activity, attention, motivation, and reward are equated between the experimental and control animals, and the stress involved in reversal of handedness is minimal. A group of 6 rats were trained during 2 x 25 minutes per day for three days. Between the first and second training session on the fourth day, the rats were injected intraventricularly on both sides with 2 x 30 µg of antiserum against S100 in 2 x 30 µl. During further training for three days after the injection, the rats did not increase in performance. It is to be noted that the rats were not affected by the S100 antiserum with respect to motor function and sensory responses.

To demonstrate the specific effect on the antiserum against the S100 protein, the following experiment was carried out. Antiserum against S100 was absorbed with S100 protein according to a procedure previously used (17).

Eight rats were trained for 4 days and injected on day 4 with 2 x 30 µg of S100 antiserum absorbed with S100, as described. The rats were then trained for further 3 days. The performance of the rats, as number of reaches per training session,(Fig. 2), increased in the same way as did the performance of the non-injected control rats shown in Fig. 1.

Since the active molecules of an antiserum have a large molecular weight, it is important to know if the antibodies injected intraventricularly will reach and can be localized to the hippocampal structures. Therefore, rats were injected intraventricularly with 2 x 30 µg of antiserum against S100, and other rats with 2 x 30 µg S100 antiserum absorbed with S100 protein.

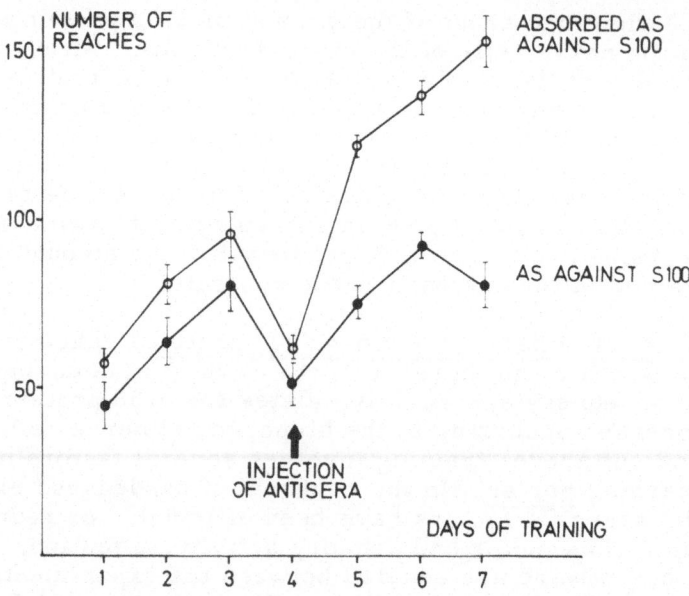

Fig. 2. Performance curves of 6 rats injected intraventricularly on day 4 with 2 x 30 µg of antiserum against
S100 protein and 8 rats injected with 2 x 30 µg of antiserum
against S100 protein absorbed with S100 protein.

One hour (2 rats) and eighteen hours (4 rats) after the
injection, the rats were decapitated and cryostat sections were
made of the hippocampus. Coons' double layer method was
applied on the two types of material to demonstrate the possible
localization of antibodies to cell structures, using a rabbit-
antirat-γ-globulin conjugated with fluorescein isothiocyanate
(Behring Werke AG, Marburg-Lahn, Germany). Specific
fluorescence was observed localized to nerve cells in the hippocampus of rats injected with antiserum against S100. No specific
fluorescence could be observed in the material from rats injected
with S100 antiserum absorbed with S100.

A pertinent question is whether the effect of the S100 antiserum on behavior is due to a S100 antibody-antigen reaction in
limbic structures since Klatzo et al. (18, 19) reported that
fluorescein-labelled globulin does not penetrate through the
ependyma into the brain tissue. In our present experiments,
the antisera are injected into the narrow lateral ventricles.
It can be suspected that the thin needle (gauge 20) may slightly

damage the walls of the lateral ventricles when inserted and
thus give free passage to globulins to enter. That such damage
occurs with the injection technique used was demonstrated with
a fluorescence technique.

At this point it can be stated that the S100 protein is specifically correlated with learning processes within training. It
does not mean, of course, that injection of S100 protein should
give rise to a spontaneous change of handedness.

It was then of interest to study a possible effect of injected
antiserum containing antibodies not directed against the S100
protein. In 14 rats various types of antisera were injected,
such as antiserum against rat γ-globulin from goat, rat γ-
globulin from rabbit, and rabbit γ-globulin from goat. As is
seen from the curves in Fig. 3, this has no impeding effect on
their performance.

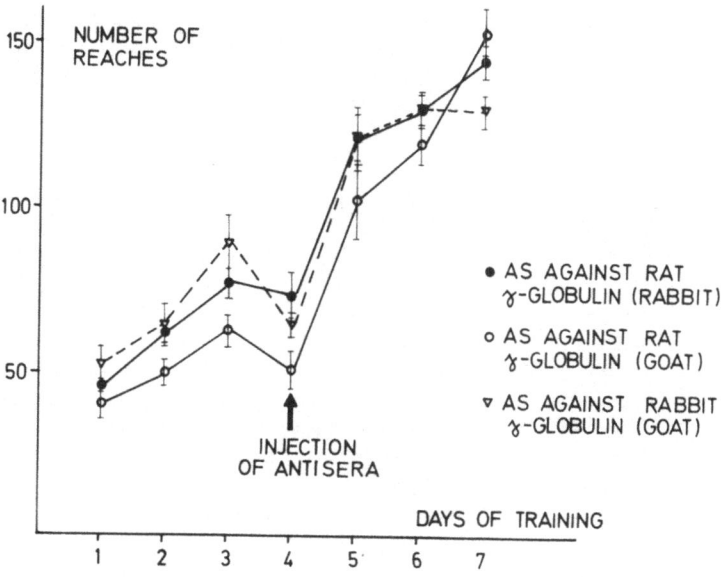

Fig. 3. Performance curves of rats injected intraventricularly on the 4th day with 2 x 30 μg of antiserum
against rat γ-globulin from rabbit (4 rats), and from goat
(4 rats), and rabbit γ-globulin from goat (4 rats).

DISCUSSION

In the present experiments, we have used a uniform nerve cell population of the hippocampus in rats. This brain region was chosen because of its importance for learning (16). In order to relate behavior to biochemical correlates, correct control experiments are essential. All factors involved in the training test were identical for experimental and control animals as far as can be with the exception of learning. The increase of the S100 protein in hippocampal nerve cells, its alteration in physical properties and the blocking effect of the S100 antiserum on behavior point to the S100 protein as a true biochemical correlate to behavior. This conclusion is backed up by the localization of the antibodies against the S100 by specific immuno-fluorescence and by the lack of effect of antisera not specifically directed to the S100 protein. Presumably, however, the S100 protein is not the only protein of importance for behavior.

It is questionable that the antiserum would influence the total brain protein synthesis in a measurable way. To confirm this assumption, the incorporation of ^{3}H-leucine into the hippo-campal nerve cell protein was studied after the last training session on the 7th day. No difference in incorporation of ^{3}H-leucine in the hippocampal protein was found between rats receiving antiserum against S100 and rats receiving rat γ-globulin antiserum intraventricularly.

In a recent paper (16), we studied the response during reversal of handedness of two protein fractions of hippocampal nerve cell (CA3). These proteins move on the acidic side at electrophoresis at pH 8.3. During the course of one month of intermittent training (day 1-4, 14, 30-33), the values for the incorporation of ^{3}H-leucine into the protein were increased after the first and second training period, but not after the last training. The rats performed well at each training period. This temporal link between behavior and protein synthesis indicated that the protein response was linked to learning processes of the neurons and not an expression of increased neural function in general.

Jankovic et al. (20) reported the in vivo effect of anti-brain protein antibodies on defensive conditioned reflexes in the cat and found significant changes in conditioned responses immediately after the administration and on subsequent days.

Mihailovic et al. (21) studied the effects of intraventricularly injected anti-brain antibodies on visual discrimination tests performance in Rhesus monkeys. The animals were injected with anti-caudate nucleus, anti-hippocampal and normal γ-globulin, respectively. The anti-caudate and anti-hippocampal animals were temporarily impaired in the performance as compared to the normal γ-globulin animals. It is also interesting to note that a high level of environmental stimulation leads to a thicker hippocampus with a higher density in both oligo- and astroglia compared to those of control rats living in isolation (22).

Acknowledgements

We thank Dr. L. Levine, Brandeis University, Waltham, Massachusetts, who kindly provided the antiserum against the S100 protein, and Dr. Blake Moore, Department of Psychiatry, Washington University, St. Louis, Missouri, who generously gave us S100 protein.

This study has been supported by the Swedish Medical Research Council, Grant B69-11X-86-05B, and a grant from Riksbankens Jubileumsfond.

REFERENCES

1. Moore,B.V., and McGregor,D. J. biol. Chem. 240:1647 (1965)
2. Hydén,H., and McEwen,B. Proc. nat. Acad. Sci. 55:354 (1966)
3. McEwen,B.S., and Hydén.H. J. Neurochem. 13:823 (1966)
4. Dannies,P.S., and Levine,L. Biochem. biophys. res. comm. 37:587 (1969)
5. Zuckerman,J., Herschman,H., and Levine,L. J. Neurochem. 17:247 (1970)
6. Moore,B.W., and Perez,V.J. in Physiological and Biochemical Aspects of Nervous Integration, (Ed. F.D. Carlson, Prentice-Hall, Inc., Englewood Cliffs, N.J., 1968, p. 343)
7. Bennett,G.S., and Edelman,G.M. J. biol. Chem. 243:6234 (1968)
8. Bogoch,S., The Biochemistry of Memory (Oxford University Press, London 1968)
9. MacPherson,C.F.C., and Liakopolou,A. Fed. Proc. 24, Part 1, Abstr. 272 (1965)
10. Kosinski,E., and Grabar,P. J. Neurochem. 14:273 (1967)
11. Warecka,K., and Bauer,H. J. Neurochem. 14:783 (1967)
12. Hydén,H., and Egyhazi,E. Proc. nat. Acad. Sci. 52:1030 (1964)
13. Wentworth,K.L. Genet. Psychol. Monogr. 26:55 (1942)
14. Hydén,H. Nature 184:433 (1959)
15. Hydén,H., and Lange,P.W. J. Chromat. 35:336 (1968)
16. Hydén,H., and Lange,P.W. Proc. nat. Acad. Sci. 65:898 (1970)
17. Mihailovic,L., and Hydén,H. Brain Res. 16:243 (1969)
18. Klatzo,I., Miquel,J., Ferris,P.J., Prokop,J.D., and Smith,D.E. J. Neuropath. exp. Neurol. 23:18 (1964)
19. Steinwall,O., and Klatzo,I. Acta Neurol. Scand. Vol. 41,

Suppl. 13 (1964)

20. Jankovic,B.D., Rakic,L., Veskov,R., and Horvat,J. Nature
    218:270 (1968)
21. Mihailovic,L., Divac,I., Mitrovic,K., Milosevic,D., and
    Jankovic,B.D. Exp. Neurol. 24:325 (1969)
22. Walsh,R.N., Budtz-Olsen,O.E., Penny,J.E., and Cummins, R.A.
    J. comp. Neurol. 137:361 (1969)

SOME ASPECTS OF ENZYME CATALYSED ASYMMETRIC REACTIONS OF SYMMETRICAL MOLECULES

A.H. ETEMADI

Institut de Chimie des Substances Naturelles - C.N.R.S.
91, GIF-SUR-YVETTE (FRANCE)

Intra- and intermolecular interactions involving the residual forces leading to the formation of bonds other than covalent ones are of paramount biological importance. If these residual forces did not exist no aggregate of organic molecules would appear, this means that at submicroscopic level a random mixture of molecules would exist; at microscopic level, well defined membrane structures would not appear ; and at macroscopic level no beings like plant,bird or man would be born. Aspects of these interactions are discussed in respect to the structure of nucleic acids, to their replication, to the conformation of proteins and the regulation of their genesis and their function, to the cohesion of membrane structures, in a word in connection with many basic processes of life ; the concept of complementarity governs the enzyme-substrate interaction.

Concept of complementarity. This is a fact long established, that enzymes generally use one of two enantiomers. Pasteur observed that the microorganism Penicillium Glaucum would accumulate (-) tartaric acid, hence use only the (+) isomer from a racemic mixture introduced in its growth medium. Emil Fischer introduced (1894) the " lock and key "image for enzyme-substrate interaction, and this image has been extended to immune (antigen-antibody, antibody-hapten) reactions, first by Paul Ehrlich (1906). Thus, the first image of complementarity was introduced at the turn of the century. In general, complementarity may exist between two geometrical forms. So far as we are concerned, the complementarity of the surface of the enzyme and that of the substrate is in question. The enzyme is asymmetric ; that is, in particular, its active site is asymmetric ; in other words the arrangement of atoms, ions, groups in its active site is such that it presents chirality or handedness.Before going further, an important point must be made : complementarity for a chiral surface does not mean only a geometrical convenience

133

but may imply all types of residual intermolecular forces : ionic interaction, hydrogen bonding, Van der Waals forces, dipole—dipole interaction. Let us now envisage in a very succinct and schematized manner
some different types of substrate molecules in connection with their
interaction with enzymes.

Complementary of Cabcd (chiral) centre and the enzyme. The term
" chiral " is used to designate a dissymmetric molecule (1), that is a
molecule which is not superimposable to its mirror image (generally one
uses the term asymmetric instead of the more correct term dissymmetric).
The chiral molecule of the enzyme uses only one of the enantiomers of
this type of molecule the dissymmetric centre of which is represented by
Cabcd. In the case both enantiomers reach the active site, discrimination between them may be explained in considering at least three interactions between the substrate and the enzyme.

The scheme shows a chiral molecule (I) and its three points of interaction with an enzyme. Its enantiomer cannot present these three interactions, thence is not a substrate.

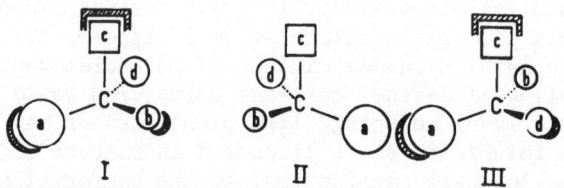

I) A chiral molecule Cabcd in its three point interactions with an
enzyme. II) The enantiomer of I. III) The same enantiomer II in its
interaction with the enzyme. It can be seen that two interactions that
of a and c groups may occur.

It seems unnecessary to give an example of this kind of specificity
as this is a rule. The reverse, the case of an enzyme being able to ca-
talyse reaction upon both enantiomers has been reported but is an exception.

Complementarity of Caabc (" meso " or " prochiral ") centre and the
enzyme. The carbon atom of Caabc type presenting two identical substituents is called a " meso " carbon (2) or to use a more recent naming
a " prochiral " carbon (3). That means that if one of the two identical
substituents is replaced by a different substituent a centre of chirality is created. Consequently, the two identical substituents of a
" prochiral " centre are sometimes called enantiotopic if the molecule
has no asymmetric centre or diastereotopic if the substitution at prochiral centre by a different substituent can bring about diastereoiso-
mers, this eventuality occuring when the molecule, besides the " prochiral " carbon, has an independent asymmetric centre.

The first emphasis on the asymmetrical use of a symmetrical molecule presenting a <u>prochiral</u> centre was brought by Ogston and dates from 1948 (4). At that time there was some reluctance to accept the idea that oxaloacetate when condensing with acetate (we know today that acetyl CoA is involved) would be converted to citric acid because the label of the carboxyl group located at α to CH_2 of oxaloacetate, appeared, in α-keto-glutarate derived enzymically, only at the carboxyl in α to the keto group and not at all at the other carboxyl. Ogston at that time explained that an enzyme, which is asymmetrical, can discriminate between the two identical groups of a symmetrical molecule. This concept is today generally accepted. Here again one partner of the complementarity is the enzyme, the factor intervening being the asymmetry of its active site. A symmetrical molecule of the type Caabc cannot present to the enzyme, indifferently, two identical substituents <u>a</u> and <u>a</u> without exchanging the positions of the two non identical substituents. In the figure we see that if the summit a_1 of the tetrahedron, at the centre of which the <u>prochiral</u> carbon is placed, is brought to a_2 and vice versa, <u>b</u> is forcibly brought to <u>c</u> and vice versa : the <u>chiral</u> enzyme cannot contract the same interaction with <u>b</u> and <u>c</u> indifferently. This may explain the stereospecificity of the catalysis.

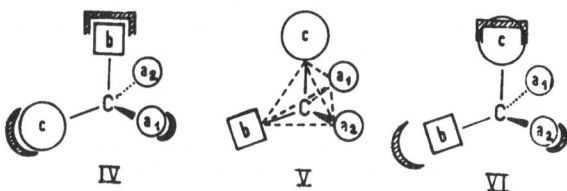

IV) A <u>prochiral</u> molecule Caabc is shown in its three point interactions with an enzyme. V) The same molecule is shown when the two identical substituents <u>a</u> and <u>a</u> (a_1 and a_2 in the fig.) are interconverted. The interconversion brings about the interconversion of the two non identical substituents <u>b</u> and <u>c</u>   . VI) As a consequence, the enzyme-substrate interaction in both points <u>b</u> and <u>c</u> is destroyed.

If the two identical substituents of a <u>prochiral</u> carbon are not hydrogen atoms certain enzyme reactions can convert it to a <u>chiral</u> centre. Thus, glycerol, a <u>prochiral</u> molecule, in presence of ATP and glycerokinase is reversibly converted to sn-glycerol 3-(dihydrogen phosphate) (5) (for the nomenclature used see (6)).

The study of the stereochemistry of enzyme catalysed reactions involving a <u>prochiral</u> centre in which the two identical substituents are H is possible only if a <u>chiral</u> analogue, stereospecifically labelled

with two isotopes of hydrogen, is used as substrate or formed as product in the course of reaction. Westheimer made the first suggestion
about the use of isotopes of hydrogen in the study of oxido-reduction
reactions using DPN or TPN as cofactors ; Westheimer and his colleagues
(for reviews see (7)) studied, by using ethanol, DPN, alcohol dehydrogenase and deuterium labelling, the stereospecificity of the hydrogen
transfer during oxido-reduction of ethanol. These authors demonstrated
that during the reaction only one of the two enantiotopic hydrogens of
the alcohol has been removed stereospecifically and showed that the hydrogen is transferred stereospecifically to one of the two sides of the
pyridine ring. In a _prochiral_ carbon of the type CH_2bc the two hydrogens
are designated as H_S and H_R and are respectively defined as the one
which, when considered as having priority over the other, will give to
the carbon an absolute S or R configuration as defined by Cahn, Ingold
and Prelog (8). In the alcohol dehydrogenase reaction the hydrogen removed from C_1 of the alcohol is H_R and is transferred to one specific
side of the pyridine ring of DPN called A side. Other studies have shown
that in other oxido-reduction reactions the opposite side of the DPN is
involved. This side has been defined as the B side. A and B sides are
likewise defined for TPN (for reviews see 9).

Owing to the work in Cornforth and Popjak's group (10, 11) we know
today that in the A group of dehydrogenases the 4-H_R of reduced pyridine
nucleotides is stereospecifically removed and that with this group of
enzymes the hydrogen introduced on the oxidized pyridine nucleotides has
the 4-H_R configuration. In the B group dehydrogenases the 4-H_S is involved during the oxido-reduction. As a consequence, if the H_R of ethanol
is replaced by deuterium the alcohol dehydrogenase catalysed hydrogen
transfer can be shown as follows :

Many examples of asymmetrical reactions involving hydrogen atoms
on meso carbons are known. In a brilliant series of researches Cornforth
and Popják's laboratory studied the substrate stereochemistry in squalene biosynthesis from mevalonate and solved all but one of the stereochemical ambiguities foreseen for the reactions involved ; they notably
solved all the stereochemical problems in respect to meso carbons. The
reader is invited to refer particularly to these author's article (12)
which constitutes the outstanding paper in the field.

Complementarity of an unsaturated molecule and the enzyme. We will
only mention the case of double bonds. The five σ bonds of the two carbons involved in a double bond are in a plane, bonds making an angle of
120°. The four atoms attached to the two carbons are in the same plane.

This situation is of stereochemical importance in the course of enzyme catalysed reactions involving double bonds.

Suppose first that the four substituents are different and consider the complementary surface of the enzyme, presented for convenience as its projection in a plane as it appears in the figure : the three point interactions between the substrate VII and the enzyme do not need to involve, as the projection seems to suggest, only one of the two faces of the molecule. Let us now turn through 180° the unsaturated molecule and present it as VIII to the enzyme, we see that no translational movement of this can satisfy the complementarity requirements. It is important to note that the direction of the molecular π orbital is perpendicular to the plane of the double bond and its form is symmetrical. Turning the substrate through 180° does not change its interaction capabilities unless the nature of the substituents changes the symmetry of distribution of the electronic cloud of the molecular π orbital as would happen, for example, if conjugation exists. The substituents may intervene in this way in enzyme-substrate interaction or directly by interacting with

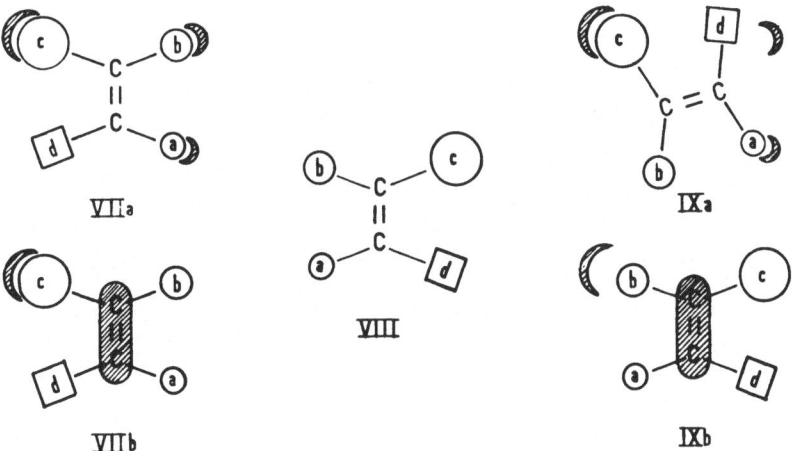

VIIa and VIIb) A molecule of C_2abcd type in its interaction with the enzyme either by three substituents (VIIa) or by involving the two carbons of the double bond and the substituent c (VIIb). VIII) The substrate molecule is turned through 180°. IXa and IXb) The substrate turned is now presented again to the enzyme. These eventualities are chosen arbitrarily for the sake of illustration.

the enzyme. Here again we can see that stereospecificity may be explained on the basis of three point interactions between the enzyme and the substrate.

In the range of different types of nonounsaturated compounds only the one of C_2a_4 and two (X and XI) of the three isomers of the $C_2a_2b_2$

can be expected to be presented to the enzyme indifferently from one
side or the other ; all the others, the third isomer of $C_2a_2b_2$, the
three isomers of C_2a_2bc and the six isomers of C_2abcd present an inequivalence of the two faces of their double bonds. This inequivalence of
the two faces or " prochirality " exists equally for keto groups. A
system of naming has been suggested for these " prochiral " carbon
atoms seen from one side or the other of the plane of the double bond.
In using the priority rule of Cahn, Ingold and Prelog (8) if the substituents of a carbon, seen from a given side, are arranged clockwise,
then it will be named re (rectus) if the substituents are seen anti-
clockwise the carbon is called si (sinister), for example if one considers fumaric acid one recognizes two faces, for one the two carbons are
re and for the other si (3).

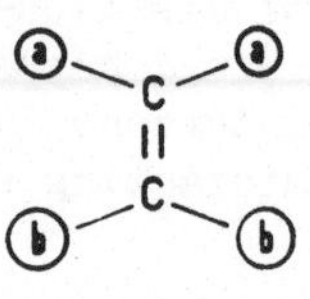

X 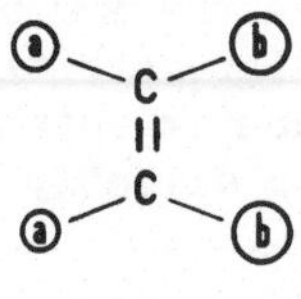   XI 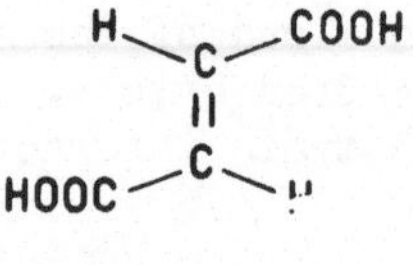   re, re face 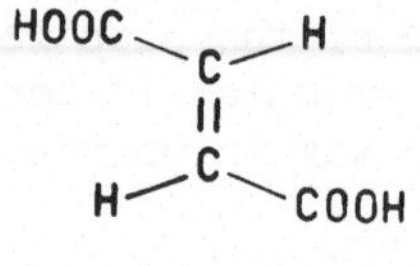   si, si face

     Let us now consider, as an example of the stereospecific enzymic
reaction involving an unsaturated molecule of the type $C_2a_2b_2$, the case
of the hydration of fumaric acid catalysed by fumarase ; here the introduction of OH and H (or their removal) proceeds stereospecifically. The
fact that the introduction of OH is stereospecific is long known (13) ;
more recent works (14, 15) have established that the addition is trans,
so that the diastereotopic hydrogen introduced has the erythro configuration vis-à-vis the OH group. If now we continue the examination of
this case, we realise that the addition of water with its known stereochemistry may still proceed in two ways A and B as shown in the figure,
in both eventualities [2S-3-H_R] malate is formed. There persists however
the question of choosing between these two ways, the answer will wait

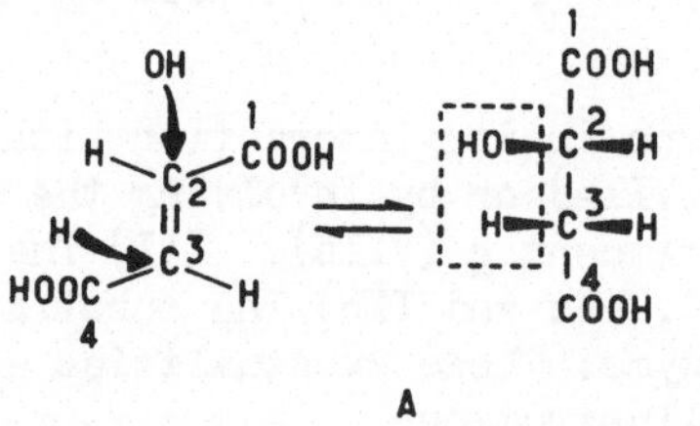

A 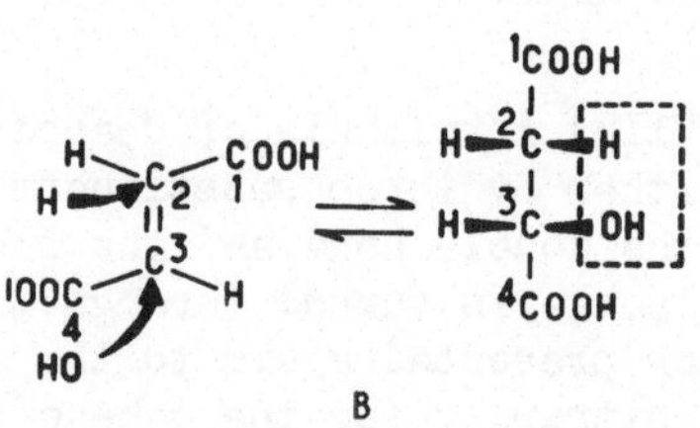   B

more knowledge about the enzyme and the exact way the enzyme-substrate
complex is formed. A similar question arises in connection with the succinate dehydrogenase reaction.

    Complementarity of Caaab centre and the enzyme. In this case no
enzyme can discriminate between the three identical substituents as the

free rotation around C—b bonding, the direction of which constitutes a ternary axis of symmetry, can interchange one of the substituents with any others. Cornforth and his colleagues in one hand (16) and Arigoni and his group on the other (17) have been able to demonstrate, by synthesizing both enantiomers of acetic acid C ^{1}H ^{2}H ^{3}H—COOH, stereospecifically labelled with deuterium and tritium, that malate synthesis from acetyl CoA and glyoxylate follows a stereospecific path as shown (to avoid entering into details the numbering of hydrogen in our figure does not refer to the isotopes) :

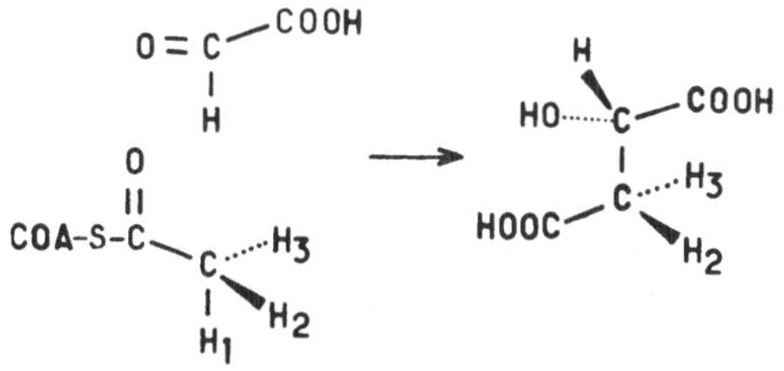

The probability being the same for anyone of the three hydrogen atoms of a non-labelled acetate sample, each of them, if involved, would be eliminated in the stereospecific path shown ; the " trick " of the synthesis of stereospecifically labelled acetate, in taking advantage of isotope effects, being used to demonstrate the inversion of configuration, otherwise indetectable, of the carbon of methyl group during the reaction (for more details see (16)and(17)).

Enzyme-substrate complementarity and the influence of other molecules. This is a broad chapter comprising many concepts of modern molecular biology as it covers among other problems that of allosteric enzymes (18), that of many enzymes present in membranous structures which are only active in presence of certain lipids (see for example (19)) and that of multienzyme complexes (see for example (20)).

It is out of the subject of this article to discuss these problems here. There is however a possibility of an asymmetrical use of a symmetrical molecule which may occur only if : i) the symmetrical molecule is an intermediate in a series of reactions and, ii) the reaction giving rise to this intermediate and that using it as the substrate for the next step are catalysed by a multienzyme system. In the laboratory of Professor Cornforth and Professor Popják at Milstead we tested such an eventuality in assaying the possible asymmetric use of squalene molecule during its oxidation and subsequent cyclisation.

The oxidation of squalene from only one end would occur if the molecule once synthesized would not dissociate from the multienzyme particle, in this case the probability exists that the two ends will not be available for the oxidation step and that the squalene might be oxidized from only one particular end. It is known that the enzymes involved in cholesterol biosynthesis are of two groups, one group of them is solubi-

lized during the homogenate preparation, the other remains retained in
the microsomal fraction. The first group comprises enzymes involved in
the conversion of mevalonate to farnesyl pyrophosphate and the second,
microsomal fraction, in the steps converting two farnesyl pyrophosphate
molecules into cholesterol. The possibility exists then for the squalene,
formed by squalene synthetase system, not to detach from the multienzyme
and to be oxidized from one particular end only. This, naturally depends
on the interaction of squalene with the enzymes synthesizing it and the
spatial arrangement of the enzyme catalysing its oxidation. We know
today that intermediates are involved in the synthesis of squalene from
farnesyl pyrophosphate the nature of which is under study in different
laboratoires (21, 22). However, it is known from the work of Popják,
Cornforth and their colleagues that the H_S of C_1 of one of the two farnesyl pyrophosphate molecules is eliminated and is replaced by the H_S
of C_4 of TPNH and that the configuration of the C_1 of the other farnesyl
pyrophosphate molecule is inverted during the condensation (for reviews
see (12)). Consequently, if the biosynthesis is carried out in the presence of $[4S-4-^3H]$ TPNH an asymmetrically labelled squalene XII (a and b)
would be formed ; then, if the oxidation proceeds specifically from the
(A) or (B) end of the squalene molecule, the ^3H will have respectively
11α- or 12β-position in lanosterol XIII, the direct product of its cyclisation. However, if no discrimination occurs, both 11α and 12β positions
would be equally labelled. These possibilities can be tested by degra-
dation of lanosterol. The acetate of 24,25-dihydrolanosterol XIV, when
submitted to chromic acid oxidation, leads to a mixture of oxidation
products and particularly to the enedione XV (23) in which the 11α-
hydrogen is lost. Control experiments showed that during this oxidation

XIIa : $R_1 = ^3H$, $R_2 = ^1H$
XIIb : $R_1 = ^1H$, $R_2 = ^3H$ XIII   XIV   XV

carried out in acetic acid medium there is no exchange of protons at
C_{12}. The oxidation of lanosterol to the 7,11 dione and the alkaline
equilibration of the latter would then permit the localisation of the
label.

From the work in K. Bloch's laboratory (24) it is known that pig
liver homogenates prepared in a Waring blendor synthesize lanosterol.
$[^3H^{14}C]$ lanosterol was prepared by incubating $[4, 8, 12-^{14}C_3]$ farnesyl
pyrophosphate with pig liver homogenates, and a system generating
$[4S-4-^3H]$ TPNH, and freed from 24,25-dihydrolanosterol which was also
formed. Then the ratio $^3H/^{14}C$ was adjusted to 7.76 by adding ^{14}C lanosterol prepared from $[2-^{14}C]$ mevalonate and the pig liver homogenates,
$[^3H^{14}C]$ lanosterol was submitted to chromic oxidation and the 7,11-
dione purified. The $^3H/^{14}C$ ratio was 4.04 ; after alkaline equilibration
the ratio fall down to 1.14. If all the ^3H atoms were localized at C-11
we would find the tritium ratio of the 11α-to 12β-position to be ∞, if
all the activity was localized at the C-12 this ratio would be 0, if
however both locations had the same proportion of ^3H the ratio would be
1. In our experiment the ratio was 1.28, that is too little deviation
from the value 1 to be attributed to an asymmetrical use of squalene
molecule in the experimental conditions described (25). The possibility
of asymmetric use of squalene is however not totally excluded as the
preparation of homogenates might have damaged the integrity and the natural organization of enzymes. Professor Cornforth is now planning a new
way of attacking the problem (26) in using, in particular, the whole rat
and the stereospecifically labelled $[5-^3H_S]$ mevalonate ; it is hoped
that we will soon learn more about this step of the cholesterol biosynthesis which has a conceptual interest.

In this brief account we envisaged different cases of asymmetrical
enzymic reactions of symmetrical molecules, all of them are explicable
on the basis of the complementarity of the enzyme and the substrate molecule. Chemists and biochemists have accomplished a most magnificent
task in establishing the stereochemistry of many different cases of this
type of reactions ; with the development of protein chemistry and X ray
crystallography and the existence of most promising precedents in the
knowledge of tridimensional structure of proteins, it is hoped that we
will learn soon about the second partner of the complementarity, the
active site of the enzymes involved. As we saw in some cases like that
of the fumarase catalysed reaction, there is still an ambiguity in defining the exact way the reaction proceeds, one important notion missing
being that of the structure of the protein partner. Interactions of different nature between molecules and, in particular, proteins and other
molecules, are of paramount biological importance. The activity of the
enzymes depends on enzyme-substrate complementarity but other molecules
may interact with enzymes and destroy or reinforce the site of the complementarity. The asymmetrical use of squalene in the sense we just discussed, if proved, would be a case in which the asymmetrical reaction
would be conditioned by an environmental factor and not by the enzyme
catalysing the reaction itself.

Finally, in connection with the urgency to learn more about the
protein partner, I wish to mention the following statement of Cornforth
and Popják in their Ciba lecture (12) : " the " three-point attachment "
is however not a unique condition for a substrate to be treated by an

enzyme in a stereospecific manner. More generally, dissymmetric treatment of substrate by an enzyme can occur whenever the enzyme imposes,
whether actively by binding or passively by obstruction, a particular
orientation on the substrate at the site of reaction ... attachment of
substrate to the enzyme by one group alone would suffice for a completely stereospecific reaction ". It is clear that the proof for this
revolutionary statement, which introduces new and additional possible
explanations for the stereospecificity of enzyme catalysed reactions,
will wait too, more knowledge about the exact enzyme-substrate interactions and particularly the knowledge of the protein partner.

1. Mislow, K. Introduction to Stereochemistry, Benjamin. New York
    (1965)
2. Schwartz, P., and Carter, H.E. Proc. Nat. Acad. Sci., Wash 40 :
    499 (1954)
3. Hanson, K.R. J. Amer. Chem. Soc 88 : 2731 (1966)
4. Ogston, A.G. Nature 162 : 963 (1948)
5. Bublitz, C., and Kennedy, E.P. J. Biol. Chem 211 : 963 (1954)
6. Bull. Soc. Chim. Biol 50 : 1363 (1968)
7. Vennsland, B., and Westheimer, F.H. in The Mechanism of enzyme
    action (Eds.W.C. Mc Elroy,B.Glass) John Hopkins Press,Baltimore(1954)
8. Cahn, R.S., Ingold, C.K., and Prelog, V. Experientia 12:81 (1956)
9. Levy, H.R., Talalay, P., and Vennsland, B. Prog. Stereochem 3 :
    299 (1962)
10. Cornforth, J.W., Ryback, G., Popjak, G., Donninger, C., and
    Schroepfer Jr.,G. Biochem. Biophys. Res. Commun 9 : 371 (1962)
11. Cornforth, J.W., Cornforth, R.H., Donninger, C., Popják, G., Ryback,
    G., and Schroepfer Jr., G. Proc. Roy. Soc.(London) B 163:436 (1965)
12. Popják, G., and Cornforth, J.W. Biochem.J. 101 : 553 (1966)
13. Straub, F.B. Z. Physiol. Chem 275 : 63 (1942)
14. Gawron, O., and Fondy, T.P. J. Amer. Chem. Soc 81 : 6333 (1959)
15. Anet, F.A.L. J. Amer. Chem. Soc 82 : 994 (1960)
16. Cornforth, J.W., Redmond, J.W., Eggerer, H., Buckel, W., and
    Gutschow, C. Nature 221 : 1212 (1969)
17. Lüthy, J., Rétey, J., and Arigoni, D. Nature 221 : 1213 (1969)
18. Monod, J., Changeux, J.P., and Jacob, F. J. Mol.Biol. 6:306 (1963)
19. Jones, P.D., Holloway, P.W., Peluffo, P.O., and Wakil, S.J.
    J. Biol. Chem 244 : 744 (1969)
20. Lynen, F. Exposés Annuels de Biochimie Médicale 1 (1967)
21. Rilling, H.C. J. Biol. Chem 241 : 3233 (1966)
22. Popják, G., Edmond, J., Clifford, K.H., and Williams, V. J. Biol.
    Chem.244 : 1897 (1969)
23. Ruzicka, L., Rey, E., and Muhr, A.C. Hel. Chem. Acta 27 : 472 (1944)
24. Tchen, T.T., and Bloch, K. J.Biol. Chem 226 : 921 (1957)
25. Etémadi, A.H., Popják, G., and Cornforth, J.W. Biochem. J 111 : 445
    1969
26. Cornforth, J.W. Quart. Rev 23 : 125 (1969)

SECTION II

BIOCHEMICAL AND MORPHOLOGICAL INTERRELATIONS

BIOCHEMISTRY OF THE DEVELOPING

AUTONOMIC NEURON

E. Giacobini

Dept. of Pharmacology

Karolinska Institutet, Stockholm, Sweden [x)]

ESTABLISHMENT OF SYNAPTIC CONNECTIONS DURING DEVELOPMENT OF THE NERVOUS SYSTEM.

According to the theory proposed by Sperry (1), (2) patterning of synaptic connections is handled by growth mechanism directly, independently of function and with very high "selectivity".According to the "selectivity theory" the establishment and maintenance of synaptic connections depends on highly specific cytochemical affinities ("similarities" ?). This mechanism can be regulated in at least three different ways: a) differentiation (genetically directed) b) induction by means of contact c) other effects, like "embryonal gradient".

One can think that growing fibers and cells in the nervous system are both marked at very early stage of development by means of individual "identification tags" ("I.D. cards" acc. to Sperry (2)).

Sperry (2) proposed that these "identification tags" may be of chemical nature. In this way the millions of nerve cells may be "recognized" by other cells through their processes (Fig. 1).

It has been subsequently suggested (3) that the "identification tags" may be identical with the specific proteins involved in the mechanism of chemical transmission, that is with those enzymes participating in synthesis, metabolism, inactivation or transport of the transmitter molecules (Fig. 1).

x) Present address: Research Laboratories, AB Draco, Fack,
                      S-221 01 Lund, Sweden

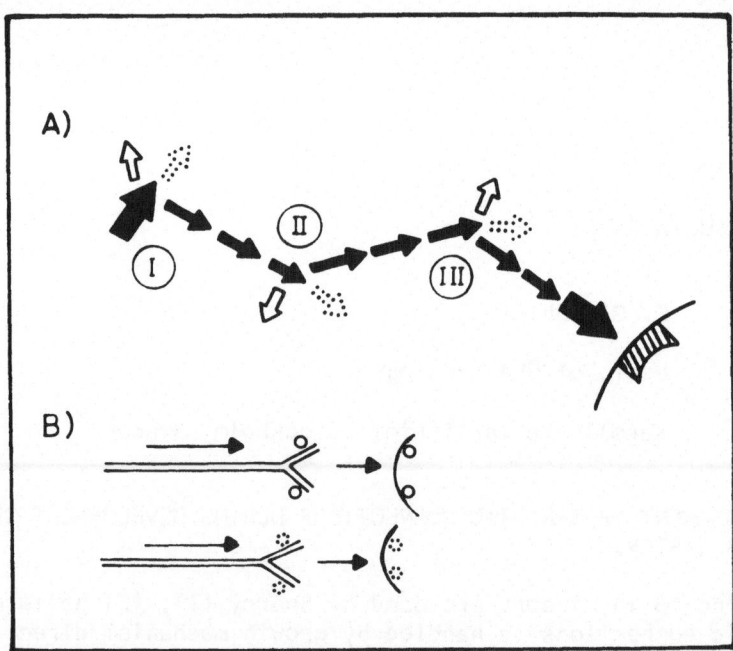

Fig. 1   Schematic diagram of growth of nerve fibers during
development: A) Inspite of various possibilities the nerve fiber
chooses selectively "one" direction of growth. B) Individual
"identification tags" of chemical nature are guiding selectively
the nerve fiber to connect with the "correct" nerve cell. See text.

THE DEVELOPMENT OF THE SYMPATHETIC NEURON

The autonomic and spinal ganglia of chick embryo are readily iden-
tifiable and separable as early as at the 4th - 5th day of incubation. The nerve cells of these ganglia undergo very marked changes,
of size and structure during the embryonic life ; during which
time the synapse is formed (only in the sympathetic ganglia).
Embryonic autonomic ganglia represent therefore an excellent material for studying the appearance and changes of specific proteins
connected with transmission.

The advantages of working with embryonic ganglia instead of the
CNS are obvious, therefore, in this paper the reported studies
will apply to this subject only.

The postganglionic transmitter, NA (noradrenaline), which is synthetized in the cell bodies of the sympathetic neurons can be demonstrated histochemically in the chick embryo of 3 - 4 days of
incubation (4). Fig. 2.

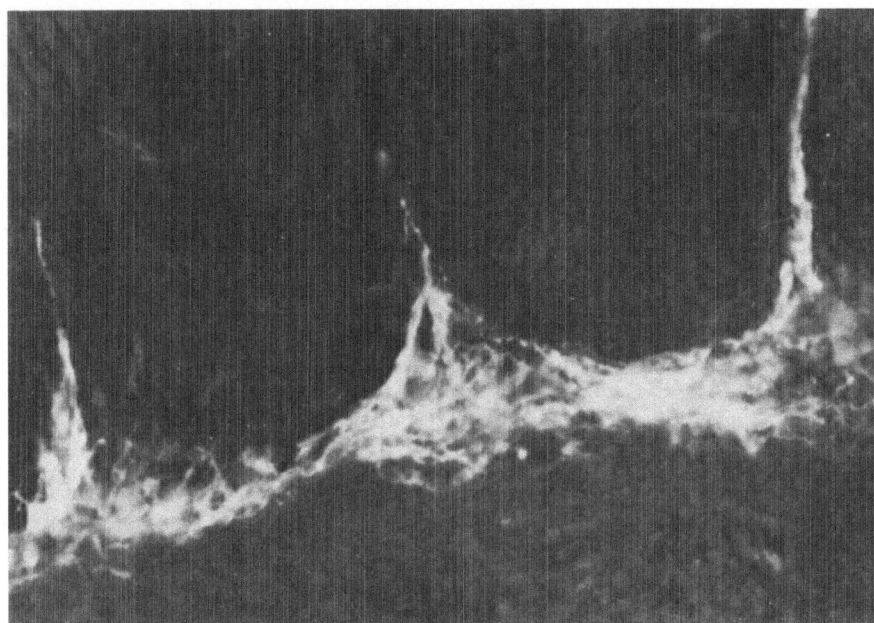

Fig. 2   Sympathetic chain in the chick embryo (4 - 5 days of
incubation). Note the strongly fluorescent segmental swellings
(primitive ganglia). (Courtesy of Dr. Enemar (4))

From the primitive sympathetic cells two different lines of development emerge (Fig. 3), the first will give rise to chromaffine
cells and the second to sympathetic neurons. The final localization
of the two cell types is multiple (Fig. 3).

The sympathetic nervous system undergoes a rapid development : in
the chick embryo the different peripheral organs receive their
sympathetic innervation during the period 3rd - 8th day of incubation (5), (6), (7).

A relationship has been described between the development of the
sympathetic synapse, the maturation of the sympathetic neuron and
the initial appearance and development of the transmitter -
metabolizing enzymes, (8).

Furthermore, a sequential appearance of NA, adrenaline (A), their
precursors and the enzymes participating in the biosynthetic
pathway of formation (dopa $\longrightarrow$ dopamine $\longrightarrow$ NA $\longrightarrow$ A) has been shown
by Burack and Badger (9) and Ignarro and Shideman (11) in homo-
genates of chick embryos and peripheral organs. In the heart,

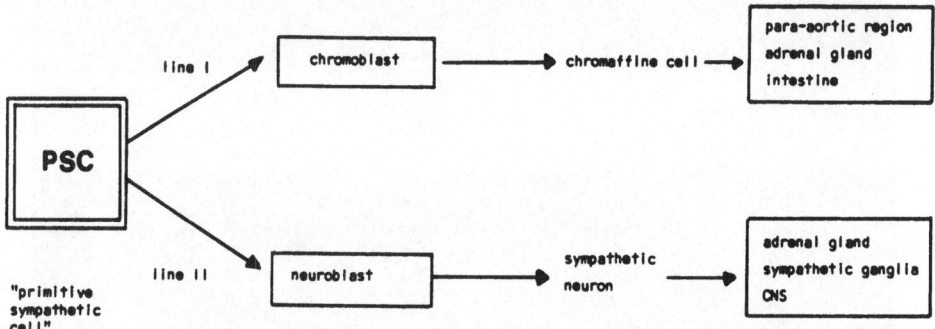

Fig. 3   Diagram of development of the sympathetic neuron.

dopa and dopamine were first detected on the 4th and the 6th day
of incubation respectively (11).

Tyrosine hydroxylase (TH), dopa-decarboxylase (dopa-DC), dopamine-
-β-oxidase and phenylethanolamine-N-methyl transferase activities
were first detected on the 1st, 2nd, 4th and 6th day of incubation,
respectively (11).

It should be mentioned that in the chicken embryo the sympathetic
nerves reach the heart as early as on the 5th day of incubation,
(5), (6), (7).

The exact time of appearance of monoamines in the cellular storage
sites inside the neuron is not known and a great deal of ultrastructural data on the development of the synapse inside the
sympathetic ganglion are still lacking.

It is known that adrenergic nerves appear in their different innervation structures at different time and a theory of "hetero-
chronism of innervation" for different organs has been suggested
(12). The study of the structural development of specialized
organelles storing and transferring the transmitter in the sympathetic neuron has just begun. However, the studies of Machado (12)
in the rat pineal gland throw some light upon a possible mode of
formation of the small granular vesicles from neurotubules or from
the smooth membranes in the endothelial reticulum. According to this
author the granule content in the vesicle is formed peripherally
and progressively increases to fill the whole vesicle.

A tentative diagram of the formation of cellular storage sites
for catecholamines in autonomic neurons is reported in Fig. 4.

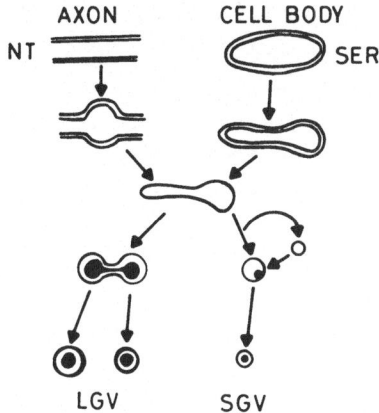

Fig. 4. Development of specific transmitter carriers (LGV = large granulated vesicles, SGV = small granulated vesicles) in the cytoplasm (SER = smooth endothelial reticulum) and in the axon (NT = neurotubules) of the sympathetic neuron (after Machado), (12).

APPEARANCE AND TREND OF ENZYME ACTIVITY DURING DEVELOPMENT IN CHICK EMBRYO SYMPATHETIC GANGLIA

In a series of investigations, cholineacetylase (ChAc), total cholinesterase (ChE), acetylcholinesterase (AChE), dopa-DC, mono-amineoxidase (MAO) activity and proteins were measured throughout the embryonic life of the chick in spinal and sympathetic ganglia, (8), (13), (14).

The enzymatic activities were determined in µg samples of the ganglia by means of the sensitive radiometric micromethods developed in our laboratory (for references, see (3)).

In chick embryo spinal and sympathetic ganglia, AChE accounts for practically all the ChE activity (8).

Acetylcholinesterase activity increases in both ganglia until the 12th day of incubation: during this period neuroblasts actively multiply and undergo differentiation. After the 12th day there is a steady decrease in spinal ganglia AChE, which drops more than 10 fold in terms of protein.

The specific activity of AChE remains at the high level in sympathetic ganglia in the second half of embryonic development. This event may be related with the morphological maturation of synapses starting around the 8th-9th days of development (15).

Acetylcholinesterase and MAO activity increases in both spinal and sympathetic ganglia almost parallel from the 6th to the 12th day of incubation. From this day on a significant divergence occurs, mainly due to a steady fall in spinal ganglia AChE.

In the sympathetic ganglia, dopa-DC does not show the same pattern of activity as other enzymes (14), but continues to increase progressively from the 5th day of incubation to the 21st and shows a tendency to increase after hatching (Fig. 5).

In the spinal ganglia the dopa-DC activity   progressively diminishes from the 5th day to the 21st (Fig. 5).

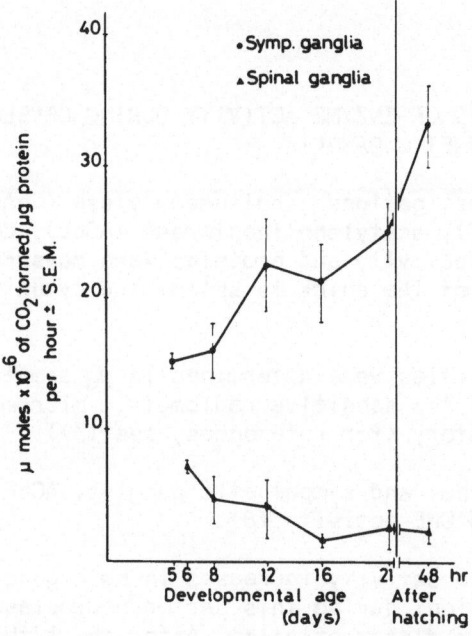

Fig. 5   Variations of dopa-DC activity in chick embryo spinal and sympathetic ganglia.

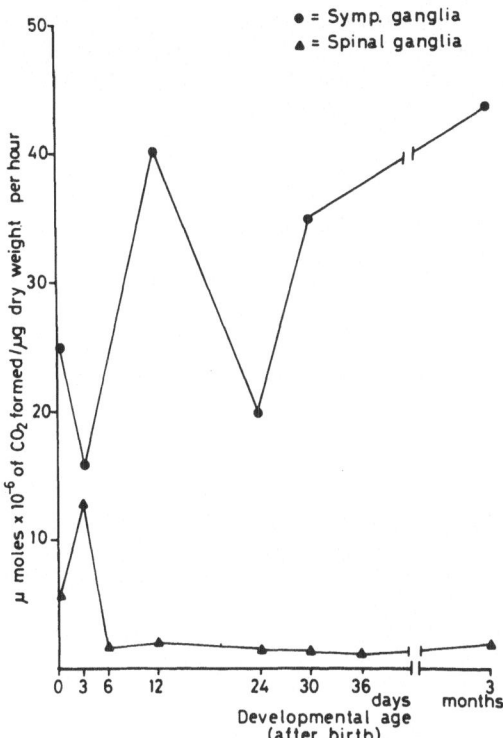

Fig. 6   Variations of dopa-DC activity in rat spinal and sympathetic ganglia.

Figure 6 shows the variation of dopa-DC activity in rat ganglia
from birth to 3 months of age (preliminary results) (16). A marked
difference between the two curves is apparent. After a rapid and
short lasting increase the dopa-DC activity in the spinal ganglia
reaches a very low level. In the sympathetic ganglia a peak is
apparent around the 12th day after birth. A trend to increase after
the 24th day can be followed up to the 3rd month.

A similar peak is noticed in the chick heart, catechol-O-methyltransferase (COMT) and MAO activities (10). These enzymes are first
detected in the embryonic heart on the 4th day and increases thereafter until the 10th day when their activity plateaux (10).
Activities of both enzymes increases on the 15th to 16th day and
attaines maximal values by the 19th day. Subsequently COMT and
MAO exhibits sharp declines in their activities. After a new,
gradual increase through the next 6 - 7 days, after hatching, COMT
activity gradually declines and its activity cannot be detected
in the adult chicken, whereas MAO activity continues to increase
throughout development (10).

The peak in dopa-DC activity found by us in the rat ganglia around
the 12th day correlates well with the peak levels of dopamine and
NA found by Ignarro and Shideman (11) in developing heart of chick
embryo in the corresponding period.

SIGNIFICANCE OF ENZYMATIC CHANGES IN SYMPATHETIC GANGLIA DURING DEVELOPMENT

In order to be able to interpret the complex changes in enzyme
activities occurring during development we must consider that the
sympathetic neuron in this period is not only growing in size but
also actively establishing functional connections with the periphery.

The variation in AChE, ChAc, MAO and dopa-DC activity in the two
types of ganglia, "synaptic" (sympathetic) and "asynaptic" (spinal),
can be better compared studying the ratio between the two ganglia
(symp./spinal) (Fig. 7). A distinct peak of activity is observed
at the 8th day with a ratio of 2, 3 and 5 for AChE, MAO and ChAc
respectively (8), (14).

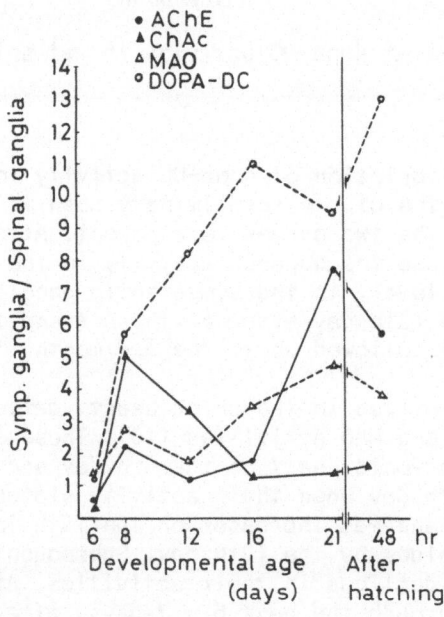

Fig. 7   Developmental changes of the ratio sympathetic / spinal
ganglia for AChE, ChAc, MAO and dopa-DC activity in chick embryo.

This indicates that maximal activity of the three enzymes is
reached during the period in which synaptic connections begin to
form in the ganglion.

As mentioned in the previous section, dopa-DC does not show in the
sympathetic ganglia the same pattern of activity as the other enzymes (14), but continues to increase from the 5th day of incubation
to the 21st (Fig. 5). The NA content of the ganglia shows a similar
pattern (4).

The localization of these enzymatic changes is principally at the
synapse for AChE and ChAc and in the cell body for MAO and dopa-DC.

In the ensuing days the ChAc and dopa-DC ratio shows a marked
difference from that of AChE and MAO and no second peak is apparent
at the time of hatching (14).

Since only AChE, MAO and dopa-DC are present (17), (18) in the postganglionic adrenergic pathway (that is in the cell body and the
efferent fibers) the second peak of enzyme activity (Fig. 7, AChE
and MAO activities) may coincide with the accomplishment of the
peripheral innervation and the starting of function at birth. The
intermediary period between the two peaks corresponds probably to
a period of migration of MAO, AChE and dopa-DC from the cell body
towards the innervated region (8), (14), similarly to that described for monoamines during growing processes.

The different subcellular localization for MAO and dopa-DC (the
first being principally mitochondrial and the second soluble and
cytoplasmic) as well as the different metabolic and functional
significance of the two enzymes may influence the rate of transport
from the cell body to the periphery and account for the different
pattern of activities during development.

The ratio between sympathetic and spinal ganglia for AChE, ChAc and
MAO activity, indicating an evident relationship between the development of the synapse in the sympathetic ganglia and the maximal
activity of these enzymes supports the hypothesis (3) that specific
enzymes related to transmission may be identical with the "identification tag" (2) molecules.

ACKNOWLEDGEMENT

The work performed in our laboratory and reported in this paper was
supported by grants from the U.S. Public Health Service, National
Institute of Health, NB-04561-06, from the Swedish Medical Research
Council project no. I2X-246-06 and B-71-I4X-246-06 and from a joint
Swedish-Italian grant from the respective Medical Research Councils
for molecular studies in neurobiology. (Swedish grant No. B70-70E-
-2774-01).

REFERENCES

(1)   Sperry, R.W. Proc. natn. Acad. Sci. U.S.A. 50:703 (1963)

(2)   Sperry, R.W. James Arthur Lecture. Publ. by The American
        Museum of Natural History New York (1964)

(3)   Giacobini, E. Int. Symp. on the Histochemistry of Nervous
        Transmission (Helsinki, 1970). In "Progress in Brain Re-
        search". (Ed. O. Eränkö), Elsevier & Co. Publ. (1970. In press.

(4)   Enemar, A., Falck, B. and Håkansson, R. Developmental Bio-
        logy, 11:268 (1965)

(5)   Huettner, A.F. Fundamentals of Comparative Embryology of
        the Vertebrates, The Macmillan Company, New York,(1949)

(6)   Hamilton, H.L. Lillie's Development of the Chick, 3rd ed.,
        Holt, Rinehart and Winston, New York (1952)

(7)   Romanoff, A.L.The Avian Embryo, The Macmillan Company, New
        York (1960)

(8)   Giacobini, G., Marchisio, P.C., Giacobini, E. and Koslow, S.
        J. Neurochem. 17:1177 (1970 a)

(9)   Burack, W.R. and Badger, A. Federation Proc. 23:561 (1964)

(10)   Ignarro, L.J. and Shideman, F.E. J. Pharmac. exp. Ther.
        159:29 (1968 a)

(11)   Ignarro, L.J. and Shideman, F.E. J. Pharmac. exp. Ther.
        159:38 (1968 b)

(12)   Machado, A.B.M. Int. Symp. on the Histochemistry of Nervous
        Transmission (Helsinki, 1970). In "Progress in Brain Re-
        search" (Ed. O. Eränkö), Elsevier & Co. Publ. (1970). In press.

(13)   Marchisio, P.C. and Consolo, S. J. Neurochem. 15:759 (1968)

(14)   Giacobini, G., Giacobini, E. and Noré, B. J. Neurochem.
        (1970 b). In press.

(15)   Wechsler, W. and Schmekel, L. Acta Neuroveg (Wien), 30:427
        (1967)

(16)   Giacobini, E., Eränkö, L. and Noré, B. (1970 b). To be
        published.

(17)   Giacobini, E., Karjalainen, K., Kerpel-Fronius, S. and
        Ritzén, M. Neuropharm. $\underline{9}$:59 (1970 a)

(18)   Giacobini, E. and Noré, B. (1970) Acta Physiol. Scand.
        In press.

METHODS FOR MEASURING INDOLEALKYLAMINE AND CATECHOLAMINE

TURNOVER RATE "IN VIVO"

E. Costa

Laboratory of Preclinical Pharmacology, National Inst.

of Mental Health, Saint Elizabeths Hospital, Washington, D.C.

Introduction

Many theoretical and practical problems are encountered in designing methods to calculate turnover rates of neuronal catecholamines and serotonin "in vivo." The great majority of these difficulties derives from our incomplete understanding of the morphology and biochemistry of monoaminergic nerves. Frequently we circumvent the barriers of our ignorance by over simplifications leading to the formulation of kinetic models. Not only are these models unsatisfactory because of the implicit assumptions but as listed in Table 1, other limitations impair the in vivo measurements of turnover rates of brain monoamines in animals. Of course, the difficulties become insurmountable if we try to measure turnover rates of brain monoamines in man.

TABLE 1

PROBLEMS CONCERNING MODELS TO MEASURE TURNOVER RATE OF
MONOAMINES "IN VIVO"

1. The cytology of monoaminergic tracts is not fully known.
2. The localization of the enzymes and substrates controlling monoamine turnover rate is not fully known.
3. The brain samples where precursor concentrations are measured may not reflect the concentration of amino acids at the sites of amine synthesis.
4. The times when measurements can be made are restricted.
5. The data contain fluctuations.
6. The number of time points that can be measured in a single animal is limited.

The contention that turnover rate of monoamines is related to the rate of neuronal activity is the basic support for studying monoamine turnover during ontogenetic development. If one accepts this relationship as a valid starting point, then by measuring turnover rate of monoamines in various brain structures at different stages of brain development one may hope to increase our understanding of the time required for the functional maturation of monoaminergic neuronal tracts. Moreover, a comparison of the changes in monoamine turnover rates elicited by either drugs or environmental conditions at various stages of brain development might be meaningful to establish when the process of postnatal development is terminated.

These considerations comprise the trying imposition of revealing the nature of the process that interlocks depolarization of neuronal membranes with turnover rates of monoamine stores. Since we attempt to understand how the function of neuronal membranes and that of cytoplasmic constituents converge into a functional interaction, it is important to keep in mind that the biochemical link between these topographically distinct structures may have to operate with a time dimension which mirrors that of neuronal depolarization. However, when turnover rates are measured during either ontogenesis or postnatal development, one should be aware that in adults turnover rate of amines in cell bodies is faster than the turnover rate of the same amines in nerve endings. Hence, it is pertinent distinguishing turnover rate changes before and after the stabilization of the synaptic field in neuronal formation.

ASSUMPTION UNDERLYING TURNOVER RATE MEASUREMENTS OF NEURONAL MONOAMINE STORES

The present inability to ascribe a functional role to several morphological and biochemical components of nerve endings is usually circumvented by representing monoaminergic nerve endings as a hypothetical model where the kinetic properties of the biological system are formulated mathematically. The assumptions which serve as a basis for mathematical models used in turnover rate measurements reported in this paper can be summarized as follows:

a) Neuronal monoamines behave kinetically as if they were contained in an open single compartment system.

b) Enzymatic hydroxylation of tryptophan and tyrosine is rate limiting for catecholamine and serotonin formation.

c) The intermediate hydroxylated amino acids are rapidly converted to their respective monoamines.

d) Neuronal function fails to distinguish between old and newly formed monoamines.

e) Newly formed serotonin and catecholamines are incorporated into a population of biosynthetic units which behave as a reliable sample of the kinetic properties of the mono-aminergic neurons contained in the tissue.

f) Monoamines retrieved from the extraneuronal spaces by the monoamine pump of the neuronal membrane rapidly mix with newly synthesized amines in the kinetically uniform compartment of nerve terminals.

g) The acidic metabolites of monoamines are synthesized in a kinetically uniform compartment and their efflux from brain is facilitated by an active transport mechanism.

ISOTOPIC METHODS TO ESTIMATE CATECHOLAMINE AND SEROTONIN TURNOVER RATES "IN VIVO"

A. Labeling Peripheral Catecholamine Stores with $Dl-NE^3H$.

Attempts to measure the turnover of endogenous monoamines stored in sympathetic nerves began with the availability of radio-active NE of high specific activity and were prompted by the discovery that an amine pump in adrenergic varicosities takes up the amine from the circulation and stores it against a concentration gradient. The turnover rate measurements were facilitated by the finding that after administering a small intravenous dose of radioactive NE, the specific radioactivity of cardiac NE declines exponentially. This characteristic of the decline indicates that within the limits of our detection ability, this storage system behaves as a single open compartment. However, we cannot exclude that the cardiac store might be composed by a two compartment open system where the size of one compartment is about 15% of the other and the transfer of monoamines from one compartment to the other is very rapid. For the purpose of turnover rate measurements we simplify our mathematical models and disregard this second pool because we cannot measure it. Hence, to measure turnover of cardiac NE after tracer labeling with ^3H-NE we consider that the change with time of cardiac NE specific radioactivity (NE_{sa}) is proportional to the NE_{sa} present at various times

$$\frac{d\ NE_{sa}}{dt} = -k_{NE}\ NE_{sa} \qquad (1)$$

where k_{NE} is the rate constant of the efflux of NE from the cardiac store. Integrating

$$NE_{sa} = (NE_o)_{sa} \; e^{-k_{NE}t} \qquad (2)$$

which expressed in common logarithms

$$\log NE_{sa} = \log (NE_o)_{sa} - \frac{k_{NE}t}{2.303} \qquad (3)$$

In practice k can be obtained by plotting in semilogarithmic paper the NE_{sa} of the heart of a population of animals killed in groups of at least 4 animals at various times after a tracer dose of ^3H NE

$$k_{NE} = \frac{0.693}{T\ 1/2} \qquad (4)$$

where T 1/2 is the half time for the decline of NE_{sa}. Turnover rate can be calculated from NE turnover = $k_{NE}NE$.

These simplifications are in order if one can prove that the equation of the line describing the decline with time of NE_{sa} is:

$$\widehat{NE}_{sa} = \overline{NE}_{sa} + b \; (x - \overline{NE}_{sa}) \qquad (5)$$

Where $\widehat{NE}_{sa}$ is the expected value for any theoretical value of x, $\overline{NE}_{sa}$ is the mean of NE_{sa} and b is the regression coefficient or the slope of the line.

B. Labeling of CNS Stores with DL-NE-^3H

Since the circulating NE reaches only limited portions of brain tissue (area postrema, infundibulum, pineal and the pituitary stalk) labeling with ^3H NE cannot be used to measure turnover rates of brain catecholamines. Several authors have proposed and used the intraventricular injection of ^3H NE to measure brain catecholamine turnover.

However, it is difficult to calculate turnover rate from these experiments because the specific radioactivity of brain NE declines as a multiphasic function due to several nonspecific binding sites of ^3H NE in brain tissue. After tracer doses of DL-^3H NE the brain tissue retains both isomers while in periphery most of the D-isomer disappears from the body very rapidly and either is excreted as such, or after metabolism by catechol-O-methyltransferase (COMT) or monoamineoxidase (MAO).

When DL-^3H NE is injected into the brain ventricles, elimination of the tracer from brain is almost exclusively controlled by MAO, but the enzymatic destruction of D-^3H NE is slower than that of L-^3H NE thus creating additional complications (see Table 2).

TABLE 2

ADVANTAGES AND DISADVANTAGES OF VARIOUS ISOTOPIC METHODS TO MEASURE TURNOVER RATE OF BRAIN MONOAMINES "IN VIVO"

METHOD	ADVANTAGE	DISADVANTAGE
Constant infusion of T* and TP*	a) Single animals b) Drug effects c) Simultaneous assay of 5HT and catecholamine turnover rate is possible	a) Immobilization stress b) Technically complex c) No brain parts d) Expensive
Pulse injection of T* and TP* (intravenously or intraventricularly)	a) Brain parts b) Minimizes stress c) Behavioral studies d) Drug effects e) Simultaneous assay of 5-HT and catecholamine turnover rate is possible	a) Measure turnover of a group of rats b) Technically complex c) Expensive
Intracerebral ^{3}H NE	a) Brain parts b) Technically simple	a) Difficult to calculate rates b) Difficult to exclude nonspecific binding
Conversion index of T* and TP* into radioactive amines	a) Suitable for short acting drug effects b) Gives measurements for single rats c) Simultaneous assay of 5-HT and catecholamine turnover rate is possible	a) Fails to correct for amine efflux b) Can measure turnover during short time intervals immediately after labeling

C. Labeling NE and Serotonin Stores with Radioactive Tyrosine (T*) and Tryptophan (TP*)

The turnover rate of endogenous NE and 5-HT cannot be estimated using principles of steady-state kinetics by injecting their
immediate precursors (labeled hydroxylated amino acids) because
aromatic amino acid decarboxylation is not a rate limiting process
for the biosynthesis of these monoamines and because in neurons
the pool of either 3,4-dihydroxyphenylalanine or 5-hydroxytrypto-
phan is exceedingly small. By measuring the conversion rate of
T* to labeled catecholamines one can estimate the turnover rate of
endogenous dopamine and NE, likewise the conversion rate of TP*
into 5-HT yields measurements of the turnover rate of endogenous
5-HT. Since the objective is that of estimating a rate, it seems
unsatisfactory to measure the specific activity of precursors and
products at one time point after the labeling with pulse injections of T* or TP*, assuming as a mathematical model a compartment
open only at one side. Obviously, the validity of such a procedure
can be questioned but we must keep in mind that since we ignore
many essential mechanisms regulating the monoaminergic function,
no procedure can be singled out as an absolutely reliable method
to measure brain monoamine turnover rates in vivo. For the purpose of calculating turnover, one can represent the monoaminergic
nerves as a kinetic compartment open at one side and express the
relationship between tyrosine and catecholamines as in equation (6).

$$\text{Conversion Index} \atop {\text{(m}\mu\text{mole of converted} \atop \text{catecholamines/g tissue)}} = \frac{\frac{\text{dpm Catecholamine/g tissue}}{\text{dpm Tyrosine/g tissue}}}{\text{m}\mu\text{mole Tyrosine/g tissue}}$$

By measuring the specific activity of tyrosine in the tissue
sample and the radioactivity incorporated in the catecholamines
stored in one gram of tissue (conversion index), one can solve
equation (6) and approximate the amount of amine formed in the
time interval between the injection of the label and the sampling
of tissues. This method has however a number of limitations: 1)
as evident from equation (6), this procedure fails to correct for
the efflux of radioactive and nonradioactive catecholamines leaving
the neuron during the time interval between the injection of the
label and the assay of the tissue. Hence, the longer this time
interval the greater the error. However, the extent of this error
can be minimized by keeping the time interval between labeling
and analysis equal to a small fraction of the turnover time of the
store measured; in the case of brain dopamine if this interval is
kept to a maximum of about 25 minutes the error may not be greater
than 15%; 2) caution must be taken to assure that measurements are
taken when the specific activity of the precursor is at least
double that of the product. When either the specific activity of
the precursor is smaller than that of the product or the two

specific activities are equal, equation (6) cannot be used to calculate turnover because the direct relationship does not hold any longer.

This method becomes a useful compromise to evaluate changes of turnover rates of short duration such as those elicited by drugs with a biological half-life of about one hour. Actually, in these experiments, equation (6) is practically the only method that seems reliable.

When T* is infused intravenously at a constant rate, the accumulation of labeled catecholamines with time is dependent on the rate of NE synthesis. Since it is impossible to measure the specific activity of tyrosine in the compartment of adrenergic neurons involved in catecholamine synthesis, one can propose as a provisional alternative that the specific activity of catechol-amines bear a relationship with the specific activity of plasma tyrosine (T_{sa}). If k_1 is the fractional rate constant for hy-droxylation of tyrosine which is only a small fraction of k_T (the fractional rate constant for the turnover of plasma tyrosine)T_{sa} would be expected to change with time as shown in (7).

$$\frac{d\,T_{sa}}{dt} = K - k_T\,T_{sa} \qquad (7)$$

where K is the rate of change of T in the plasma compartment. On integration and imposing the condition that $T_{sa} = 0$ when the duration of infusion (t) is = 0 equation (7) becomes

$$T_{sa} = \frac{K}{k_T}\,(1 - e^{-k_T t}) \qquad (8)$$

At steady-state in an open, single compartment system, the change in concentration of radioactive NE (NE*) would be related to T* as follows

$$\frac{d\,NE*}{dt} = k_1\,T* - k_{NE}\,NE* \qquad (9)$$

and dividing by tissue levels of NE, equation (9) will be pre-sented in terms of changes of NE specific activity (NE_{sa})

$$\frac{d\,NE_{sa}}{dt} = \frac{k_1\,T*}{NE} - k_{NE}\,NE_{sa} \qquad (10)$$

At steady-state the following relationship must be true

$$k_1\,T = k_{NE}\,NE \qquad or \qquad NE = \frac{k_1\,T}{k_{NE}} \qquad (11)$$

substituting for NE in equation (10)

$$\frac{d\,NE_{sa}}{dt} \;=\; k_{NE}\,(T_{sa} - NE_{sa}) \qquad (12)$$

Equation (8) can now be substituted in equation (12) for T_{sa}

$$\frac{d\,NE_{sa}}{dt} \;=\; k_{NE}\left[\frac{K}{k_T}\,(1 - e^{-k_T t)} - NE_{sa})\right] \qquad (13)$$

Since $NE_{sa} = 0$ at $t = 0$ equation (13) integrates into

$$NE_{sa} \;=\; \frac{K}{k_T}\left[1 + \frac{1}{k_{NE} - k_T}\,(k_T\,e^{-k_{NE}t} - k_{NE}\,e^{-k_T t})\right] \qquad (14)$$

Various values of NE_{sa} can be calculated from equation (14) when k_{NE} is equal to either 0.1, 0.2, 0.3, etc. and $t = 40$ minutes. In fact, k_T and k can be calculated from the rate of infusion of T* and the plot of the increase of T_{sa} during the infusion of T*. The results of the various NE_{sa} calculated from equation (14) solved for different sequential and hypothetical values of k_{NE} will allow a plot of k_{NE} versus NE_{sa}. Obtaining the NE_{sa} of each sample, one can read from this plot the appropriate value of k_{NE} and calculate turnover rate multiplying k_{NE} BY NE.

   The turnover rate of NE can also be measured after a pulse injection of T* given intravenously or intraventricularly. The latter procedure should be preferred when the turnover rate is measured in discrete brain areas but it cannot be used to measure turnover rates of peripheral NE stores. To obtain uniformity in the results and to avoid the interferences brought about by anesthesia, a small polyethylene cannula is implanted in the lateral cerebral ventricle a few days before the experiments in a group of twenty rats. Groups of 4 animals each are sacrificed at 40 minute intervals beginning two hours after the injection of the labeled amino acid dissolved in 10 µl of saline. Specific activity of amine and amino acid is measured and turnover rate can be measured from equation (15).

$$k_{NE} \quad \frac{\dfrac{NE_{sa\ t_2} - NE_{sa\ t_1}}{t_2 - t_1}}{\dfrac{(T_{sa} - NE_{sa})_{t_1} + (T_{sa} - NE_{sa})_{t_2}}{2}} \qquad (15)$$

After plotting in semilogarthmic paper the specific activities
of amino acid and amine, the best fit line is drawn by eye. The
graph is divided in consecutive 20 minute intervals and k_{NE}
calculated for each time interval according to equation (15).
These k_{NE} values are averaged and the SE of the mean is calculated.
The method is essentially identical when turnover is calculated
after a single intravenous injection of the labeled amino acid.
The advantages and disadvantages of these isotopic methods are
reported in Table 2.

NONISOTOPIC METHODS TO MEASURE SEROTONIN AND CATECHOLAMINE TURNOVER RATES

A. Inhibition of Catecholamine Biosynthesis

When tyrosine hydroxylase is instantaneously blocked by
appropriate intravenous doses of α-methyltyrosine (MT), catecholamines are no longer formed and the amine levels decline at a
rate that is proportional to the concentration of the remaining
amine. Mathematically, this relationship can be described with
equation (1). Integrating and converting to common logarithms
this relationship is

$$\log NE = \log NE_0 - 0.434 \, k_{NE} t \qquad (16)$$

where NE_0 is the initial steady-state amine concentration and
NE is the concentration at any time t after administration of MT.
A plot of log NE versus time yields a straight line, the slope
of which is 0.434 times the rate constant of NE efflux (k_{NE}).
Since MT inhibits tyrosine hydroxylase by competing with tyrosine
an optimal tissue level of the inhibitor should be maintained
throughout the experiment to assure maximal inhibition of the
process that limits catecholamine biosynthesis. In rats, maximal
in vivo inhibition of tyrosine hydroxylase can be obtained if
tissue concentrations of MT are maintained at about 5×10^{-4}M.
However, caution should be taken to assure that experimental
conditions do not change steady-state levels of brain tyrosine.
If these were to increase concentrations of MT greater than

5×10^{-4} M are required. It is important to calculate the slope
of the decline of tissue NE from at least three experimental
points which should be selected keeping in mind that they have
to include a decline of brain NE concentrations equal at least
to one half-life of endogenous NE.

This method cannot be extended to 5-HT because p-chloro-
phenylalanine, the nonequilibrium, noncompetitive inhibitor of
tryptophan hydroxylation does not act promptly. Until fast
acting inhibitors of 5-HT synthesis are found, 5-HT turnover can
not be measured by this method.

B. Accumulation of Serotonin and Catecholamines After

Inhibition of Monoamine Oxidase

After the intraperitoneal injection of pargyline hydrochloride
(75 mg/kg), a nonreversible MAO inhibitor, brain concentrations
of 5-HT increase linearly for about 2 hours and then plateau at
about three-fold their normal steady-state concentrations. The
turnover rate of brain 5-HT is calculated from the rate of increase
of 5-HT concentration.   When 5-HT turnover rate is measured by
this method it is assumed that: 1) pargyline acts almost instantaneously; 2) the increase of other monoamines elicited by
pargyline does not interfere with the rate of firing and therefore
the rate of production of 5-HT (see Table 3); 3) the initial 5-HT
accumulation reflects the rate of production because at this time
no 5-HT is lost from brain by diffusion; 4) the metabolism of
5-HT by pathways other than MAO is negligible; 5) pargyline does
not interfere with the concentration of TP at the site of 5-HT
synthesis. The latter assumption should be checked whenever 5-HT
turnover rate is estimated after drug injections because the
concentration of TP in brain are below the k_m of tryptophan
hydroxylase (Table 3). Moreover, TP is the only essential amino
acid which is largely bound to plasma proteins, hence any drug
could interfere with this binding and increase the free concentrations of plasma tryptophan. The latter change can reflect
itself in the concentrations of brain TP which are kept by an
active transport mechanism operating on the concentrations of
free TP in plasma. This method cannot be used to estimate turnover rates of brain NE or DM because after MAO inhibition the
concentration of these two amines does not increase linearly.
Perhaps, this is due to product inhibition on the rate limiting
enzyme (Table 3).

TABLE 3

PRESENT UNDERSTANDING OF THE MECHANISMS CONTROLLING
TURNOVER RATE OF MONOAMINES IN MONOAMINERGIC NERVES

FACTORS INFLUENCING BRAIN AMINE TURNOVER RATE	IMPORTANCE OF THE FACTOR IN NERVES STORING		
	NE	DM	5-HT
Tissue concentration of substrates of rate limiting enzymes	above 2X k_m	above 2 X k_m	below 2X k_m
Rate of nerve activity	yes	yes	yes
End product inhibition	yes	yes	no
Tissue O_2 tension	?	?	yes
Blood level of cortico-steroids	no	yes	yes
Cations	yes	yes	yes

C. Decline of Concentrations of Acidic Monoamine Metabolites After Inhibition of Monoamine Oxidase

Assuming that brain 5-HT is metabolized solely by MAO to form 5-hydroxyindoleacetic acid (5-HIAA), then the steady-state rate of 5-HIAA efflux from brain must be equal to the rate of. formation of 5-HT. Following an intraperitoneal injection of pargyline (75 mg/kg) brain concentrations of 5-HIAA decline exponentially. By using the same rationale employed for the calculation of NE biosynthesis from the exponential decline of the NE concentrations after blockade of synthesis (2,a), the equation for the decline of 5-HIAA can be derived as follows

$$\log 5\text{-HIAA} = \log 5\text{-HIAA}_o - 0.434 \, k \, _t \qquad (17)$$

where 5-HIAA$_o$ is the steady-state or extrapolated zero time value for the acid and 5-HIAA is the level of acid in brain at any time t after inhibition of MAO and the slope is 0.434 times k. If the assumption that 5-HIAA is the main catabolite of brain 5-HT is correct then the synthesis rate of 5-HIAA must be equal to the synthesis rate of 5-HT. Therefore, the latter can be approximated

by multiplying the molar concentration of 5-HIAA at the steady-
state by k (equation 16), the fractional rate constant of 5-HIAA
decline. The assumptions made for the method described in (2,b)
hold for this procedure also. Because of the multiple pathway
regulating the metabolism of brain catecholamines this method is
not suitable to measure turnover rate of these brain amines.

D. Accumulation of 5-HIAA after Probenecid Administration

Removal of 5-HIAA from brain by an acid transport is supported
by the accumulation of this acid in spinal fluid and brain tissue
following probenecid injections. Rats receiving intraperitoneally
200 mg/kg of probenecid exhibit a linear increase of 5-HIAA levels
lasting longer than ninety minutes. When the formation of 5-HIAA
is inhibited by pargyline, probenecid treatment will not longer
induce a rise of 5-HIAA; however, this inhibitor of the acid
transport prevents the usual decline of 5-HIAA levels elicited by
pargyline.

The accumulation of 5-HIAA is related to the dose of probenecid administered, maximal rates of accumulation are observed
after 200 mg/kg i.p. of probenecid. The concentrations of 5-HIAA
in brain increase by about 6-folds when rats receive probenecid
(200 mg/kg i.p.) every two hours during an eight hour period.
Applying the rationale of amine accumulation after MAO inhibition,
turnover rate of 5-HT can be calculated from the molar rate of
5-HIAA accumulation in brain of rats receiving probenecid.

Although accumulation of catechol acid elicited by probenecid
can be a suitable index of NE turnover, this method fails to yield
absolute measurements of NE turnover rates because alcohols derived
from MAO action on catecholamines leave the brain and spinal fluid
before they are converted into the respective acids by alcohol
dehydrogenase. Hence, the measurements of catechol acid accumulation in spinal fluid of man receiving probenecid cannot be readily
interpreted in terms of catecholamine turnover rates.

Table 4 lists the problems on the advantages deriving from
calculations of monoamine turnover according to the nonisotopic
methods listed above.

CONCLUSIONS

As stated at the beginning of this presentation, none of the
methods reported are absolutely reliable to measure in vivo the
turnover rates of brain catecholamines and serotonin. Table 5
lists the results obtained in laboratory experiments when monoamine turnover rates were measured with different methods. The
data reported indicate that within the limits of usual biological

Table 4

ADVANTAGES AND DISADVANTAGES OF VARIOUS NONIOSOTOPIC METHODS TO MEASURE TURNOVER RATE
OF BRAIN 5-HT, NE, and DM "IN VIVO"

METHOD	ADVANTAGES	DISADVANTAGES
Amine accumulation after MAO inhibition	None for NE and DM Inexpensive for 5-HT	Unreliable for NE and DM because of product inhibition. For 5-HT, confirmation by another method should be obtained
Decline of deaminated metabolites after MAO	None for NE and DM Inexpensive for 5-HT	Methodological problems for NE and DM. For 5-HT, confirmation by another method should be obtained
Accumulation of deaminated metabolites after probenecid	None for NE and DM Inexpensive for 5-HT	Unreliable for NE and DM. Changes TP levels and it might be a sourse of error when measuring turnover of brain 5-HT
Initial decline of monoamine after blockade of synthesis	Inexpensive for NE. We do not have a suitable inhibitor of 5-HT synthesis	Toxicity of α-methyltyrosine Not suitable for short acting drug effects

TABLE 5

RATE CONSTANT AND SYNTHESIS RATE OF NE, DM, and 5-HT FOR WHOLE RAT BRAIN CALCULATED USING VARIOUS METHODS

AMINE	METHOD	REFERENCE	k_m Hr^{-1}	SYNTHESIS RATE $m\mu mol/g/hr$
5-HT	Constant rate infusion of ^{14}C TP	Lin et al., 1969	0.72	1.5
	Pulse injection of radioactive TP	Neff et al., in press	0.65	1.6
	Accumulation of 5-HT after inhibiting monoamine oxidase	Lin et al., 1969	0.71	1.7
	Decline of 5-hydroxyindoleacetic acid after inhibiting monoamine oxidase	Lin et al., 1969	0.74	1.5
	Accumulation of 5-hydroxyindoleacetic acid after probenecid	Lin et al., 1969	0.76	1.4
NE	Constant rate infusion of ^{14}C T	Neff et al., 1969	0.25	0.71
	Pulse injection of radioactive T	Neff et al., in press	0.27	0.71
	Decline of NE after α-methyltyrosine	Neff et al., in press	0.17	0.56
DM	Constant rate infusion of ^{14}C T	Neff et al., in press	0.26	2.4
	Pulse injection of radioactive T	Neff et al., in press	0.34	1.9
	Decline of DM after α-methyltyrosine	Brodie et al. 1966	0.37	2.6

variability the methods discussed in this paper are in fairly good agreement. Despite this agreement and because of various problems involved in turnover rate measurements, we believe that one should measure turnover rates with at least two different methods before reaching any conclusions on a given result.

The data in the literature present discrepancies greater than those listed in Table 5 when results obtained with different methods are compared. A possible explanation for this greater discrepancy is offered by the data listed in Table 6 concerning turnover rate differences in Sprague-Dawley rats purchased from different suppliers.

Table 6

COMPARISON OF RATE CONSTANT OF AMINE LOSS AND TURNOVER RATE OF SPRAGUE-DAWLEY RATS FROM DIFFERENT SUPPLIERS
(Method of Inhibition of Catecholamine Synthesis)

SOURCE OF RATS	TISSUE	AMINE	RATE CONSTANT OF AMINE LOSS hr^{-1}	TURNOVER RATE $m\mu mol/g/hr$
Marland Farms	Brain	NE	0.12 ± 0.006	0.21
N.I.H.	Brain	NE	0.17 ± 0.008	0.41
N.Y. Breeding Farms	Brain	NE	0.17 ± 0.04	0.56
Marland Farms	Brain	DM	0.28 ± 0.09	1.3
N.I.H.	Brain	DM	0.37 ± 0.05	2.6
Marland Farms	Heart	NE	0.076 ± 0.005	0.28
N.Y. Breeding Farms	Heart	NE	0.10 ± 0.01	0.54

The data reported in this table suggest that when measuring turnover rate in vivo one should carry on simultaneously experimental and control studies because of the possibility that environmental and seasonal factors influence turnover rates of brain monoamines.

Following is a list of bibliographical references concerning various techniques to measure in vivo the turnover rate of neuronal monoamines.

Algeri, S. and Costa, E., Biochem. Pharmacol., in press

Azmitia, E. C., Jr., Algeri, S. and Costa, E., Science 169:
    201-203, 1970.

Bapna, J., Neff, N. H. and Costa, E., Neuropharmacol. 9:
    333-340, 1970.

Beaven, M. A., Costa, E. and Brodie, B. B.: Life Sci. 2: 241, 1963.

Brodie, B. B. and Costa, E., Nervosa Superior 5: 264, 1963.

Brodie, B. B., Costa, E., Dlabac, A., Neff, N. H. and Smookler,
    H. H., J. Pharmacol. Exp. Ther. 154: 493, 1966.

Cheney, D. L., Goldstein, A., Algeri, S. and Costa, E.,
    Science, in press

Costa, E., In: Proceedings of VI International C.I.N.P. Meeting,
    Excerpta Medica Foundation, Amsterdam, pp. 11-35, 1969.

Costa, E., In: Advances in Biochemical Psychopharmacology, Vol.
    II (Eds. E. Costa and E. Giacobini), Raven Press, New York,
    pp. 169-204, 1970.

Costa, E., Boullin, D. J., Hammer, W., Vogel, W. and Brodie, B.B.,
    Pharmacol. Rev. 18: 577, 1966.

Costa, E. and Groppetti, A., In: International Symposium on
    Amphetamines and Related Compounds; Proceedings of the Mario
    Negri Institute for Pharmacological Research (Eds. E. Costa
    and S. Garattini), Raven Press, New York, pp. 231-255, 1970.

Costa, E. and Neff, N. H., in: Proceedings of the V. International
    C.I.N.P. Meeting, Excerpta Medica International Congress
    Series 129: 959, 1966.

Costa, E. and Neff, N. H.: In: Biochemistry and Pharmacology of the
    Basal Ganglia (Eds. E. Costa, L. Cote, M. D. Yahr), Raven
    Press, New York, pp. 141-156, 1966.

Costa, E. and Neff, N. H., In: Interactions between pharmacological
    activity of MAO and their effects in man, Pergamon Press,
    Oxford, p. 15, 1968.

Costa, E. and Neff, N. H., In: Topics in Medicinal Chemistry,
    Vol. 2 (Eds. J. L. Rabinowitz and R. M. Myerson), John
    Wiley and Sons, Inc., New York, pp. 363-407, 1968.

Costa, E. and Neff, N. H. In: Importance of Fundamental Principles in Drug Evaluation (Ed., D. Tedeschi and R. E. Tedeschi), Raven Press, New York, pp. 185-202, 1968.

Costa, E., Neff, N. H. and Ngai, S. H., Brit. J. Pharmacol. 36: 153-160, 1969.

Costa, E. and Neff, N. H. In: Handbook of Neurochemistry, Vol. 4 (Ed. A. Lajtha), New York, Plenum Publ. Co., p. 45, 1970.

Costa, E., Spano, P. F., Groppetti, A., Algeri, S. and Neff, N.H. Atti. Acc. Med. Lomb. 23: 1100, 1968.

Diaz, P., Ngai, S. H. and Costa, E.: Amer. J. Physiol. 214: 591, 1968.

Diaz, P., Ngai, S. H. and Costa, E. In: Advances in Pharmacology Vol. 6 (Eds.S. Garattini, P. Shore, E. Costa and S. Sandler) New York, Academic Press, pp. 75-92, 1968.

Eakins, K. E., Costa, E., Katz, R. L. and Reyes, C.L., Life Sci. 7: 71, 1968.

Groppetti, A., Misher, A., Cattabeni, F., Naimzada, M., Revuelta, A. and Costa, E., J. Pharmacol. and Exp. Therap., in preparation.

Herr, B. and Costa, E., Int. J. Neuropharmacol. 8: 463, 1969.

Lin, R. C., Costa, E., Neff, N. H., Wang, C. T. and Ngai, S. H., J. Pharmacol. Exp. Ther. 160: 232-238, 1969.

Lin, R. C., Neff, N. H., Ngai, S. H. and Costa, E., Life Sci. 8: 1077-1084, 1969.

Montanari, R., Costa, E., Beaven, M. A. and Brodie, B. B., Life Sci. 2: 232, 1963.

Ngai, S. H., Neff, N. H. and Costa, E., Life Science 7: 847, 1968.

Ngai, S. H., Neff, N. H. and Costa, E., J. Anesthesiology 31: 53, 1969.

Neff, N. H., Barrett, R. E. and Costa, E., Eur. J. Pharmacol. 5: 348-356, 1969.

Neff, N. H. and Costa, E. In: Proc. International Symposium on Antidepressant Drugs, Excerpta Medica International Congress Series 122: 242, 1966.

Neff, N. H. and Costa, E., Life Sci. 5: 951, 1966.

Neff, N. H. and Costa, E., J. Pharmacol. Exp. Ther. 160: 40, 1968.

Neff, N. H., Lin, R. C., Ngai, S. H. and Costa, E. In: Advances in Biochemical Psychopharmacology, Vol. 1 (Eds., E. Costa and P. Greengard) Raven Press, New York, p. 91, 1969.

Neff, N. H., Ngai, S. H., Wang, C. T. and Costa, E., Molecular Pharmacol. 5: 90-99, 1969.

Neff, N. H., Spano, P. F., Groppetti, A. and Costa, E., J. Pharmacol. Exp. Ther., in press

Neff, N. H. and Tozer, T. N. In: Advances in Pharmacology, Vol. 6 (Eds., S. Garattini, P. Shore, E. Costa, and M. Sandler), New York, Academic Press, pp. 97, 1967.

Neff, N. H., Tozer, T. N. and Brodie, B. B.: J. Pharmacol. Exp. Ther. 158: 214, 1967.

Neff, N. H., Tozer, T. N., Hammer, W., Costa, E. and Brodie, B.B., J. Pharmacol. Exp. Ther. 160: 48, 1968.

Spano, P. F., Connor, J. D. and Neff, N. H.: Atti. Acc. Med. Lomb. 23: 1110, 1968.

Stern, D. N., Five, R. R., Neff, N. H., and Costa, E., Psychopharmacologia 14: 315-322, 1969.

Tozer, T. N., Neff, N. H. and Brodie, B. B., J. Pharmacol. Exp. Ther. 153: 177, 1966.

DEVELOPMENT OF MONOAMINERGIC TRANSMISSIONS IN THE RAT BRAIN[*]

Dr. Jean Renson

Laboratory of Physiopathology

University of Liège, Belgium

The neuronal machinery necessary for precise monoaminergic transmissions in the central nervous system is quite complex. The present state of knowledge about monoaminergic mechanisms in synaptic transmissions has been adequately reviewed recently by Phyllis (1), Iversen and Callingham (2) and Renson (3).

We shall limit this chapter mostly to the development of monoamines in the rat brain since a great deal of information is already available in this species.

It should be pointed out immediately that, in addition to their hypothetical roles in the central nervous system, monoamines could have important functions during the first steps of embryogenesis in the biochemical mechanisms of cellular differentiation. Experimental evidences for such roles have been collected by Buznikov and his school(4, pp. 274-276). In several species of invertebrates and a teleost, 5-hydroxytryptamine (5-HT) is found in early embryos and even in some mature but unfertilized eggs. Soon after fertilization, 5-HT levels rise sharply and the first four cleavage divisions are accompanied by cyclical rises and declines of 5-HT. These observations suggest that 5-HT is an important participant in the processes of early embryogenesis in these species. Another example of such a role for amines in differentiation is a recent report showing that in fetal rat brain, histamine and spermidine concentrations peak at 17 days gestation, then decrease sharply at 21 days. They achieve their highest levels about 5 to 10 days after birth and then steadily decline to low adult levels by the time of weaning (5).

[*]This work was supported by the F.R.F.C. contract No. 755, Belgium.

Significantly, no histidine decarboxylase activity is found in either fetal or neonatal brain tissues, which indicates that in the absence of a blood-brain barrier, histamine is taken up from peripheral sources.

This developmental pattern of histamine and spermidine is quite different than the gradual increase with age observed for other monoamines and correlates best with periods of rapid cell proliferation and growth during brain maturation (Stages 1 and 2, see below). All these data support the idea that amines may play important roles in some basic cellular processes during early development.

The rat brain during development goes through four characteristic stages of unequal length (6):
1. From the appearance of the neural crest until birth, during which period cell divisions take place until the complete number of neurons is reached.
2. During the first ten days after birth, neurons grow and develop axons and dendrites.
3. Between the tenth and twentieth days of life, myelination occurs and the growth rate of the brain decreases. At this stage electrical activity is easily demonstrable and some neuromuscular controls begin to function.
4. Until complete maturation of the brain which is reached six weeks after birth. During this last stage, myelination is still active but the increase in size is slow and the chemical composition is fairly constant.

Some other physiological functions will eventually reach complete maturity later on; i.e., sexual maturity in the rat occurs only three months after birth.

A.   DEVELOPMENT OF AMINE LEVELS

Several physiological factors can influence brain amine levels (species, strain, sex, age, seasons, time of day, environmental factors). Striking differences in amine levels according to age have been repeatedly demonstrated, despite some quantitative discrepancies in the intensity or the onset of these phenomena. 5-HT levels have been shown to follow an irregular pattern of development in the rat (4). However, all the observations available indicate periods of rapid increase in the brain of fetal and newborn rats until adult levels are reached (4, 7-12, 14). Norepinephrine (NE) also seems to follow an irregular pattern (12-14); the levels of dopamine (DA) and normetanephrine levels have also been measured (12), and show, at least for DA, a similar increase in the developing rat brain.

Karki et al. (9) correlated the content of brain 5-HT and NE
to the degree of physiological maturity at birth. Their studies
demonstrated that a species like the guinea pig, which is fully
mature at birth does not follow the same pattern of monoaminergic
development, for this species almost has adult levels at birth.
The validity of such a behavorial correlation has been questioned
by Baker and Quay (4) and more comparative studies are certainly
needed before making any definite generalization. More recently,
regional studies have demonstrated a caudal to rostral progression
of some amine levels (5-HT, NE) since they derive from nuclei located in the lower brain stem (13-14).

B. MECHANISMS OF DEVELOPMENT

There exist two ways to explain the development of monoamines
with age:

a) A sequential development of the successive enzymatic steps
in completely mature neurons with their fully developed axonal and
dendritic systems as seen in the chick embryo (23). Axons are known
to furnish millions of so-called varicosities which are highly
fluorescent in the adult. The nerve endings receive, by axonal
flow, all the subcellular organelles with the molecular machinery
necessary for biosynthesis, storage, release and uptake of monoamines in order to carry adequate synaptic transmissions. In this
hypothesis, full anatomical development of neuronal cells would
precede a sequential synthesis of all the molecular tools necessary
for neurochemical transmissions by reading in a sequential fashion
the genetic message designed for these highly specialized neuronal
cells.

b) A second explanation would tend to visualize a parallel
development of neuroanatomical connections with all the molecular
mechanisms necessary for synaptic transmission already present in
growing cell bodies and their axons during propagation toward their
target areas. Thus, the biosynthetic machinery would be synthesized
during cellular differentiation. The quantitative increase in
monoamines would merely represent a progressive spreading of these
fully competent cells while innervating the rest of the brain from
their original locations.

This second possibility is well-supported by recent histochemical evidences. In the rat, it seems that cell divisions of
monoaminergic neurons are still taking place at birth and axonal
development is in full operation during the first ten days of life.
The first EEG manifestations in the rat can only be clearly demonstrated after this period. Let us now examine more closely the
biochemical and neuroanatomical hypotheses.

C. BIOCHEMICAL DATA

Steady amine levels represent merely a dynamic equilibrium between their rates of biosynthesis and catabolism. Low amine levels in the young rat, as compared to adults, could arise either from limited biosynthesis, low storage capacity, or higher rates of disappearance due to some deficiency of the normal mechanisms protecting amines from enzymic destructions. A combination of these factors is also conceivable.

Several biochemical factors could be incriminated for the early transitory low values of monoamines:

1. Aromatic amino acid precursors in the central nervous system are in limited supply due to low plasma levels with poor uptake into the brain.

2. The two specific hydroxylases, which are the rate limiting steps in the adult, are not fully developed at birth.

3. The common decarboxylating system, aromatic L-amino acid decarboxylase is absent or deficient.

4. After biosynthesis, storage granules may be absent or momentarily physiologically incompetent.

5. Catabolic enzymes, monoamine oxidase (MAO) and catechol-O-methyl transferase (COMT) are excessively active.

We shall review in turn the known facts about these various factors and show that none of the above by themselves can explain completely the phenomenon, although some of these factors can probably afford a partial explanation (see table).

1. Availability of Aromatic Amino Acid Precursors in the Newborn Rat Brain

Biosynthesis of monoamines, mostly 5-HT, is highly dependent upon an adequate concentration of tryptophan, an essential amino acid furnished by the plasma. It can be seen in the table that tyrosine and tryptophan plasma levels in the newborn rat plasma are greater than in adults. The active transport system which takes up aromatic-L-amino acids into the brain is also more developed in the young rat than in the adult. Consequently, pools of both amino acids in the newborn rat are significantly greater. It can therefore be assumed that a deficiency of both precursors cannot be held responsible for a low biosynthetic rate in the newborn rat.

COMPARISON OF BRAIN MONOAMINERGIC MECHANISMS BETWEEN NEWBORN AND ADULT RATS

LEVELS OR ACTIVITIES	NEWBORN	ADULT	UNITS	REFERENCES
5-Hydroxytryptamine	0.24	0.52	μg/g	10
Norepinephrine	0.12	0.54		9
Dopamine	0.28	0.67	μg/g	12
Normetanephrine	0.00	0.01		12
5-Hydroxyindolacetic acid	0.38	0.41		10
Tryptophan in plasma	19.0	12.0	μg/ml	10
Tyrosine in plasma	24.6	15.5	μg/ml	10
Uptake of tryptophan in brain (in vivo)	51.7*	28.2*	μg/g/30'	10
Uptake of tyrosine in brain (slices)	300.	200.	μg/g/hour	15
Tryptophan in brain	21.6	6.7	μg/g	10
Tyrosine in brain	36.6	22.8		10
Tryptophan-5-hydroxylase (pineal body)	40.	100.	% adult activity	16
Tryptophan-5-hydroxylase (brain stem)	3.0	3.6	nM/g/hour	18
Tyrosine hydroxylase (kitten)	3.4	3.1	nM/g/hour	13
Aromatic-$\underline{\text{L}}$-amino acid (whole brain) decarboxylase	0.31	0.38	mg 5-HT formed/g/h	9
(pineal body)	0.	100.	% adult activity	16
Storage	80	78	% bound	11
Monoamine oxidase	0.59	1.75	mg 5-HT oxidized/g/h	9
Catechol-O-methyl transferase	20	100	% adult activity	20
Neuronal development 5-HT	+	+++	Fluorescence intensity	22
NE	+	+++		
DA	++	+++		

*TRY (66 μg/g) injected - I.P.

2. Tyrosine and Tryptophan-5-Hydroxylase

These two enzymes represent the first steps leading to the biosynthesis of catecholamines and 5-hydroxytryptamine, respectively. They share some common properties, such as the same cofactor (tetrahydrobiopterin) and probably have a similar mechanism since they are both monoxygenases. They are the rate limiting steps in the conversion of L-tyrosine to L-DOPA and in the biosynthesis of 5-HTP from L-tryptophan. Their deficiency, due to a relative lack of apoenzymes, their cofactors, or dihydrobiopterin reductase, could possibly explain low levels of monoamines in the young rat brain.

Unfortunately, no data on tyrosine hydroxylase in the developing rat brain is presently available, but one study on the kitten (13) shows no differences between young and adult animals in several brain areas.

Tryptophan-5-hydroxylase has been measured in the brain stem of the newborn rat with a highly sensitive and rapid assay using $5-^3H$-tryptophan (17). This enzyme in the newborn rat has 80% of the adult level (18). In the pineal gland, which is the richest source of tryptophan-5-hydroxylase, Håkanson et al. (16) also found a significant activity (40% of adult values). Even more interesting in this case is the observation that no 5-HT could be detected either chemically or by histofluorescence. Measurements of aromatic-L-amino acid decarboxylase demonstrated a total absence of the enzyme in this particular organ. This would represent the only case where the decarboxylating step could be the rate limiting factor around birth.

3. Aromatic-L-Amino Acid Decarboxylase

This enzyme is common to L-DOPA and L-5-HTP and will decarboxylate both amino acids into their corresponding amines, dopamine (DA) and 5-HT. At birth, decarboxylase activity in the brain is almost as high in the young rat as in the adult animals (see table). Furthermore, this enzymatic activity in the brain is generally present in large excess and is capable of decarboxylating, in vitro, 2 μM DOPA/g/hour. 5-HTP is probably decarboxylated at a slightly lower rate. No deficiency of this enzyme is responsible for low levels of monoamines which are observed in the young rat brain, except maybe in the pineal gland.

4. Catabolic Enzymes

Direct measurements of MAO activity in the brain have repeatedly shown less activity (see table) (8,9,11,16). 5-HIAA

levels are almost normal in the young rat brain and even somewhat
superior during the first few weeks of life (10). The presence of
high levels of 5-HIAA indicate adequate activity of MAO or poor
transport of 5-HIAA by an active process carrying acidic metabolites
out of the brain into the blood stream.

No data is available about the development of COMT in the rat
brain during development but the barely detectable amounts of normetanephrine found in the newborn rat brain (12) certainly indicate
that this enzyme is not more efficient in the central nervous system than in peripheral tissues where it represents only 20% of
adult activity (20).

5. Uptake and Storage of Monoamines

Bennett and Giarman (11) investigated the possibility that
the storage mechanism could be partially deficient but a careful
study of the ratio of free to bound 5-HT revealed that the relative binding of 5-HT was the same in the newborn as in the adult
rat. Furthermore, administration of 5-HTP resulted in accumulation of bound 5-HT, demonstrating that storage sites were far from
being saturated (11).

A study made on peripheral storage and uptake mechanisms of
NE by Iversen et al. (19) have shown very different patterns of
behavior according to the organs examined. These differences have
been attributed to the differences of sympathetic innervation,
which varies with age.

D. HISTOCHEMICAL DATA

Correlations between biochemical data and histochemical data
are rare. Using the fluorescence histochemical technique of Falck
and Hillarp, Loizou (22) made a study of the development of monoamine-containing neurons in the brain of the normal albino rat
from birth to adulthood. He observed that nearly all the monoaminergic neurons described by Dahlstrom and Fuxe (21) in the adult
brain could be visualized at birth. The intensity of fluorescence
was lower in neonates than in adult animals but it increased gradually to attain the adult pattern by the fifth week. Virtually no
5-HT or NE-containing terminals were present at birth, but dopamine-
containing terminals were present in the neostriatum and other
areas of minor importance. The dopamine, NE and 5-HT-containing
terminals were found to proliferate at different speeds and at
various times, so that adult patterns were reached at various
periods. Up to about two weeks of age, monoamines could cross
the blood-brain barrier and be taken up by monoamine neurons,
while their precursors, DOPA and 5-HTP, were taken up rapidly

at all ages (22).

These histochemical findings demonstrate that at birth, cell bodies of monoaminergic neurons are present, but terminals are sparsely distributed, particularly in the rostral part of the brain. There follows a gradual increase in the fluorescence of monoamine-containing cell bodies and a proliferation of terminals throughout the brain with the complete adult pattern being attained at the fourth to fifth postnatal week in all areas of the brain.

It can be concluded that the increase in the levels of mono-amines in the rat brain is a consequence of the progressive pro-liferation of nerve terminals with their varicosities where bio-synthesis and storage of intraneuronal amines take place. This explains the concomitant increase of the synthesizing enzymes and storage vesicles.

E. CONCLUSIONS

More work is needed before deciding the exact sequence of appearance of the different subcellular components necessary for monoaminergic transmissions in the young animals. More refined techniques such as electron microscopy, histofluorescence, single cell analysis and regional biochemical studies are necessary to correlate biochemical development with electrophysiological and behavorial data.

We know, for instance, that young mammals spend most of the time sleeping with alternating periods of slow wave sleep and paradoxical sleep (24). Since 5-HT and NE are involved in this complex pattern of sleep, monoaminergic transmissions might be more fully developed than hitherto suspected. We still do not know exactly how much monoamines are sufficient to perform ade-quately their neurophysiological functions. The so-called mature behavior might be more dependent on myelination of the cholinergic pathways and primitive pattern of behavior, like sleep, might be functional even with low amine levels such as those found at birth.

REFERENCES

(1) Phyllis, J. W. in The Pharmacology of Synapses, Pergamon Press (1970).

(2) Iversen, L. L., and Callingham, J. in Fundamentals of Biochemical Pharmacology (Eds. Z. M. Bacq, R. Paoletti, R. Csapek, J. Renson) Pergamon Press (1971).

(3) Renson, J. in Fundamentals of Biochemical Pharmacology (Eds. Z. M. Bacq, R. Paoletti, R. Csapek, J. Renson) Pergamon Press (1971).

(4) Baker, P. C., and Quay, W. B. Brain Research 12:273 (1969).

(5) Pearce, L. A., and Schanberg, S. M. Science 166:1301 (1969).

(6) McIlwain, H. in Biochemistry and the Central Nervous System, 3rd Edition, J. A. Churchill, London, p. 244 (1966).

(7) Kato, R. J. Neurochem. 5:202 (1960).

(8) Nachmias, V. J. Neurochem. 6:99 (1960).

(9) Karki, N., Kuntzman, R., and Brodie, B. B. J. Neurochem. 9: 53 (1962).

(10) Tyce, G., Flock, E. V., and Owen, Ch. A. Progress in Brain Research 9:198 (1964).

(11) Bennett, D. S., and Giarman, N. J. J. Neurochem. 12:911 (1965).

(12) Agrawal, H. C., Glisson, S. N., and Himwich, W. A. Biochim. Biophys. Acta 130:511 (1966).

(13) McGeer, E. G., Gibson, S., Wada, J. A., and McGeer, P. L. Canadian J. Biochem. 45:1943 (1967).

(14) Loizou, L. A., and Salt, P. Brain Research 20:467 (1970).

(15) Guroff, G., and Udenfriend, S. Progress in Brain Research 9:187 (1964).

(16) Håkanson, R., Lombard Des Gouttes, M.-N., and Owman, Ch. Life Sciences 6:2577 (1967).

(17) Renson, J., Daly, J., Weissbach, H., Witkop, B., and Udenfriend, S. Biochem. Biophys. Res. Commun. 25: 504 (1966).

(18) Renson, J. unpublished observations.

(19) Iversen, L. L., DeChamplain, J., Glowinski, J., and Axelrod, J. J. Pharmacol. Exptl. Therap. 157:509 (1967).

(20) Glowinski, J., Axelrod, J., Kopin, I. J., and Wurtman, R. J., J. Pharmacol. Exptl. Therap. 146:48 (1964).

(21) Dahlström, A., and Fuxe, K., Acta Physiol. Scand. 62:1, suppl. 232 (1964).

(22) Loizou, L. A. J. Anatomy (London) <u>104</u>:588 (1969).

(23) Ignaro, L. J., Shideman, F. J. Pharmacol. Exptl. Therap. <u>159</u>: 38 (1968).

(24) Jouvet, M. Physiological Rev. <u>47</u>:117 (1967).

ACKNOWLEDGEMENTS

I would like to thank Drs. Sidney Udenfriend, Herbert Weissbach and Sidney Spector for their kindness and helpful advice. The skillful technical help of Mr. Victor Bourdon and the kind secretarial assistance of Miss Elena Smith were also extremely appreciated.

CONTROL MECHANISMS IN THE SYMPATHETIC NERVOUS SYSTEM

Rita Levi-Montalcini

Laboratory of Cell Biology, CNR, Rome; and

Dept. of Biology, Washington University, St.Louis, USA

Ever since the science of neurobiology emerged from the more general field of biological sciences, the main effort has been directed to gaining information on the laws which control growth and differentiation of its 10 billion nerve cells and on the factors which operate in the building of the immensely complex wiring system between neurons and between these cells and their highly diversified end organs. It is fortunate that the first students in this field perhaps did not even realize the magnitude of the problem which they were facing and the impossibility of solving it with the primitive tools and techniques which were then available, for otherwise they would have invested no further efforts along this line.

The first attempts to gain some knowledge on the control mechanisms in the nervous system date back to the beginning of this century. They took their start from, and were in fact part of, the more general area of investigation known as experimental embryology, which explores growth and differentiation of organs and tissues of the developing embryo under normal and experimental conditions. The experimental approach consisted in ablation and implantation of end organs, permutation and translocation of different segments of the central nervous system and exchange of end organs between embryos of closely related species, with microsurgical instruments devised for this purpose. The results of a long series of investigations performed at first in amphibian larvae and then in chick embryos, gave evidence for the all-important role played by peripheral end organs in the differentiation of the associated nerve centers: these centers would in fact fail to differentiate in the absence of their peripheral field of

innervation and, conversely, some nerve cell populations such as
those of the spinal ganglia would increase in size when faced with
a larger than normal peripheral territory (1, 2, 3, 4, 5, 6).
After about four decades of intense work, it was realized that
little more knowledge could be gained by further experimenting
along these lines; a feeling of frustration and scepticism pre-
vailed and discouraged young biologists from entering the field of
experimental neuroembryology.

It was at that time--two decades ago--that sheer change
rather than calculated search opened an entirely new and far more
promising approach to the problem of control mechanisms in the
nervous system. Too many times has the history of this accidental
discovery been reported and I will therefore not recount it (7, 8,
9). Suffice it to mention that the first evidence for the existence of a diffusible nerve growth promoting factor came with the
discovery that fragments of mouse sarcomas 180 or 37, implanted
onto the chorio-allantoic membrane of 4 to 6 day chick embryos,
call forth a marked size increase of the entire sympathetic chain
ganglia of the host embryo and the production by these hypertrophic and hyperplastic ganglia of an extraordinary number of
nerve fibers which invade the viscera and even form large neuromas
inside the blood vessels (10). From a retrospective viewpoint, it
appears that the discovery that nerve cells are responsive to the
action of humoral factors was the main, but not the only contribution. The other contribution, which has rarely been mentioned, is
to have shifted the attention from the billions of nerve cells
lodged in the central nervous system to the well circumscribed and
fairly homogeneous nerve cell population of the sympathetic
ganglia. It is of interest to note in passing that this cell
population had already played a fundamental role in the fields of
neurophysiology and neuropharmacology, two decades earlier, by
revealing the existence and mechanism of action of neurotransmitters (11). Now these same cells played a different and no less
important role in the field of neurogenesis, which deals with
growth control mechanisms of developing nerve cells. Although it
remains to be seen to what extent the findings to be reported in
the following pages apply also to other nerve cell populations,
the results of this investigation offer a promising lead to the
approach to this problem.

In 1953 (12), the growth effect elicited by fragments of
mouse sarcomas 180 or 37 was explored in vitro by culturing fragments of these tumors in a semisolid medium in proximity to sensory or sympathetic ganglia dissected out from 7 to 9 day chick
embryos. The production by these ganglia in a six to twelve hour
period of a dense fibrillar halo gave definite evidence for the
release of a humoral nerve growth promoting factor from these
tumors (13) and at the same time afforded the possibility of

exploring the chemical nature of the growth promoting factor
released by neoplastic cells.

In 1954, this factor was identified in a nucleoprotein particle isolated from the ribosomal fraction of the tumor (14).
Subsequent investigations identified this factor, which became
known as the Nerve Growth Factor (NGF), in a protein rather than
nucleoprotein tumoral fraction. In 1956 and in 1958 respectively,
a protein particle endowed with identical growth promoting properties was isolated from snake venoms and from the salivary glands
of adult male mice (15, 16, 17, 18). The presence of this protein
molecule at a very high concentration in both sources made possible its isolation in fairly large quantities from snake venom
and even more from the mouse salivary glands (19, 20). Ever since
these investigations were first pursued in our laboratory, the NGF
isolated from the mouse salivary glands and, to a lesser extent,
the same factor isolated from snake venoms, became, in our and
other laboratories, the object of extensive investigations directed to elucidating the chemical nature of this protein. Space
restrictions do not allow consideration of the extensive and con-
flicting data reported by different research groups nor comment on
the regrettable amount of confusion which resulted from most of
these none too accurate evaluations of the biological and chemical
properties of the NGF. Experiments still in progress in our
laboratory have succeeded in establishing the remarkable similarity of the NGFs isolated from different snake venoms and from the
mouse salivary gland. Suffice it to mention that the amino acid
composition of the venom and salivary NGFs are strikingly similar,
the molecular weights being respectively 27,000 and 29,000 (21).
Results reported in detail elsewhere support the conclusion that
the native salivary NGF is composed of two identical or very
similar subunits of 14,500 molecular weight. The two units, each
of which contains three intramolecular disulfide bonds, are associated by non-covalent forces in the native molecule (22). The
half molecules are biologically active and are immunochemically
identical to the native 29,000 molecular weight salivary NGF (23).
It remains to be established whether the NGF acts in vivo in the
monomeric or dimeric form. As an approach toward clarifying the
molecular basis for its mechanism of action, determination of the
primary structure of mouse NGF has been initiated.

The growth response of the target nerve cells was studied in
newborn and adult mice at the structural and ultrastructural
levels. While during early developmental stages both sensory and
sympathetic nerve cells are receptive to the NGF effects, only the
sympathetic nerve cells retain this prerogative during later
developmental phases and in adult life. These differences between
the two cell populations, already apparent from earlier studies in
the chick embryo, became more evident when the NGF effects were

explored in newborn and adult mice. In neither the newborn nor
the adult mouse does the NGF elicit overgrowth of the sensory
ganglia. The sympathetic ganglia on the contrary are strikingly
increased in size.

         The NGF effects in newborn mice will be considered first.
The NGF was injected subcutaneously soon after birth in the amount
of 0.05 ml per g of body weight. The purified NGF protein is
present in this preparation at the concentration of about 300 µg/
ml. The injections were repeated daily and the amount of NGF
injected increased with the size and weight of the animal. Experimental and control mice were sacrificed every other day till
the ninth day. The sympathetic ganglia were dissected out from
both groups and compared. The superior cervical ganglia were then
sectioned serially, stained and used for volume determinations,
mitotic counts, and counts of the entire nerve cell population.
A number of experimental mice were also examined between the ninth
and the thirtieth day of continuous NGF treatment and compared to
controls. The mitotic activity as determined between the third
and the ninth day after birth is considerably increased in the
NGF-treated mice. The total nerve cell population is between two
and three times as large as that of controls. At 9 days, the
mitotic activity comes to an end in control and NGF-treated animals. In the subsequent developmental periods the NGF calls forth
increase in size of individual neurons but no further increase in
their number (24). The hyperplastic and hypertrophic response of
the sympathetic ganglia of newborn mice and rats varies from one
case to another and depends on the strength and the purity of the
preparation. In a first series of experiments, a four- to sixfold
volume increase was obtained in the superior cervical ganglia of
treated mice as compared to littermates. In subsequent experiments it was found that the growth effect could still be considerably enhanced by injecting a more highly purified NGF protein
molecule. The superior cervical ganglia of mice treated with this
preparation for a 9-day period are nearly twelve times larger than
controls (7).

         The baby mice injected with this highly purified NGF do not
differ from controls in somatic size and vitality. The size
differences which were reported in a first experimental series
performed soon after the discovery of the NGF (24) were then
traced to contaminants present in the injected NGF fraction. The
histological analysis of serially sectioned specimens shows a
marked hyperinnervation of the viscera and of hair bulbs but no
otherwise abnormal distribution of the sympathetic nerve fibers.
In young and adult mice the NGF evokes hypertrophic, but no hyperplastic, response from the sympathetic ganglia. The volume increase of the superior cervical ganglia is two to three times that
of the controls in a 5-day period of treatment. Similar results
were obtained in newly hatched chickens. In a 3-day period the

superior cervical ganglion undergoes a threefold increase of its normal size (8). In mice, as in chicks, the hypertrophic sympathetic nerve cells differ from controls, not only in size but also in the intense basophilic reaction to toluidine stain which is a reflection of the increased RNA in the cell cytoplasm. Also there is a marked increase in size and number of nucleoli. While control sympathetic nerve cells have an average of three to four nucleoli, the NGF-treated sympathetic nerve cells have an average of five nucleoli. Recent electronmicroscopic studies gave evidence of the massive production of neurofilaments and neurotubules in the target nerve cells (25). This extraordinary increase in neurofibrillar material may well result from the stimulation of some more basic and highly specific processes in the sequence of events which take place in the cells, triggered by the NGF. In line with this concept are the results of an extensive series of metabolic studies which demonstrated the marked enhancement of protein, lipids and RNA synthesis and the utilization of carbohydrates by the target nerve cells incubated in presence of the NGF (26, 27).

The growth control exerted by the NGF on sympathetic neurons and the vital role played by this molecule in the life of these cells, came in sharp relief in 1959 with the discovery that a specific antiserum to the NGF injected in newborn mammals causes the selective and permanent destruction of sympathetic chain ganglia of the treated animals (19, 28). This process which became known as "immunosympathectomy" (29) has been the object of extensive investigations in these past years, directed to the study of the morphological, physiological, pharmacological and behavioral changes taking place in immunosympathectomized animals. Here we shall consider only the sequence of the degenerative events caused by the antiserum to the nerve growth factor in sympathetic nerve cells. Twelve hours after the first injection given a few hours after birth, a decrease of mitotic activity becomes evident in sympathetic ganglia. Between the second and third day, the mitotic activity continues to decrease and is practically nil in the subsequent two days. At the same time, the process of differentiation, which is very active in control ganglia, ceases in experimental animals and a progressively larger number of neuroblasts undergo disintegration and are removed by macrophages which surround these necrotic cells. Between the fourth and the ninth days, the process of cell atrophy and death continues at a slower pace because of the drastic reduction in the cell population produced in the first days of the treatment. While the nerve cells show all these signs of disintegration, satellite cells are spared such effects during the early phases of the treatment, but in subsequent stages they also undergo atrophy and the ganglia are reduced to exceedingly small sclerotic nodules. Histological studies show that the ganglia consist of a small

population of glial cells and a few sparse nerve cells of a size
smaller than in controls (28). Electronmicroscopic studies of
antiserum treated ganglia showed that the first lesions are apparent in the nuclear compartment (30). Actually the earliest
lesions are seen in the fine structure of the nucleoli, which
undergo marked disaggregation. Distinct dense bodies form from
the separation of the nucleolar material. Eventually the nuclear
envelope breaks down and only in later stages does marked disorganization of ribosomes become apparent in the cell cytoplasm
together with alterations of mitochondria (31). Incorporation
experiments show that RNA synthesis is already considerably reduced in the experimental ganglia 12 hours after the first injection of the antiserum (32).

A new facet of the growth control mechanisms which operate
on immature and mature sympathetic neurons recently came to light
with the discovery by Dr. P. Angeletti in our laboratory of the
dramatic destructive effects caused by a dopamine analog,
6-hydroxydopamine, in sympathetic neuroblasts of newborn mammals.
Previous studies (33) had shown that this compound produces a
long-lasting depletion of norepinephrine storage in sympathetic-
ally innervated organs. Subsequent electronmicroscopic studies
showed that this effect is due to the selective destruction of
adrenergic synaptic vesicles by this agent (34, 35). No lesions
were observed in the cell perikarya. A time sequence study gave
evidence for the reversibility of the damages inflicted by this
agent on the synaptic vesicles. Regeneration of the synaptic
terminals takes place after discontinuation of the treatment and
parallels the gradual and progressive restoration of the function.
Space restrictions do not allow detailed description of the effects of 6-hydroxidopamine (6-OHDA) on sympathetic ganglia of
newborn mammals. The reader is referred to recent publications
in which this effect was first reported (36, 37). Here it is of
interest to consider the sequence of degenerative events in sympathetic nerve cells of newborn mice and rats injected daily for
the first week after birth with 50 µg/g of body weight. At the
light microscope, two days after the beginning of the treatment,
a number of neuroblasts show clear degenerative marks such as
cytoplasm vacuolization, nuclear pyknosis and complete cell lysis.
Glial cells do not appear to be affected. At the same time, the
nerve fibers emerging from the ganglia are considerably larger
than in controls and show irregular thickenings or beadings,
along the course of the nerves. In subsequent days, the degenerative process extends to all neuroblasts until, toward the end of
the first week, practically all nerve cells have disappeared from
the field. A systematic histological study of all para- and
prevertebral ganglia in treated mice 2 to 3 months of age, shows
that all ganglia are equally affected and the residual cell population amounts to less than 2 per cent. No degenerative effects

are apparent in the adrenergic ganglionic complexes innervating the sex organs. It is of interest to note that this ganglionic complex was likewise resistant to the effect elicited by the antibody to the nerve growth factor.

An electronmicroscopic study of sympathetic ganglia taken at different time intervals during the first 24 hours after the beginning of the treatment shows the precocity of cytoplasmic alterations which become evident within a few hours from injection of 6-OHDA. They consist of shrinking of the mitochondria, appearance of large and numerous lacunar spaces and disorganization of the endoplasmic reticulum. In the same section, the severity of degenerative marks varies in different neurons. The cytoplasmic lesions seem to precede the alterations in the fine structure in the nuclear compartment. The latter become more apparent at later stages. The glial cells do not show degenerative signs in any of these early stages. At 7 days only glial and connective cells are seen in most sections (36, 37).

Concluding Remarks

Evidence has been presented that growth and differentiation of sympathetic nerve cells can be dramatically altered in opposite ways by injecting a protein molecule endowed with specific nerve growth promoting activity or by an antiserum to this protein which became known as the NGF.

Several lines of evidence indicate that in both instances the reaction must take place in the cell perikarya. The recent discovery that sympathetic nerve cells in newborn mammals are selectively destroyed by injections of a dopamine analog, 6-OHDA, points to the existence of a still different control mechanism which is exerted in a centripetal fashion, from the axon to the cell body and from the cytoplasm to the nuclear compartment. Thus it is apparent that the growth and indeed the life of sympathetic nerve cells can be regulated by at least two different mechanisms: in one case by interfering with a specific growth factor which plays a key role in the life of these cells; in the other case by diverting the specialized function of the same cells from the uptake of a necessary metabolite to that of a noxious substitute which results in cell death.

Acknowledgments

This work was supported by grants from the USPHS (NS-03777), from the National Science Foundation (GB-7304 and GB-16330), and from the Consiglio Nazionale delle Ricerche (Rome).

References

1. Harrison, R. G. Proc. Roy. Soc. London Ser. B. $\underline{118}$:155 (1935).

2. Weiss, P. J. Exptl. Zool. $\underline{113}$:397 (1950).

3. Hamburger, V. J. Exptl. Zool. $\underline{68}$:449 (1934).

4. Hamburger, V. Physiol. Zool. $\underline{12}$:268 (1939).

5. Hamburger, V., and Levi-Montalcini, R. J. Exptl. Zool. $\underline{111}$: 457 (1949).

6. Levi-Montalcini, R. Science $\underline{143}$:105 (1964).

7. Levi-Montalcini, R. Harvey Lectures Ser. $\underline{60}$:217 (1966).

8. Levi-Montalcini, R. Ann. N. Y. Acad. Sci. $\underline{118}$:149 (1964).

9. Levi-Montalcini, R., and Angeletti, P. U. Physiol. Rev. $\underline{48}$: 534 (1968).

10. Levi-Montalcini, R. Ann. N. Y. Acad. Sci. $\underline{55}$:330 (1952).

11. Dale, H. H. Proc. Roy. Soc. Med. $\underline{28}$:319 (1935).

12. Levi-Montalcini, R. Proc. XIV Intern. Cong. Zool. Copenhagen, 1953, p. 309.

13. Levi-Montalcini, R., Meyer, H., and Hamburger, V. Cancer Res. $\underline{14}$:49 (1954).

14. Cohen, S., Levi-Montalcini, R., and Hamburger, V. Proc. Natl. Acad. Sci. U. S. $\underline{40}$:1014 (1954).

15. Cohen, S., and Levi-Montalcini, R. Proc. Natl. Acad. Sci. U. S. $\underline{42}$:571 (1956).

16. Levi-Montalcini, R., and Cohen, S. Proc. Natl. Acad. Sci. U. S. $\underline{42}$:695 (1956).

17. Cohen, S. in Chemical Basis of Development (Eds. W. D. McElroy, B. Glass), Johns Hopkins Press, p. 665 (1958).

18. Levi-Montalcini, R. in Chemical Basis of Development (Eds. W. D. McElroy, B. Glass), Johns Hopkins Press, p. 646 (1958).

19. Cohen, S. Proc. Natl. Acad. Sci. U. S. $\underline{46}$:302 (1960).

20. Bocchini, V., and Angeletti, P. U. Proc. Natl. Acad. Sci. U. S. $\underline{64}$:787 (1969).

21. Angeletti, R. Hogue Proc. Natl. Acad. Sci. U. S. $\underline{65}$:668 (1970).

22. Angeletti, R. H., Bradshaw, R. A., and Wade, R. D. Biochem. in press.

23. Zanini, A., Angeletti, P. U., and Levi-Montalcini, R. Proc. Natl. Acad. Sci. U. S. $\underline{61}$:835 (1968).

24. Levi-Montalcini, R., and Booker, B. Proc. Natl. Acad. Sci. U. S. $\underline{46}$:373 (1960).

25. Levi-Montalcini, R., Caramia, F., Luse, S. A., and Angeletti, P. U. Brain Res. $\underline{8}$:347 (1968).

26. Angeletti, P. U., Liuzzi, A., Levi-Montalcini, R., and Gandini Attardi, D. Biochim. Biophys. Acta $\underline{90}$:445 (1964).

27. Levi-Montalcini, R., and Angeletti, P. U. in Organogenesis (Eds. R. L. DeHaan, H. Ursprung), Holt Rinehart & Winston, p. 187 (1965).

28. Levi-Montalcini, R., and Booker, B. Proc. Natl. Acad. Sci. U. S. $\underline{46}$:384 (1960).

29. Levi-Montalcini, R., and Angeletti, P. U. in Regional Neurochemistry (Eds. S. Kety, J. Elkes), Pergamon, p. 362 (1961).

30. Sabatini, M. T., De Iraldi, A. P., De Robertis, P. J. Exptl. Neurol. $\underline{12}$:370 (1965).

31. Levi-Montalcini, R., Caramia, F., and Angeletti, P. U. Brain Res. $\underline{12}$:54 (1969).

32. Sabatini, M. T., Levi-Montalcini, R., and Angeletti, P. U. Ann. Ist. Super. Sanita $\underline{2}$:349 (1966).

33. Porter, C. C., Totaro, J. A., and Stone, C. A. J. Pharmacol. Exp. Therap. $\underline{140}$:308 (1963).

34. Tranzer, J. P., and Thoenen, H. Experientia (Basel) $\underline{24}$:155 (1968).

35. Tranzer, J. P., Thoenen, H., Snipes, R. L., and Richards, J. G. in Prog. in Brain Res., vol. 31 (Eds. K. Akert, P. G.

Waser), Elsevier, p. 33 (1969).

36. Angeletti, P. U., and Levi-Montalcini, R. Proc. Natl. Acad.
     Sci. U. S. 65:114 (1970).

37. Angeletti, P. U., and Levi-Montalcini, R. Arch. ital Biol.
     108:213 (1970).

DRUGS INTERFERING WITH CENTRAL

CHOLINERGIC MECHANISMS

G. PEPEU

Dept. of Pharmacology - School of Pharmacy

Cagliari University - Italy

In this paper the mechanisms by which many drugs could affect the acetylcholine (ACh) cycle in the central nervous system are listed and examples taken from the literature and from previous work of the author are discussed. No attempt has been made to cover all the vast literature concerning the effects of drugs on cerebral cholinergic mechanisms and data on the determination of the changes either in brain ACh output or content induced by drugs are mainly presented.

Interference with brain ACh cycle can be either the main central action of a drug, for example atropine, tremorine, or a secondary effect of a wider action which only the determination of the level of stored and released ACh may reveal as with general anaesthetics.

ACh is present in various concentrations in the brain of mammalian and non mammalian animal species (1) with a non uniform distribution as shown in Table 1 for the cat brain. A closely related distribution has been reported for choline acetyltransferase (2), the enzyme forming ACh and for acetylcholinesterase (3). The caudate nucleus, lateral geniculate bodies and superior colliculus show the highest ACh content. Similar results were also obtained in the rat brain (4).

Table 1

Acetylcholine distribution in some regions of cat brain

Region	ACh level µg/g	Region	ACh level µg/g
Cerebral cortex:		Lateral genicula	
frontal	0.84	te bodies	3.34
parietal	0.93	Basal ganglia	2.66
occipital	1.06	Hypothalamus	2.31
Cerebellum	0.26	Pons	2.78
Caudate nucleus	3.43	Medulla oblongata	2.20
Superior colliculus	4.50	Medulla	0.38

The values for ACh level were determined in unanasthetized cats transected at midpontine pretrigeminal level (5, 6 and unpublished results).

On the basis of the classical and widely accepted description of the ACh cycle in the nervous system made by BIRKS and Mc INTOSH (7), it appears that the drugs could act at the following steps:
1) ACh synthesis:
    inhibition of a) choline acetyltransferase; b) choline
    uptake
    decrease of available acetate
    stimulation of choline acetyltransferase
    increase of precursors.
2) ACh storage
3) ACh inactivation: inhibition of acetylcholinesterase
                     inhibition of uptake mechanism
4) ACh release from the nerve endings.
5) Postsynaptic cholinoceptive receptors.

Extraction of ACh from the brain was usually carried out by means of relatively simple procedures (4, 6, 8, 9) and the content of the extracts was determined by bioassay on frog rectus abdominis muscle, leech dorsal muscle, guinea pig ileum. Recently reliable gas chromatographic methods have been introduced (10) and the content thus obtained are in good agreement with the previous determinations. The gas chromatographic methods have definitely confirmed the identity of the bioassayed material with true ACh. ACh output was investigated by bioassay but recently attempts to label cerebral ACh and to determine its output by chromatographic and radiochemical methods have been reported (11). Direct quantitative studies of ACh turnover in the brain are only beginning (12).

In the following paragraphs the drugs acting on ACh cycle will be discussed in some detail.

DRUGS WHICH BLOCK ACETYLCHOLINE SYNTHESIS

Inhibition of choline acetyltransferase. A variety of types of compounds have been reported to be weak inhibitors of choline acetyltransferase (13). Few of these have I_{50} values below 10^{-3} M. Recently a group of styrylpiridine analogs which actively inhibit "in vitro" this enzyme was synthetized (14). The most active, hexamethylene-1-4 (1 naphthylvinyl) pyridinium 6-trimethylammonium dibromide, has an I_{50} value of 9×10^{-7} M. The most specific is 4 (1 naphthylvinyl) pyridine. The pharmacological actions observed "in vitro" and "in vivo" (15) demonstrate an impairment of the cholinergic peripheral mechanisms. However no direct measurment of ACh content and of ACh release either at peripheral level or in the brain has been, to our knowledge, as yet reported.

Competition with choline uptake. It has been shown that ACh synthesis can be inhibited through a competition with choline for a carrier mechanism necessary to transport extracellular choline to the intracellular sites at which it is acetylated. The most active compounds available for this purpose are hemicholinium (HC-3) (16) and triethyl choline (TEC) (17). Both possess a quaternary ammonium structure and they do not readily penetrate into the brain from the peripheral circulation. Injected intraperitoneally HC-3 caused a marked respiratory depression, muscular weakness but no changes in brain ACh content (18). In order to overcome the blood brain barrier these two drugs can be injected directly into some brain structures or intraventricularly. By this route a marked reduction in ACh content of the structures surrounding the ventricles, associated with EEG alteration was observed in the dog.

Both HC-3 and TEC caused a rapid decrease in the ACh content of the entire brain in the rat (19).

<u>Decrease of active acetate</u>. The decrease in brain
ACh observed in rats rendered thiamine deficient or
treated with thiamine antagonists could find an explanation
in the block of the pyruvate-acetylCoA pathway (20). Other
metabolic alterations could affect the ACh cycle through
this mechanism.

DRUGS WHICH STIMULATE ACETYLCHOLINE SYNTHESIS

<u>Possible precursors</u>. Attempts have been made to
increase ACh formation in the brain by administering
choline or dimethylaminoethanol (DMAE), a precursor of
choline in tissue of the rat. However neither DMAE nor
choline acutely and chronically administered caused an
increase in total brain ACh in the rat (21). Choline also
administered into the cerebral ventricles did not affect
the levels of brain ACh in the rat (19) but only antagoni-
zed the effects of HC-3 and TEC on ACh content. On the
other hand it was shown that the endogenous formation of
acetyl CoA and not choline is rate limiting in the synthesis
of ACh (22).

<u>Stimulation of choline acetyltransferase.</u> According
to the observation of STERN and GASPAROVICH (23) recently
confirmed (24), tremorine and its active metabolite oxotremorine have an activating effect on brain choline
acetyltransferase which brings about an increase in brain
ACh level (refs. in 24). Both drugs cause in several
animal species tremors and a strong peripheral muscarinic
stimulation which can be prevented by pretreatment with
anticholinergic drugs. A possible causal relationship
between the rise in brain ACh level and the onset of the
tremors is still matter of investigation. However it
should be pointed out that the increase seems to take
place in the areas of the extrapyramidal system involved
in the control of movements and posture as shown in Table
2. Furthermore arecoline, another muscarinic drug, also
causes tremors associated with an increase in brain ACh
level (25). It also appears that the regions in which
oxotremorine causes the most evident increase of ACh
content are those with highest choline acetyltransferase
activity (2).   Nevertheless the reasons for the non
uniform effect of this drug are not yet explained.

Table 2

The effect of oxotremorine (OT), 1 mg/Kg i.p., on ACh content of some brain regions of young cats.

Regions	ACh content μg/g		% Changes
	Controls	After OT	
Cerebral cortex	1.12 (4)	1.43 (4)	+ 27
Cerebellum	0.26 (2)	0.27 (2)	+ 3
Diencephalon and upper brain stem	2.53 (4)	4.79 (4)	+ 89
Caudate nucleus	1.86 (2)	4.52 (2)	+143
Lower brain stem	3.32 (4)	3.45 (4)	+ 4

In parenthesis the number of determinations. Part of the data have already been reported (26).

DRUGS WHICH AFFECT ACETYLCHOLINE STORAGE

Subcellular fractionation procedures and electronmicroscopic investigations demonstrated that brain ACh is stored in three neuronal compartments: 1) cytoplasmatic, 2) vesicular and 3) on the external membrane or in vesicles close to it (27). These ACh pools would correspond to the so called "free" and "bound" ACh (the latter present in stable and labile forms) which are obtained with simple extraction procedures.

Drugs can affect the pattern of ACh subcellular distribution. It was shown that anaesthetic increase equally "bound" and "free" ACh while eserine and tremorine increase the "free" fraction to greater extent than the "bound" (28).

By subcellular fractionation it was observed (29) that anaesthetics increase the "free" and the labile fraction of the "bound" form while eserine affects mainly the "free" fraction. On the other hand ACh depleting agents, such as scopolamine and metrazole, mainly decrease the labile and stable forms of the bound fraction. The changes in ACh subcellular distribution seem to be caused by effects of the drugs more on the release from the nerve endings and on the inactivation than on the binding mechanisms in the vesicles.

DRUGS WHICH AFFECT ACETYLCHOLINE INACTIVATION

<u>Acetylcholinesterase inhibition.</u> Brain ACh is
inactivated by acetylcholinesterase and to a lesser
extent by other cholinesterases, called also pseudoesterases,
which are present in the glial cells and in some neurons
(30). These esterases are inhibited by a score of compounds
among which the most common are eserine and its analogs
and the organophosphorus derivatives used as pesticides,
combat gases and rarely for medical purposes. The
administration of cholinesterases inhibitors is followed
by a marked increase in ACh content, as shown in table 3,
and by severe peripheral and central cholinergic symptoms.
A direct relationship between inhibition of brain acetyl-
cholinesterases, increase in brain ACh and some behavioural
effects was demonstrated in the rat after the administration
of eserine (31).

Table 3

Effects of some acetylcholinesterase inhibitors on brain
ACh content in the rat.

Inhibitors	Dose mg/Kg i.p.	ACh content $\mu g/g \pm$ S.E.		% increase	Rfs.
		Controls	Treated		
Eserine	1.0	2.44 ± 0.07	3.70 ± 0.12	51	32
TABUN	0.16	2.60 ± 0.29	4.17	60	33
TEPP	1.0			117	18
Paraoxon	0.5	2.90 ± 0.3	5.10 ± 0.4	75	34

<u>Inhibition of ACh uptake</u>. It has been shown that
uptake of ACh against a concentration gradient into
slices of mouse and rat brain occurs in the presence of
suitable cholinesterase inhibitors (35, 36). A number
of drugs, among them eserine, HC-3, atropine, oxotremorine,
morphine, cocaine, nicotine, are competitive inhibitors
with K_i values in the order of 10^{-5} M. The importance
of the uptake in the physiological inactivation of ACh
has not yet been defined.

DRUGS WHICH AFFECT THE RELEASE OF ACETYLCHOLINE

Table 4
List of drugs affecting ACh release from the cerebral cortex.

Drug	Route Administ.	Dose mg/Kg	Animal species	Effect on ACh output	Rfs
Ether	respir.	–	sheep	decrease	37
Cyclopropane	"	–	"	decrease	37
Halothane-N_2O	"	–	cat	decrease	38
Pentobarbital	i.v.	20	rabbit	decrease	39
Morphine	i.v.	5	"	decrease	39
Chloralose	i.v.	60	cat	decrease	37
Metrazol	i.v.	80	"	increase	40
Strychnine	i.v.	5	"	increase	40
Nicotine	i.v.	0.004	"	increase	41
Scopolamine	i.v.	0.75	"	increase	42
	local	0.001/ml	"	increase	
Atropine	i.v	1	"	increase	43
	local	0.0002/ml	"	increase	
Cocaine	local	10/ml	"	decrease	42
Tetrodotoxin	local	0.01/ml	"	decrease	38

Table 4 summarizes some of the available data on the changes of ACh release from the cortex induced by many drugs. It can be seen that drugs which either depress the nervous activity or prevent the depolarization of the nerve endings, like cocaine and tetrodotoxin, bring about a decrease of ACh output. On the other hand all the drugs which stimulate nervous activity increase ACh output from the brain. The changes in ACh output are followed by variations in ACh content the brain as shown in Table 5.

DRUGS BLOCKING THE POSTSYNAPTIC RECEPTORS

The anticholinergic drugs, particularly the anti-muscarinic agents, exert a score of behavioral and EEG effects which have been widely investigated (46). From

table 4 and 5 it appears that both atropine and scopolamine also increase ACh output from the cortex and decrease
brain ACh content.

Table 5
Drug induced changes of brain ACh in the rat.

Drug	Dose mg/Kg	% change	References
Pentobarbital	30	+ 21	18
Morphine	50	+ 47	18
Chloralose	60	+ 47	44
Metrazol	75	- 22	18
Nicotine	1	- 55	44
Scopolamine	0.6	- 31	45
Atropine	5	- 33	45

It has been shown (45) that ACh reduction after
scopolamine was restricted to the cerebral hemispheres
and did not appear in subcortical regions of the brain.
In the cat the effect of atropine on ACh output was completely abolished by mesencephalic lesions or by tetrodotoxin (38). These findings support the hypothesis that
anticholinergic drugs increase ACh output by blocking
cortical cholinergic synapses which are part of a circuit
inhibiting cholinergic neurons. However the observation
(24) that atropine prevents in vivo and in vitro the increase in brain ACh caused by oxotremorine suggests the
possibility that anticholinergic drugs might also affect
ACh level by a different mechanism.

CONCLUSIONS

In this short review it has been shown that many
drugs influence the amount of neurotransmitter available
at the central cholinergic nerve endings. Direct effects
on ACh formation have been discussed, but a number of
evidences demonstrate that ACh synthesis in the brain is
regulated by the concentration of ACh in the vicinity of
the sites of synthesis (47). Therefore drugs which cause
ACh accumulation in the nerve endings bring about a decrease of ACh formation while drugs which stimulate ACh

output enhance choline acetyltransferase activity (39). By this mechanism ACh metabolism seems to mantain rather efficiently the level of neurotransmitter within relatively narrow limits. For example drugs stimulating ACh output cause a maximum decrease of 50% while it is not uncommon to find a total depletion in the brain monoamines content after such drugs as reserpine.

1) Whittaker, V.P. in Cholinesterases and Anticholine-sterases Agents. (Koelle, G.B. ed.) Springer Berlin (1963).

2) Hebb, C.O. and Silver, A. J. Physiol. 134,: 718 (1956)

3) Burgen, A.S.V. and Chipman, L.M. J. Physiol. 114: 296 (1951).

4) Takahashi, R. and Aprison, M.H. J.Neurochem. 11: 887 (1964).

5) Pepeu, G. Sett. Med. 54; 57 (1966).

6) Deffenu, G., Bertaccini, G. and Pepeu, G. Expl. Neurol. 17: 203 (1967).

7) Birks, R. and McIntosh, F.C. Canad. J. Biochem. 39: 787 (1961).

8) Dren, A. and Domino, E. J. Pharmacol. Exp. Therap. 161: 141 (1968).

9) Beani, L. and Bianchi, C. J. Pharm. Pharmac. 15: 281 (1963).

10) Hanin, I. in Adv. Biochem. Psychopharmacol. Vol. 1 (Eds. E. Costa and P. Greengard) Raven Press. N.Y. (1969).

11) Chakrin, L.W. and Shideman, F. E. Int. J. Neuropharmacol. 7: 337 (1968).

12) Schubert, J., Sparf, B. and Sundwall, A. J. Neurochem. 16; 695 (1969).

13) Nachmansohn, D. in Cholinesterases and Anticholine-esterases Agents. (Ed. G.B. Koelle) Springer Berlin (1963).

14) Crispin Smith, J., Cavallito, C.J. and Foldes, F.F.

Biochem. Pharmacol. 16: 2438 (1967).

15) Hemsworth, B.A. and Foldes, F.F. European J. Pharmac.
11: 187 (1970).

16) MacIntosh, F.C., Birks, R.I. and Sastry, P.B. Nature
178: 1181 (1956).

17) Bull, G. and Hemsworth, B.A. Nature 199: 487 (1963)

18) Giarman, N.J. and Pepeu, G. Br. J. Pharmac. 19: 226
(1962)

19) Slater, P. Int. J. Neuropharmacol. 7: 421 (1968).

20) Cheney, D.L., Gubler, C.J. and Jaussi, A.W. J. Neurochem. 16: 1283 (1969).

21) Pepeu, G., Freedman, D.X. and Giarman, N.J. J. Pharmacol. Exp. Therap. 129: 291 (1960).

22) Smallman, B.N. J. Neurochem. 2: 119 (1958).

23) Stern, P. and Gasparovic, I. in First Internat. Pharmacol. Meeting 8: 149 Pergamon Press Oxford (1961)

24) Holmstedt, B. and Lundgren, G. in Int. Symposium on
the effect of drugs on Cholinergic mechanisms in the
C.N.S. Skokloster, Sweden (1970).

25) Holmstedt, B. in Cholinergic Mechanisms. Ann. N. Y.
Acad. Sci. 144: 433 (1967).

26) Bartolini, A., Bartolini, R. and Pepeu, G. J. Pharm.
Pharmac. 22: 60 (1970)

27) Barker, L.A., Dowdall, M., Essman, W.B. and Whittaker,
V.P. in Int. Symposium on cholinergic mechanisms in
The C.N.S. Skokloster, Sweden (1970).

28) Crossland, J. and Slater, P. Br. J. Pharmac. 33: 42
(1968)

29) Beani, L., Bianchi, C., Megazzini, P., Ballotti, L.
and Bernardi, G. Biochem. Pharmacol. 18: 1315 (1969).

30) Phillis, J.W. The pharmacology of Synapses, Pergamon
Press, Oxford (1970).

31) Rosencrans, J.A., Dren, A.T. and Domino, E. Int. J.
Neuropharmacol. 7: 127 (1968).

32) Pazzagli, A. and Pepeu, G. Int. J. Neuropharmacol. 4: 291 (1964).

33) Giarman, N.J. and Pepeu, G. Boll. Soc. Ital. Biol. Sper. 37: 128 (1961).

34) Milosevic, M.P. J. Pharm. Pharmac. 21: 469 (1969).

35) Schubert, J. and Sundwall, A. J. Neurochem. 14: 807 (1967).

36) Liang, C.C. and Quastel, J.H. Biochem. Pharmacol. 18 1169 (1969).

37) Mitchell, J.F. J. Physiol. 165: 98 (1963).
38) Dudar, J.D. and Szerb, J. C. J. Physiol. 203: 741 (1969).

39) Beani, L., Bianchi, C., Santinoceto, L. and Marchetti, P. Int. J. Neuropharmacol. 7: 469 (1968).

40) Beleslin, D., Polak, R.L. and Sproull, D.H. J. Physiol. 181: 308 (1965).

41) Armitage, A.K., Hall, G.H. and Seller, C.M. Br. J. Pharmac. 35: 152 (1969).

42) Bartolini, A. and Pepeu, G. Br. J. Pharmac. 31: 66 (1967).

43) Szerb, J.C. Canad. J. Physiol. Pharmac. 42: 303 (1964)

44) Pepeu, G. Arch. Ital. Sci. Farmacol. III 15: 146 (1965)

45) Giarman, N.J. and Pepeu, G. Br. J. Pharmac. 23: 124 (1964).

46) Bradley, P.B. and Fink, M. Anticholinergic drugs and brain function in animals and man. Elsevier, Amsterdam (1968).

47) Sharkawi, M. and Schulman, M.P. Br. J. Pharmac. 36: 373 (1969).

THE GABA SYSTEM IN BRAIN DEVELOPMENT

Eugene Roberts

Director, Division of Neurosciences

City of Hope National Medical Center, Duarte, Calif. 91010

INTRODUCTION

All normal or adaptive activity is a result of coordination of excitation and inhibition in the nervous system within and between neuronal subsystems in a particular organism. The underlying principle of information-processing is a coordinated interplay of excitatory and inhibitory influences. Most communication that takes place between receptor and neuron, neuron and neuron, and neuron and effector probably occurs via the presynaptic liberation of substances that have either excitatory or inhibitory influences on postsynaptic membranes (see ref. 1 for review and some pertinent general references). Several known naturally-occurring substances have been implicated as potential excitatory or inhibitory transmitters. Acetylcholine and glutamic and aspartic acids may be excitatory transmitters. γ-Aminobutyric acid (GABA), glycine, the catecholamines, histamine, and serotonin may be inhibitory transmitters. Knowledge of the properties and distributions of the enzymes which form some of these substances and degrade them and of the neural circuits in which they exist is only now becoming available. Mechanisms of presynaptic release of transmitters and their modes of action on postsynaptic membranes are being studied. The transport mechanisms which have been identified in neuronal membranes and which, in most instances, may be the chief mechanisms for removal of the active substances from synapses also are becoming known only now.

Biochemical studies of development are difficult because there is superimposition of developmental variables onto already difficultly interpretable experimental situations. Some of the develop-

mental processes which take place during the early period are growth
and arborization of dendritic neuronal processes; myelination of
axons; growth and branching of the capillary bed; differentiation
and proliferation of glial elements; and the differentiation, movement, and multiplication of the short-axoned granule cells or micro-
neurons. In addition, the types and extent of sensory input during
active postnatal cerebral growth might selectively enhance neural
pathways that are used extensively; and, therefore, such pathways
might have a greater probability of becoming important information-
processing units than those which are used less or not at all.
Thus, in a developing nervous system there is a mutual interplay of
developmental and stimulatory variables. Minimally, all of the
above factors must be kept in mind when one is studying a developing nervous system. Certainly all of the cellular structures, as
well as the extracellular compartment, must be considered when an
assessment is being made of the existence of different metabolic
"pools" and their changes during development and in different
normal and abnormal functional states. To complicate matters even
more, there are obvious differentiations even within a single given
neuron. The morphology, chemistry, and function of the dendrites,
soma, axon, and synaptosome differ from each other in the adult
and change in different ways at varying rates during development.

     In developmental studies, even more than in studies of fully
developed structures, one must face the problem of obtaining chemical information that is meaningful with regard to the information-
processing in a particular structure, and not just related to its
maintenance metabolism. Most questions still remain to be answered
about the transmitter biochemistry of the nervous system. The
conclusion from the work done to date is inescapable that specific
chemical, enzymatic, or immunocytochemical probes will have to be
developed for the localization of the transmitters and the enzymes
of their metabolism at an ultrastructural level in functionally
meaningful loci before completely conclusive data can be obtained
with regard to their exact relationships. Until the latter is
achieved, an experimenter who would study only the changes in some
biochemical parameter of a whole brain or a dissected part of it
as a function of development in a particular species often would be
likely to be wasting his time, since the specificity of the information obtained would be of a low order and there would be little
probability of eventual detailed correlation with morphological and
physiological parameters. The question might be asked about what
might be expected to be learned from studies during development of
a variety of biochemical variables in brain homogenates, slices,
or subcellular particulates from grossly dissected areas of brain.
It appears to me that by and large such work will have to be redone
when the neuronal systems studied become better known from physiological and morphological points of view.

THE GABA SYSTEM IN THE ADULT CEREBELLUM

The GABA system in nervous tissue is relatively well known from a biochemical point of view, and neuroanatomical, neurophysiological, and neurochemical studies of the vertebrate cerebellum have made this a relatively well-known neuronal system (see references 2 and 3). GABA is formed in the CNS of vertebrate organisms to a large extent, if not entirely, from L-glutamic acid. The reaction is catalyzed by an L-glutamic decarboxylase (GAD), an enzyme found in mammalian organisms only in the CNS, largely in gray matter. This enzyme is inhibited by anions and carbonyl-trapping agents. The α-decarboxylation of L-glutamic acid now has been shown to occur in kidney and several other non-neuronal tissues (4-6). The enzyme catalyzing the latter process requires high concentrations of anions for maximal activity and also is activated by carbonyl-trapping agents. In the brain the latter enzyme is present in glial cells and in cerebral blood vessels. The reversible transamination of GABA with α-ketoglutarate is catalyzed by an aminotransferase, GABA-T, which in the CNS is found chiefly in the gray matter, but also is found in other tissues. The products of the latter reaction are succinic semialdehyde and glutamic acid. Nervous tissue also contains a dehydrogenase which catalyzes the oxidation of succinic semialdehyde to succinic acid, which in turn can be oxidized via the reactions of the tricarboxylic acid cycle.

The overall function of the cerebellum probably is entirely inhibitory, the only output cells of the cerebellum, the Purkinje cells, inhibiting in Deiters' and intracerebellar nuclei monosynaptically, and also inhibiting each other through axon collaterals. The basket, stellate, and Golgi cells are believed to play inhibitory roles within the cerebellum. The basket cells make numerous powerful inhibitory synapses on the bottom region of the somata of the Purkinje cells and on their basal processes or "preaxons." The superficial stellate cells form inhibitory synapses on the dendrites of Purkinje cells. The Golgi cells make inhibitory synapses on the dendrites of the granule cells. Afferent excitatory inputs reach the cerebellum via the climbing and mossy fibers, which excite the dendrites of the Purkinje and granule cells, respectively. The granule cells are believed to be the only cells lying entirely within the cerebellum which have an excitatory function.

A reasonable interpretation of all of the data obtained to date on adult cerebellum is that GABA is formed and stored in the presynaptic endings of the basket cells and upon stimulation is released from them onto membranes of Purkinje cells. The inactivation process probably consists of removal of GABA from the synaptic gap by transport processes. A part of the released GABA could be transported into the presynaptic endings, where most of it

would be retained for subsequent release; or it could be taken up
by the Purkinje cell bodies, where it could be transported down the
axon or metabolized via the GABA-T pathway. The possibility of
glial uptake and metabolism also must be considered. The Purkinje
cells appear to be GABA neurons, and both GAD and GABA are present
in their axons, possibly in a state of transport from the cell body
to the presynaptic endings of these cells on Deiters' neurons in
the vestibular nucleus and the cells of the intracerebellar nuclei
(7,8). It also is possible that the Golgi cells may employ GABA
to mediate their inhibition, but information on this point still
is inconclusive.

THE GABA SYSTEM IN THE CEREBELLUM OF THE DEVELOPING CHICK

Because of its conveniently layered structure and the known
progression of development of individual cell types within the
cerebellum, it appeared that cerebellum of developing chick embryo
might be a favorable site in which to begin to examine the development of the GABA system in correlative manner with the morphological development (9). The time-sequence of the development of
components of the GABA system was correlated with development as
observed at both light microscopic and electron microscopic levels.
Study also was made of the changes of histochemically visualized
GABA-T activity and of the subcellular distribution pattern of the
GABA system with the age of the embryo.

Characteristic synaptic structures were noted in the chick
cerebellum at 11 days of incubation, the degree of synaptogenesis
at this early stage being small by comparison with that observed
at subsequent stages of development. From the histological observations it was suggested that the synapses observed might be between
Purkinje cell axons and the somata of cells of the intracerebellar
nuclei and between Purkinje cell axon collaterals and the somata
and dendrites of other Purkinje cells. Much further detailed work
will be required to settle definitively the location of the electron
microscopically observed synapses at this stage of development.

The chief enzymes of the GABA system, GAD and GABA-T, began to
increase much later in development than did the weight and protein
content of the cerebellum (Fig. 1). Likewise, the amount of material recovered in the synaptosome-rich fraction, P_2-B, also began to
increase rapidly only at 17 days of incubation (Fig. 2). It appears
that the development of the key components of the GABA system is
temporally better correlated with the development of recognizable
synaptic structures than with accretion of the total mass of the
cerebellum. The fractionation data were consistent with the
supposition that both GAD and GABA are present at least in some
inhibitory nerves and are distributed throughout the neuron, the

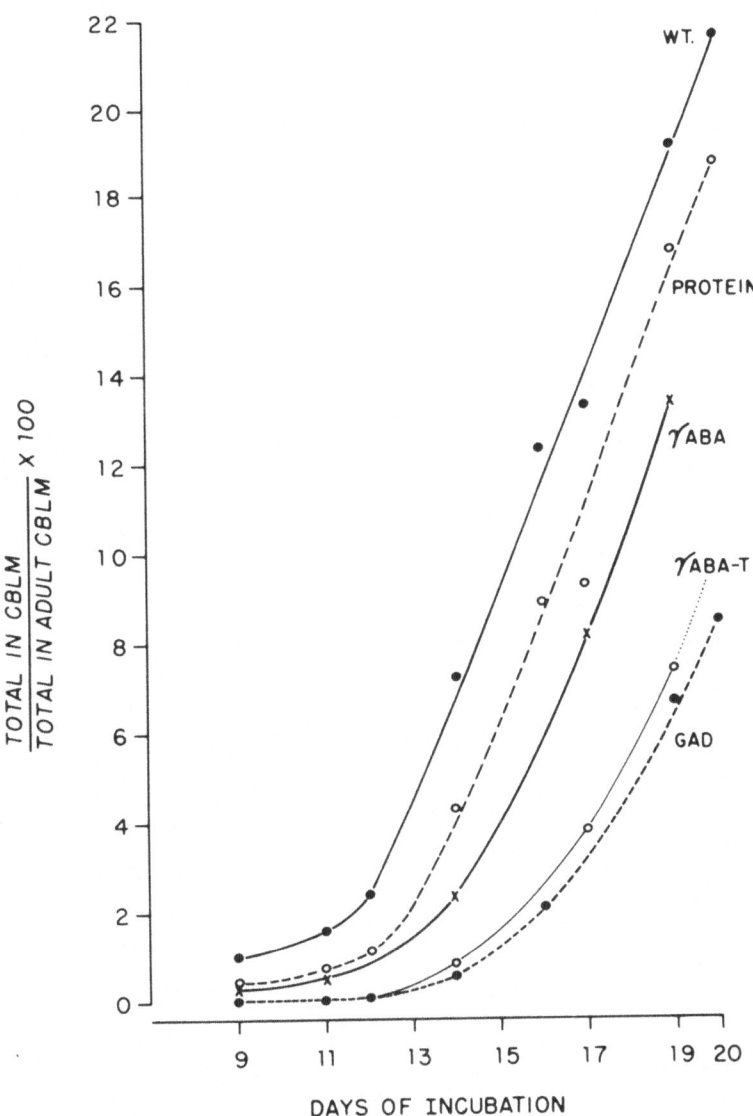

FIG. 1. Increases of weight, protein and GABA contents, and GAD
and GABA-T activities in cerebellum of chick embryo expressed
as percentages of adult values (9).

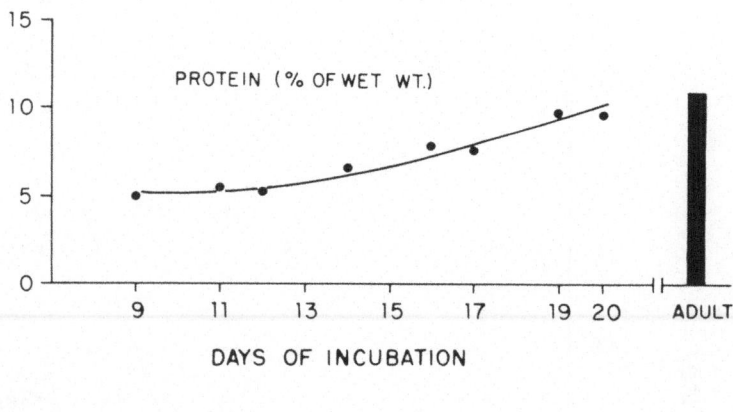

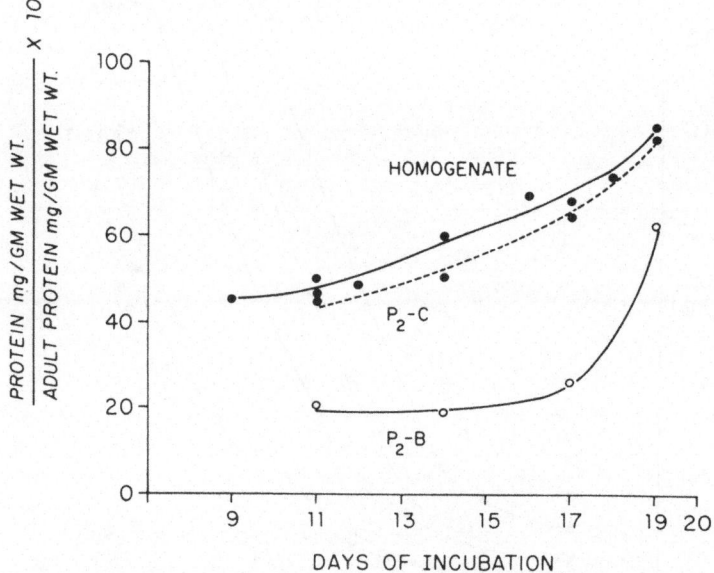

FIG. 2. Protein content (per cent of wet weight) of chick embryo
cerebellum and adult chicken cerebellum; and protein contents
in whole homogenates of chick embryo cerebellum and of the
P_2-B (synaptosome-rich) and P_2-C (mitochondrial) fraction, as
a function of age (9).

GAD being somewhat more highly concentrated in the presynaptic
endings than elsewhere. The GABA-T was found to be particularly
high in the free mitochondria, which probably come largely from
those neuronal sites onto which GABA might be liberated, such as
perikarya and dendrites that receive inhibitory inputs, and also
from glial and endothelial cells in the vicinity of inhibitory
synapses. GABA-T activity also was found to a small extent in the
nerve endings, and this probably is attributable to the nerve ending
mitochondria (see ref. 2 for discussion and pertinent references).
A centrifugally derived preparation of presynaptic nerve endings
is a mixture from inhibitory and excitatory nerves. In some brain
areas there are inhibitory presynaptic endings on nerve endings
which, themselves, serve as excitatory presynaptic inputs on other
postjunctional neuronal sites. It is, therefore, possible that
the GABA-T found in the nerve ending fraction is located in those
presynaptic endings that liberate excitatory transmitter, and which
receive an input of GABA from inhibitory neurons; while the true
inhibitory presynaptic endings may not contain any GABA-T at all.
The results fit the suggestion that in the chick cerebellum GABA
largely is formed at presynaptic sites and metabolized at postsynaptic sites onto which it is liberated. However, proof of this
idea will only come when it is possible to visualize GAD, GABA-T,
and GABA at an ultrastructural level in specific neuronal sites of
ultrathin sections of the cerebellum.

     Thus, in the above case, results obtained by methods at the
limits of our current technical capabilities give us a modicum of
meaningful information about the development of a much studied
chemical transmitter system in the best known vertebrate neuronal
system. Further advances will depend, in part, on the development
of quantitative methods for measurement of synaptogenesis, and of
methods for the identification of synapses that are inhibitory
and excitatory and the determination for each type of synapse of
the major transmitter system with which it deals.

                              COMMENT

     Many studies have been made in the past of changes in GABA and
GAD levels with development in whole brains or in specific brain
regions of a variety of species (see ref. 10). In all instances
there have been found to be progressive increases in the levels of
components of the GABA system during development which correlated
well with other developmental features, such as increases in degree
of dendritic arborization and synaptogenesis. Studies similar to
those for the GABA system in the developing chick cerebellum could
be performed for other biochemical systems and in other conveniently layered neuronal structures such as retina, cortex, hippocampus,
etc. However, in none of the latter instances has the degree of

knowledge with regard to structural and physiological organization begun to approach that available for the cerebellum, and years of combined effort might be required before a comparable degree of sophistication is attained. We have chosen to put our own efforts into attempting to increase the acuity and specificity of the information about the cerebellum, by some of the means suggested above, rather than to increase our empirical knowledge about many other vertebrate neural systems not so well known.

REFERENCES

(1) Roberts, E., and Matthysse, S. Ann. Rev. Biochem. 39:777 (1970).

(2) Roberts, E., and Kuriyama, K. Brain Research 8:1 (1968).

(3) Eccles, J. C., Ito, M., and Szentagothai, J. The Cerebellum as a Neuronal Machine, Springer, Berlin, 1967, 335 pp.

(4) Haber, B., Kuriyama, K., and Roberts, E. Biochem. Pharmacol. 19:1119 (1970).

(5) Haber, B., Kuriyama, K., and Roberts, E. Science 168:598 (1970).

(6) Kuriyama, K., Haber, B., and Roberts, E. Brain Research 23: 121 (1970).

(7) Obata, K., Ito, M., Ochi, R., and Sato, N. Exp. Brain Research 4:43 (1967).

(8) Fonnum, F., Storm-Mathisen, J., and Walberg, F. Brain Research 20:259 (1970).

(9) Kuriyama, K., Sisken, B., Haber, B., and Roberts, E. Brain Research 9:165 (1968).

(10) Baxter, C. F. in Handbook of Neurochemistry, Vol. 3 (Ed. A. Lajtha) pp. 289-353 (1970), Plenum Press, New York.

HORMONES AND BRAIN DEVELOPMENT

F. Caviezel and L. Martini

Department of Pharmacology, University of Milan

20129 Milan, Italy

I. INTRODUCTION

Recent experimental and clinical evidence indicates that the development and the maturation of the Central Nervous System (CNS) of mammals is deeply influenced by hormones; these interact with several other factors which are also of importance for the physiological maturation of the brain (state of nutrition, temperature, etc.). Hormones participate in the process of myelination, in the organization of neuroendocrine phenomena and in the establishement of behavioral patterns. It is apparent from many data that, in all species examined so far, the period of highest sensitivity and vulnerability of the CNS corresponds to that in which the brain grows fastest.

Only a selection of the most significant results regarding the effects of hormones on brain development will be presented in this paper. Special consideration will be given to the information obtained in the authors' laboratory.

II. METHODS USED FOR EVALUATING THE EFFECTS OF HORMONES ON THE DEVELOPING BRAIN

Several procedures have been adopted for analyzing the effects of endocrine factors on the development of the brain and of its functions. These include: the study of the histological appearance and of the biochemical composition of the brain (changes in cerebral

DNA, RNA, enzymes, etc.), the study of the incorporation of labelled aminoacids into CNS proteins, the recording of neural activity
using electrophysiological techniques, the evaluation of the alterations of several endocrine functions induced by hormones administered during the neonatal period, etc.

It is generally recognized that changes in the composition of
the brain parallel the process of maturation. For instance, the concentration of some lipids (e.g. cerebrosides and cholesterol)   , of
proteins of DNA and RNA, are very low at birth and increase
progressively with advancing age (1,2, 3). The changes in brain
lipids are particularly evident in the rat and in the rabbit, animals
which are born with an immature brain, while they are very limited
in the guinea pig, which shows a high degree of neurological maturity at birth. The incorporation of labelled aminoacids into subcellular fractions obtained from mouse brain is higher in immature
animals (3-10 days of age) and decreases to the adult level starting from the 15th day of life (4). It is interesting that, in the immature brain, aminoacids are preferentially incorporated into the
crude mitochondrial fraction as well as in synaptosomes and/or
nerve endings, while, in the adult brain, the uptake of aminoacids
occurs mainly in microsomes (4,5). Amongst the electrophysiological
techniques which have been particularly useful for evaluating the
process of brain maturation, one may quote the "maximal electroshock seizure pattern" and the "transcallose responses" introduced
by Timiras and her co-workers (6,7).

III. EFFECT OF HORMONES ON THE DEVELOPING BRAIN

A. Sex Hormones and Other Steroids

Accumulating evidence indicates that sex steroids, when injected during a "critical" stage of maturation, may influence the sexual
differentiation of the CNS of both male and female mammals and lead
to modifications of the secretory patterns of gonadotropic hormones
and to changes in sexual behavior. The following conclusions may
be derived from the classic findings by Pfeiffer, Barraclough, Harris
and many others (see 8 for references):
    1) animals of both sexes are born with a CNS which is potential-
ly of the female type;
    2) in male animals the final patterns of sexual behavior and of
gonadotropin secretion are organized during the first few days of life

by the presence of the small amounts of testosterone secreted by the immature testis;

3) in female animals the absence of androgens during this "critical" neonatal period permits the permanent organization of the brain in the original feminine pattern.

In support to these conclusions one may quote the following data. Castration, performed in male rats immediately after birth, alters the development of the CNS, so that after puberty the hypothalamic-pituitary axis becomes capable of releasing the Luteinizing Hormone (LH) in a female way, i.e. cyclically, and in amounts sufficient to induce ovulation. This is clearly demonstrated by the observation that ovulation and the formation of corpora lutea occur in the ovarian tissue transplanted into the anterior chamber of the eye of male rats castrated when newborn (9). It is relevant that only castration performed immediately after birth is effective in inducing ovulation and luteinization in the transplanted ovary; castration performed in sexually-mature rats is ineffective. These results indicate that a "critical" period of brain sensitivity to androgens exists (9). Castration of male rats immediately after birth is followed also by severe alterations of sexual behavior; male rats castrated at birth display the lordosis response (i.e., the most typical pattern of female sexual behavior) when estrogen and progesterone are implanted in their hypothalamus; these animals never display male sexual behavior, even if androgens are implanted in their brains (10).

An additional proof of the physiological role played by androgens in the sexual differentiation of the brain has been obtained using the very effective antiandrogen Cyproterone (1,2 alpha-methylene-6-chloro-delta-4,6-pregnadien-17 alpha-ol-3,20 dione) as an experimental tool. The administration of this compound to pregnant female rats induces the development of a vagina in a high proportion of male fetuses. At the same time, the hypothalamic-pituitary axis of male fetuses acquires the ability of secreting gonadotropins according to a female pattern. After orchidectomy and transplantation of ovarian tissue, male animals born from mothers treated with Cyproterone develop corpora lutea and show cyclic changes in their rudimental vagina (11). Like males castrated at birth, male animals born from mothers treated with Cyproterone during pregnancy (or given the antiandrogen immediately after birth) have a sexual behavior of the female type; following orchidectomy and transplantation of an ovary, they show lordosis responses when stimulated by a normal male (11).

It has also been shown that estrogens injected into newborn male
rats (prior to the 10th day of life) induce an inhibition of spermato-
genesis, a reduction in the weight of sex organs, and permanent
sterility. Convincing data indicate that the primary lesion induced
by estrogens is located in the CNS and probably in the hypothalamus
(12,13).

The observation that female rats exposed during neonatal life
to the influence of high doses of androgens, administered either
subcutaneously (8) or directly into the hypothalamus (14,15), develop a syndrome characterized by precocious vaginal opening,
sterility, constant vaginal cornification, decreased ovarian weight
and an alteration of ovarian steroidogenesis (conversion of pre-
gnenolone into androgens with no aromatization to estrogens) (16)
is usually taken as an additional proof of the role played by androgens in directing the differentiation of the immature brain toward a
male pattern. Female animals treated with testosterone when newborn are usually referred to as being "androgenized". However,
the writers feel that these experiments, while proving a demonstra-
tion of a conspicuous effect of testosterone on the developing brain,
do not conclusively prove that androgens play a physiological role
in directing the sexual differentiation of the brain toward the male
type. There are two arguments that support this statement:
    1) a very similar syndrome can be obtained by treating neonatal
female rats with either estrogens (17, 18, 19, 20) or antiestrogens
(e.g. clomiphene) (21, 22); this means that one can paradoxically
obtain "androgenized" animals by administering estrogenic compounds;
    2) the proof that female rats neonatally treated with testosterone
secrete gonadotropins in a fashion similar to that characteristic of
males is still lacking; in particular, it has never been established
whether "androgenized" females exhibit the circadian cycle of synthesis and release of Follicle-Stimulating Hormone (FSH) and of LH,
which is usually found in normal male rats (23). Preliminary data
obtained in the authors' laboratory seem to indicate that "androgenized" females do not have such a circadian rhythm (Caviezel, Carraro and Colombo, unpublished observations).

The results quoted above have been obtained mainly in rats;
there is evidence that mice and hamsters respond in a similar manner.
Information available in other species (rabbits, cats, beagle puppies,
rhesus monkeys, etc.) is so scanty that final conclusions cannot

be drawn (24).

Relatively little is known on the effects of neonatal injections of other steroids, either natural or synthetic (progesterone, 6-alpha-methyl-17 alpha-hydroxyprogesterone, corticosteroids, etc.) on neuroendocrine mechanisms; only prolonged treatments in the neonatal period seem to produce some alterations of the estrous cycle and of fertility (25).

The possibility has been recently considered that maternal progesterone might protect the hypothalamus of the female fetus from the effects of endogenous androgens (26). Recent data obtained in the authors' laboratory suggest that the protective effect of progesterone might be linked to the ability of this steroid to compete with the 5-alpha reductase. The latter one transforms testosterone into dihydrotestosterone which is believed to be the metabolite of androgens active at tissue level (27).

Relevant effects of hormonal steroids on the developing brain have been recently reported by Heim and Timiras (28). They have shown that all phases of the response to an electric shock appear at a significantly earlier date in estradiol-treated animals. Estradiol also induces the precocious appearance of the adult "maximal electroshock seizure pattern" (see Section II). This method of study has emphasized that the " critical" period in which the brain of the rat is particularly sensitive to estradiol occurs between the 6th and the 7th day of life (29). Cortisol also markedly accelerates the appearance of the " maximal electroshock seizure pattern"; however, the " critical" period for this effect of cortisol occurs much later than for estradiol (6). In this context it is interesting to recall that Schapiro et al. (30) have shown that neonatal treatments with cortisol may delay, rather than hasten, the appearance of the swimming behavior. The apparent discrepancy between the data obtained using the electroshock technique and those obtained using the evaluation of integrated behavioral patterns proves how complex is the mode of action of hormonal steroids on the developing brain.

B. Growth Hormone

It has been repeatedly suggested that Growth Hormone may regulate not only organ and body growth, but also cerebral deve-

lopment. Growth Hormone, administered to pregnant rats, seems
to influence the development of the fetal brain in such a way that
when adulthood is reached, several types of behavioral responses
are improved. This phenomenon is coupled with a hypertrophy of
the neurons, with an enlargement of the perikarya and with an expansion of the protoplasmic processes. This results in an increased interaction among the cerebral neurons.The electroencephalo-
gram is scarcely affected (31). Krawiec et al. (32) have recently
found that Growth Hormone, when administered to neonatally-thy-
roidectomized rats (within the 10th day of age) is able to restore
the cerebral and the cerebellar weight reduced by the operation.
In this instance Growth Hormone behaves like thyroid hormones
themselves (see Section III,D). Under these experimental conditions the concentrations of cerebral DNA and RNA and of various
types of cerebral enzymes are also restored by Growth Hormone
(32).

C. Hormones of the Pineal Gland

Studies on the regulatory influence exerted by the pineal
gland and by its secretory products (melatonin, 5-methoxytrypto-
phol, 5-hydroxytryptophol, etc.) on brain maturation are scanty
(33). Prenatal and early postnatal treatments with melatonin delay puberty in female rats; however, the estrous cycle is not significantly influenced (34). These data suggest that neonatally-
administered melatonin can distort only partially the neuroendocrine mechanisms which control gonadotropin secretion. Surpri-
singly enough, in neonatal female rats, the simultaneous administration of melatonin and of testosterone reduces the disturbing
effect exerted by the steroid on the estrous cycle (see Section
III,A) (35); an action similar to that of melatonin has also been reported for the endogenous activation of melatonin biosynthesis induced by blinding (35). Pinealectomy does not reverse the effects
brought about by early postnatal androgen treatment (35).

D. Thyroid Hormones

There is ample evidence based on anatomical, biochemical,
electrophysiological, and behavioral studies showing the participation of thyroid hormones in the processes of brain maturation.

Neonatal thyroidectomy (surgical or radiochemical) is followed by a reduction in the size and a change in the shape of
the brain (36), a retardation of myelination (37), a reduction

in the amount of the neuropils surrounding the cortical neurons and the appearance of abnormal membranous bodies in the most superficial layers of the visual cerebral cortex; consequently, the number of synapses is reduced (38). In hypothyroid rats, the cerebral concentrations of DNA and RNA, the efficiency of brain enzymes (e.g. GABA-transaminase, etc.) and the incorporation of aminoacids are modified. These parameters are all restored to normal when physiological amounts of thyroxine are administered (32,39,40). These data support the concept that thyroid hormones exert a direct influence on the maturation of the CNS by way of their interference with the synthesis of proteins within the cerebral tissue (41,42).

The influence of the thyroid on the developing mammalian nervous system can be demonstrated also by administering an excess of thyroid hormones during infancy. In these conditions, thyroid hormones slightly enhance the maturation of the cerebellar folia in the rat (43); moreover, they significantly advance the time of appearance of several types of behavioral patterns, like, for instance, the swimming behavior (30). Neonatal treatments with thyroid hormones also cause persistent changes in pituitary cytology (44), delayed puberty, and mild hypothyroidism (45), without producing significant modifications in the hypothalamic thyreostatic servomechanism (46).

The ability of thyroid hormones to influence the maturation of neuroendocrine phenomena has been recently studied in the authors' laboratory. It has been previously reported that the early postnatal administration of 50 μg of reserpine in female rats induces a syndrome characterized by a delayed puberty, by prolonged diestrous periods and by a reduction in pituitary stores of LH, FSH and Thyrostimulating Hormone (TSH) (47,48,49). The T/S ratio and the concentrations of the FSH-Releasing Factor (FSH-RF) and of the Thyrotropin-Releasing Factor (TSH-RF) in the hypothalamus have also been shown to be significantly reduced following neonatal treatments with this tranquilizer (48,49).

It was deemed of interest to study whether the administration of thyroxine in the neonatal period concurrently with the injection of reserpine might overcome the damaging effects exerted by this compound on the maturation of the neuroendocrine phenomena described above. The results so far obtained indicate that the injec-

tion of 1-thyroxine (5μg/day on day 4,6,8 of life) is able to counte-
ract almost completely the effects of reserpine (50μg/rat on day 4). In
the thyroxine-reserpine-treated animals puberty occurs at normal age,
estrous cycles are regular, ovarian and body weights are not significantly different from those of untreated controls. The T/S ratio (Table 1),
the stores of LH and FSH in the pituitary (Fig. 1), and the concentrations of FSH-RF in the hypothalamus (Fig. 2) of animals given reserpine and thyroxine simultaneously, are significantly higher than in animals given reserpine alone.

Table 1. Effect of Neonatal Treatment with Reserpine, Thyroxine
and with Reserpine+Thyroxine on the T/S ratio of Female Rats, killed at 120 days of age.

Groups		T/S ratio
Controls	(26)	13.88±1.18
Reserpine	(22)	4.80±1.06
Reserpine + Thyroxine	(16)	13.36±1.70
Thyroxine	(24)	15.43±1.81

Number of rats in parentheses

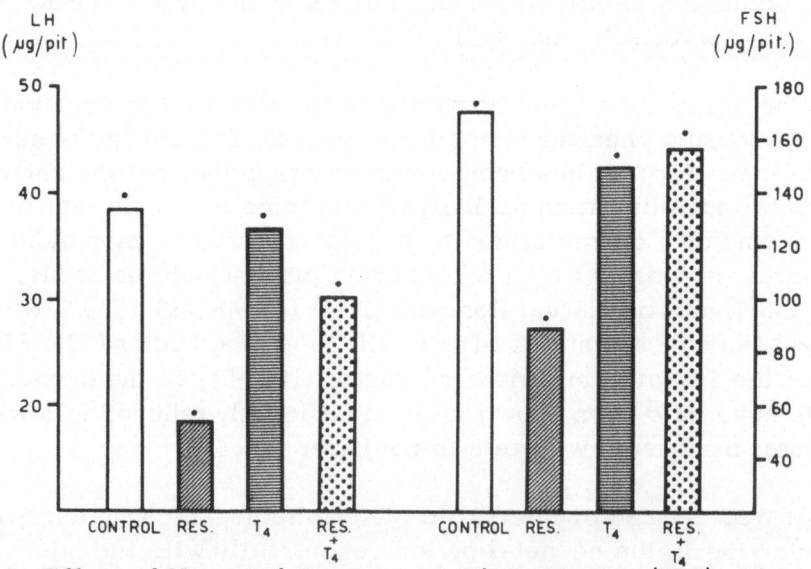

Fig. 1. Effect of Neonatal Treatment with Reserpine (RES), Thyroxine (T_4) and with Reserpine+Thyroxine on Pituitary LH and FSH
content of Female Rats, killed at 120 days of age.

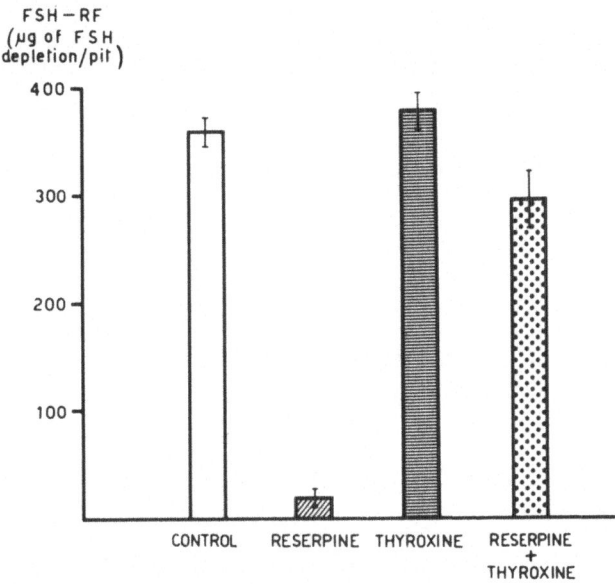

Fig.2. Effect of Neonatal Treatment with Reserpine,Thyroxine and
with Reserpine+Thyroxine on Hypothalamic FSH-RF Content of Female
Rats, killed at 120 days of age (FSH-RF was measured according to the
procedure of David et al.) (50).

In order to explain these results one might formulate the
following two hypotheses:

1) the damaging effect of reserpine is specifically connected with the ability of the drug in reducing thyroid function in
neonatal animals; according to this hypothesis exogenous thyroxine would act as a "substitution therapy"; and

2) the damage induced by reserpine is not due to hypothyroidism, but to some unknown mechanism; if this hypothesis were
correct thyroxine would act aspecifically by facilitating brain maturation as previously described.

ACKNOWLEDGEMENTS

The work described in this paper has been supported by the
following Grants: 97-530 of the Ford Foundation, New York; and
64-64,65-118 and 66-121 of the Population Council, New York.
Standard preparations of FSH, LH and TSH were generously given
by the National Institutes of Health, Bethesda, Maryland. All such
support is gratefully acknowledged.

REFERENCES

1. Galli, C., and Re Cecconi, D.   Lipids 2:76 (1967).

2. Dalal, K.B., and Einstein, E.R.   Brain Res. 16:441 (1969).

3. Dobbing, J., and Sand, J.   Brain Res. 17:115 (1970).

4. Ito, K., and Arimatsu, Y.Sci.Pap. Coll. Gen Educ., Univ.
   Tokyo 18:41 (1968).

5. Klee, C.B., and Sokoloff, L. J. Neurochem. 11:709 (1964).

6. Vernadakis, A., and Timiras, P.S. in Hormonal Steroids (Eds.
   L. Martini, F. Fraschini, M. Motta), p. 908, Excerpta Med.,
   Intern. Congr. Ser. n. 132, Amsterdam (1967).

7. Hatotani, N., and Timiras, P.S. Neuroendocrinology 2:147
   (1967).

8. Barraclough, C.A. in Neuroendocrinology (Eds. L. Martini,
   W.F. Ganong), Vol. 2, p.62, Academic Press, New York (1967).

9. Harris, G.W. Endocrinology 75:627 (1964).

10. Lisk, R.D., and Suydam, A.J. Anat. Record 157:181 (1967).

11. Neumann, F., and Kramer, M. in Hormonal Steroids (Eds. L.
    Martini, F. Fraschini, M. Motta), p. 932, Excerpta Med.,
    Intern. Congr. Ser. n. 132,Amsterdam (1967).

12. Kincl, F.A., Pi, A.F., and Lasso,L.H. Endocrinology 72:966
    (1963).

13. Jacobsohn,D., and Norgren,A. Acta Endocrin. 49:453 (1965).

14. Wagner, J.W., Erwin,W., and Critchlow,V. Endocrinology
    79:1135 (1966).

15. Nadler, R.D. Excerpta Med. Intern. Congr. Ser. 111:363
    (1966).

16. Goldzieher, J.W., and Axelrod, R.L. Fertil. Steril. 14:631
    (1963).

17. Gorski, R.A. Am. J. Physiol. 205:842 (1963).

18. Swanson, H.H. J. Endocrin. 36:327 (1966).

19. Arai, I., and Kusama, T. Anat. Record 157:207 (1967).

20. Presl, J., Jirasek, J., Horsky, J., and Henzl, M. Experientia
    23:374 (1967).

21. Leavitt, W.W., and Meismer, D.M. Nature 218:181 (1968).

22. Kato, J., Kobayashi, T., and Villee, C.A. Endocrinology 82:
    1049 (1969).

23. Fraschini, F., and Motta, M. Program 49th Meet. Endocrine
    Society, Miami, p. 128 (1967).

24. Neumann, F., Steinbeck, H., and Dahn, J.D. in The Hypotha-
    lamus (Eds. L. Martini, M. Motta, F. Fraschini), p. 569,
    Academic Press, New York (1970).

25. Saunders, F.J. Physiol. Rev. 48:601 (1968).

26. Lloyd, C.W., and Weisz, J. in Hormonal Steroids (Eds. L.
    Martini, F. Fraschini, M. Motta), p. 917, Excerpta Med.,
    Intern. Congr. Ser. n. 132, Amsterdam (1967).

27. Kniewald, Z., Massa, R., and Martini, L. Excerpta Med.
    Intern. Congr. Ser. 210:59 (1970).

28. Heim, L.M., and Timiras, P.S. Endocrinology 72: 598 (1963).

29. Heim, L.M. Endocrinology 78:1130 (1966).

30. Schapiro, S., Salas, M., and Vukovich K. Science 168:147
    (1970).

31. Clendinnen, B.G., and Eayrs, J.T. J. Endocrin. 22:183 (1961).

32. Krawiec, L., García Argiz, C.A., Gómez, C.J., and Pasquini,
    J.M. Brain Res. 15:209 (1969).

33. Fraschini, F., and Martini, L. in The Hypothalamus (Eds. L.
    Martini, M. Motta, F. Fraschini), p. 529, Academic Press,
    New York (1970).

34. Vaughan, M.K., O'Steen, W.K., and Vaughan, G.M. Neuro-
    endocrinology 6:10 (1970).

35. Reiter, R.J. Fed. Proc. 27:319 (1968).

36. Eayrs, J.T., and Taylor, S.H. J. Anat. (Lond.) 85:350 (1951).

37. Balász, R., Brooksbank, B.W.L., Davison, A.N., Eayrs, J.
    T., and Wilson, D.A. Brain Res. 15:219 (1969).

38. Cragg, B.G. Brain Res. 18:297 (1970).

39. Geel, S.E., and Timiras, P.S. Brain Res. 4:135 (1967).

40. Geel, S.E., Valcana, T., and Timiras, P.S. Brain Res. 4:143 (1967).

41. Tata, J.R. Acta Endocrin. 124:141 (1968).

42. Sokoloff, L. Proc. Nat. Acad. Sci. U.S. 60:652 (1968).

43. Tusques, J. Biol. Méd. (Paris) 45:395 (1956).

44. Eayrs, J.T., and Holmes, R.L. J. Endocrin. 29:71 (1964).

45. Bakke, J.L., Gellert, R.J., and Lawrence, N. Program 51st Meet. Endocrine Society, New York, p. 72 (1969).

46. Schreiber, V., and Kmentová, V. Neuroendocrinology 1:121 (1965/66).

47. Carraro, A., Corbin, A., Fraschini, F., and Martini, L. J. Endocrin. 32:387 (1965).

48. Martini, L., Carraro, A., Caviezel, F., and Fochi, M. Excerpta Med. Inter. Congr. Ser. 111:366 (1966).

49. Caviezel, F., Carraro, A., Colombo, L., and Martini, L. Atti Accad. Med. Lomb. 22:512 (1967).

50. David, M.A., Fraschini, F., and Martini, L. Experientia 21: 483 (1965).

THE ROLE OF THE ENDOCRINE GLANDS IN MAMMALIAN BRAIN DEVELOPMENT

N. B. MYANT

Medical Research Council Lipid Metabolism Unit

Hammersmith Hospital, London, W.12

The influence of the thyroid on the development of the brain and, indeed, on that of the whole animal, is far greater than that of any other hormone. Hence, this chapter will be concerned mainly with the influence of the thyroid on brain development. The effects of other hormones on brain development will only be considered briefly.

THYROID HORMONE

Brain Development in Man

In human beings, thyroid deficiency during the neonatal period causes mental retardation. If retardation is slight and if treatment with thyroid hormone is begun at an early age, mental capacity may eventually become normal. In severe cretinism, however, the intelligence quotient may remain subnormal throughout life, despite early and continuous treatment with thyroxine (1). This suggests that there is a critical stage in the development of the human brain during which deficiency of thyroid hormone may, if sufficiently marked, cause irreversible damage.

Little is known about the effects of cretinism on the physical development of the human brain, largely because opportunities to study the brains of young cretins so seldom arise. Lotmar (2) has described abnormalities in the brains of adult cretins, including general hypoplasia of the cerebral cortex and a decrease in the size of the Betz cells, particularly in area 4. However, since the youngest of Lotmar's cretins was aged 19, the

changes he observed were not necessarily due to an effect of
thyroid deficiency on brain development, but could have arisen
during adult life.

There are several descriptions of the electroencephalograms
of hypothyroid children and of cretinous adults.   A characteristic finding is a slowing or disappearance of the alpha rhythm and
a diminished response to light stimuli (3).   Schultz et al. (4)
have also observed a retardation of the development of the normal
electroencephalographic pattern in sleep in some hypothyroid
infants.

The question as to whether the human brain can be damaged by
thyroid deficiency in utero is extremely difficult to answer.
Thyroid hormone is certainly required for normal skeletal development in the human foetus (5, 6), which might suggest that it is
also necessary for the normal development of the foetal brain.
However, it is not possible to assess the level of development of
the central nervous system of a living new-born infant in the way
that one can so easily assess the maturity of the skeleton.   It
is therefore impossible to diagnose mental retardation at birth
unless it is extreme.   The question of the susceptibility of the
human foetal brain to damage by thyroid deficiency has a practical
aspect which should not be ignored.   If irreversible damage to
the brain can be caused by thyroid deficiency in utero, then it
is obviously important to begin treatment before birth whenever
there is presumptive evidence of foetal thyroid deficiency, as in
pregnant women who have previously borne an athyreotic infant or
who have been given large doses of ^{131}I after conception.   The
hormone secreted by the maternal thyroid is evidently unable to
cross the placenta in sufficient quantities to replace that
normally secreted by the foetal thyroid (7).   However, Carr and
co-workers (8) have shown that very large doses of thyroid
hormone given to pregnant women can compensate for the lack of
thyroid hormone in an athyreotic foetus.

Brain Development in Animals

The effects of thyroid deficiency on brain development in man
are so striking that many attempts have been made to reproduce
them in experimental animals.   With this object in mind, Eayrs
and his colleagues (see 9, 10) have studied the behaviour of rats
made hypothyroid soon after birth.   The rat is particularly
suitable for experiments of this type because several of the
critical stages in the development of its central nervous system
take place after birth.   It is therefore possible to study the
effects of thyroid deficiency on developmental changes in its
central nervous system which in human beings take place in utero.

Effects on Behaviour.   If thyroid deficiency is induced in new-born rats, the time at which certain complex reflexes first appear is markedly delayed.   These reflexes include sudden movement in response to a loud noise and the righting reflex (the ability to land upright when the animal is dropped).   The development of these automatic responses presumably depends upon the maturation of sense organs and of subcortical neural connections in the brain which integrate the response as a whole.   Impairment of these responses is not, therefore, comparable with the mental retardation that is so conspicuous a feature of cretinism in man, because the physical basis of diminished mental capacity must lie, at least partly, in the higher regions of the brain concerned in learned behaviour.   Moreover, the innate responses whose development is delayed by thyroid deficiency in the rat do eventually appear, even if the hypothyroid state is maintained, whereas the mental retardation seen in cretinism persists in the absence of treatment and, as we have seen, may be irreversible if treatment is delayed.   Some of the effects of neonatal thyroid deficiency on learned or "adaptive" behaviour in rats seem to resemble more closely the effects of cretinism on human behaviour.

When rats are thyroidectomized at birth and are subsequently tested for their capacity to behave adaptively, they perform less efficiently than normal rats of the same age.   When trained to escape from a T-shaped maze they take longer to learn and they make more errors than normal rats.   They also show impaired performance in a closed-field test which has been shown to provide a very sensitive index of those aspects of the rat's behaviour that depend upon the integrity of the cerebral cortex.   Eayrs (11), using this test, has shown that rats thyroidectomized early in life tend to make more errors than normal rats and to repeat the same errors more frequently.   Furthermore, the degree of impairment is related to the age at which the rat is thyroidectomized and to the age at which treatment with thyroid hormone is begun.   If thyroidectomy is delayed until after the 24th day, the subsequent performance of the rats is indistinguishable from that of their normal litter-mates.   If the rats are thyroidectomized at birth and are then treated from the 10th day onwards, their performance is normal.   If, however, they are thyroidectomized at birth and are not treated until the 24th day, their performance never reaches that of normal rats.   The fact that delayed thyroidectomy does not impair performance in the Hebb-Williams test shows that the defect observed in animals thyroidectomized during the first ten days of life is not simply a consequence of the metabolic effects of thyroid deficiency. On the contrary, it must reflect a disturbance of development. Moreover, the fact that the impairment cannot be reversed unless treatment is begun before the 10th day shows that there is a critical early period during which the damage to the central nervous system is irreversible.   In this respect, the behavioural

abnormality in rats thyroidectomized at birth resembles that in
severely affected human cretins.

In contrast to the effects of thyroid deficiency, excess of
thyroid hormone during the neonatal period causes an acceleration
of general development. When new-born rats are treated with
thyroid hormone, their eyes open prematurely, the skull assumes
the adult form before the normal time and the animals become
physically active and run about their cage before their untreated
litter-mates. These changes suggest precocious development of
the central nervous system. In keeping with this, treatment
with thyroid hormone accelerates the time of appearance of
certain innate responses in immature rats (12) and mice (13).

Effects on Electrical Activity. Changes in the electrical
activity of the brain have been observed in rats thyroidectomized
early in life. If rats are thyroidectomized at birth and tested
at 15 days, the amplitude of the waves recorded from the skull is
diminished and the response to auditory and visual stimuli is
reduced or absent. These effects are not observed if the thyroidectomy is delayed until after the 24th day, indicating that they
are not due to the metabolic effects of thyroid deficiency but,
like the behavioural changes described above, are caused by an
impairment of the development of the brain.

Histological Effects. Thyroid deficiency has been shown to
affect the development of the rat's cerebellum. During the
course of post-natal development of the rat, the cells of the
external granular layer of the cerebellar cortex migrate to a
deeper layer where they form the internal granular zone of the
mature cerebellum. During their migration they pass through the
molecular layer, which contains the dendrites of the Purkinje
cells. Under normal conditions, the external granular layer
disappears completely by the 21st day. If, however, rats are
thyroidectomized before the 14th day, this layer persists until
the 28th day (14) and the cells continue to synthesize DNA in
preparation for mitosis (15). If the hypothyroid state is maintained indefinitely, the external granular layer eventually disappears, but the dendrites of the Purkinje cells in the molecular
layer never develop to their full extent. In an attempt to
explain these observations, Hamburgh and Burkart (15) suggest that
the presence of thyroid hormone is necessary for the proper timing
of the change from proliferation to migration of the cells of the
external granular layer, and that these cells exert an inductive
influence on the Purkinje cells as they pass through the molecular
layer.

Of greater relevance to the behavioural and electroencephalographic effects of thyroid deficiency are the histological
changes that have been observed in the rat's cerebral cortex (10).

During the final stages of the development of the cerebral cortex the cell bodies of the neurones enlarge and there is an increase in the size and amount of branching of their axons and dendritic processes.   In rats thyroidectomized at birth, the rate of growth of the cell bodies is diminished and the density of the axonal and dendritic network is reduced, particularly in layer IV.

Effects on Chemical Composition.   Balázs et al. (16) have shown that the concentration of cerebroside in the brains of rats thyroidectomized at birth is significantly less than that of normal rats.   Since cerebroside is predominantly a myelin lipid, this suggests that thyroid deficiency has a selective effect on myelination.   In keeping with this, Balázs et al. (16) found that the concentration of myelin in the brains of their thyroidectomized animals was less than that in normal rat brain.

Experiments on the effect of excess of thyroid hormone on the myelin content of developing brain have not given such clear results.   Hamburgh (17) found that thyroxine accelerated the appearance of myelin in cultures of cerebellum from new-born rats. Myant and Cole (18), on the other hand, could not detect any selective increase in the myelin phospholipids of the brains of rats treated with thyroxine during the most active period of myelination.

The nucleic acid content of the developing brain is influenced by thyroid hormone.   During the normal development of the rat's cerebral cortex the amount of DNA/g wet weight decreases and the amount of RNA/g of DNA increases.   In rats thyroidectomized at birth, these changes are less marked, with the result that the DNA concentration in the brain is higher, and the RNA/DNA ratio is lower, than in normal rats (19, 20).

Metabolic Effects.   The brain of an adult animal differs from almost all other tissues in being unresponsive to the metabolic influence of thyroid hormone.   The O_2 consumption of developing brain, on the other hand, is affected by thyroid hormone.   Thus, the O_2 consumption of slices of cerebral cortex from immature rats is increased if the rats are treated with thyroxine (21) and is diminished by thyroidectomy at birth (22).

Thyroid hormone also influences protein synthesis in developing brain.   Gelber et al. (23) have shown that thyroxine stimulates the incorporation of amino acid into protein in cell-free preparations from adult animals.   Incorporation of amino acid into the brains of immature rats in vivo is stimulated by treatment with thyroid hormone (24) and is depressed by neonatal thyroidectomy (20).   The stimulatory effect of thyroid hormone on protein synthesis in the immature brain appears to be mediated by an effect on the mitochondria which leads indirectly to an enhancement of the rate of transfer of amino acids from sRNA into microsomal protein (25, 26).

Many of the enzymes in brain increase in activity during development, and in some cases the increase is modified by thyroidectomy in the neonatal period.   Hamburgh and Flexner (27), for example, have shown that the rise in the activity of succinic dehydrogenase which takes place in the cerebral cortex of the immature rat is diminished by neonatal thyroidectomy.   Succinic dehydrogenase activity eventually reaches the normal level if treatment with thyroxine is begun on the 10th day, but if treatment is delayed it remains permanently subnormal.

The effect of thyroid deficiency on enzyme activity in the developing brain is to some extent selective, in that only certain brain enzymes are affected by thyroidectomy.   The effect of thyroidectomy at birth on the activities of several enzymes in rat brain is shown in Table 1.

This Table shows that in the case of over half the brain enzymes tested, activity fails to increase to the normal level during development if thyroid hormone is deficient at birth. It is possible, therefore, that the stimulatory effect of thyroxine

TABLE 1

Effect of Neonatal Thyroidectomy on the Activity of Enzymes
in the Cerebral Cortex of Developing Rats

Enzyme	Age of rat[*] (days)	Effect	Reference
Succinic dehydrogenase	30	Decrease	27
Acetylcholinesterase	22	Decrease	20
Glutamate decarboxylase	40	Decrease	28
GABA transaminase	40	Decrease	28
ATPase	40	Decrease	28
Aspartate aminotransferase (latent)	40	Decrease	19
Aldolase	30	None	27
Cytochrome oxidase	30	None	27
Glutamate dehydrogenase	46	None	29
Alanine aminotransferase	46	None	29
Lactate dehydrogenase	46	None	29

[*] Age at which brain was examined after neonatal thyroidectomy.

on the incorporation of amino acid into the protein of developing brain is due partly to increased enzyme formation.

Discussion of the Rôle of the Thyroid

We know very little about the way in which structure is related to function in the mammalian brain. For this reason, discussion of the relation between the behavioural, histological and biochemical effects of thyroid deficiency on brain development must be largely speculative.

The diminished density of the axo-dendritic network in the cerebral cortex might well contribute to some of the impairment of adaptive behaviour seen in rats thyroidectomized in the neonatal period, since Eayrs (10) has shown that this histological abnormality results in a reduction in the probability of interaction between different neurones. However, thyroxine treatment may completely reverse the histological changes in the brains of rats thyroidectomized at birth, without a corresponding reversal of the impairment of behaviour (10). It is likely, therefore, that changes not visible under the microscope contribute to the behavioural effects of neonatal thyroid deficiency.

The reduced size of the cell bodies in the developing cerebral cortex of thyroid-deficient rats may be due in part to the diminished rate of amino acid incorporation into brain protein, but there must be many factors other than protein synthesis that limit the growth of a cell. Balázs et al. (29) have pointed out that two of the brain enzymes (succinic dehydrogenase and glutamate decarboxylase) affected by neonatal thyroidectomy are largely confined to the mitochondria of nerve terminals. This may have a bearing on the nature of the effect of thyroid deficiency on the growth of the axo-dendritic network in the cerebral cortex.

It is worth noting that the cells of the rat's cerebral cortex have ceased multiplying and have migrated to their final positions when the animal is born. Any influence of thyroid hormone on the post-natal development of the cerebral cortex must therefore be limited to effects on the growth and differentiation of its cells. The question as to how far thyroid hormone acts as a physiological differentiating agent in the developing mammal has been discussed elsewhere (30). The very striking differentiating effect of thyroid hormone in amphibia is associated with increased RNA synthesis and a change in the pattern of enzyme synthesis towards that characteristic of the adult form. Some of the effects of thyroid deficiency discussed in this chapter are consistent with the possibility that the rôle of thyroid hormone in developing mammalian brain is similar in essence to its rôle in

amphibian metamorphosis.   In particular, they suggest that a
normal output of thyroid hormone is necessary for the full
development of the adult pattern of enzyme synthesis in the brain.
On the other hand, this may indicate merely that thyroid hormone
is an essential constituent of the environment in which the
mammalian brain develops.   If it is a true differentiating agent
in mammals, one would expect to find that it had some influence on
RNA synthesis in developing brain, but attempts to demonstrate
such an effect have not been successful;   the decreased RNA content of the brains of neonatally thyroidectomized animals seems to
be due to an increase in utilization, rather than to a decrease
in synthesis, of RNA (31, 32).   More information about the ability
of thyroid hormone to evoke the premature appearance of the adult
pattern of enzyme synthesis in mammalian brain would undoubtedly
help us to understand more about the way in which the thyroid
influences brain development.

OTHER HORMONES

Growth Hormone

Growth hormone given to pregnant rats increases the DNA content of the brains of the new-born young (33), but this effect
must be very indirect because the placenta is more or less
impermeable to growth hormone (34).

Corticosteroids

Schapiro (35) has observed histological and biochemical
changes in the brains of rats given single injections of cortisol
at birth.   However, this observation does not necessarily throw
any light on the physiological role of the adrenal cortex in
brain development, since the doses of cortisol administered in
these experiments were large.

Sex Hormones

The sex hormones have a profound effect on the development
of the brain (36, 37, 38).   In the female rat, the sexual cycle
of oestrus followed by ovulation is regulated by the cyclic output
of gonadotrophin from the pituitary and this, in turn, is regulated by the hypothalamus.   In the male, on the other hand, the
output of gonadotrophin by the pituitary is not cyclic.   One may
therefore speak of a "female" hypothalamus, capable of maintaining
a cyclic pattern of sexual activity, and a "male" hypothalamus
whose regulatory pattern is acyclic.   If a female rat is given a
single injection of testosterone during the first few days of

life, its hypothalamus becomes masculinized, in the sense that it never becomes capable of cyclic control of the pituitary.   When the female becomes mature, it remains permanently in oestrus, it never ovulates and its mating behaviour is abnormal.   Male sex hormone, therefore, has an inductive influence on the differentiation of the hypothalamus at an early and limited stage of post-natal development.   That the inductive influence of testosterone is exerted on the hypothalamus, and not on the pituitary, is shown by the results of implantation experiments.   A pituitary taken from a testosterone-treated female can maintain a normal ovulatory cycle when implanted beneath the hypothalamus of a hypophysecto-mized female not previously treated with testosterone.

Attempts to demonstrate an analogous influence of male sex hormone on the functional differentiation of the hypothalamus in primates have not been entirely successful.   In female monkeys exposed to androgens in foetal life, a menstrual cycle may occur (39).   Nevertheless, the sexual behaviour of these animals may come to resemble that of the male.

REFERENCES

(1) Lewis, A. Lancet $\underline{i}$:1505 and $\underline{ii}$:5 (1937).

(2) Lotmar, F. Ztschr. ges. Neurol. u. Psychiat. $\underline{146}$:1 (1933).

(3) Bertrand, I., Delay, J., and Guillain, J. C.R. Soc. Biol., Paris, $\underline{129}$:395 (1938).

(4) Schultz, M.A., Schulte, F.J., and Parmelee, A. H. Mschr. Kinderheilk. $\underline{115}$:284 (1967).

(5) Dorff, G. B. Am. J. Dis. Child. $\underline{48}$:1316 (1934).

(6) Wilkins, L. The Diagnosis and Treatment of Endocrine Disorders in Childhood and Adolescence, 2nd ed., p. 99. Blackwell Scientific Publications, Oxford (1957).

(7) Myant, N.B. in The Thyroid Gland, Vol. 1 (Eds. R. Pitt-Rivers, W. R. Trotter), pp. 283-302. Butterworths, London (1964).

(8) Carr, E. A., Jr., Beierwaltes, W. H., Raman, G., Dodson, V. N., Tanton, J., Betts, J. S., and Stambaugh, R. A. J. clin. Endocr. $\underline{19}$:1 (1959).

(9) Eayrs, J. T. Br. med. Bull. $\underline{16}$:122 (1960).

(10) Eayrs, J. T. Sci. Basis Med. p. 317 (1966).

(11) Eayrs, J. T.   J. Endocr. $\underline{22}$:409 (1961).

(12) Eayrs, J. T., and Lishman, W. A.   Br. J. Anim. Behav. $\underline{3}$:17 (1955).

(13) Hamburgh, V., and Vicari, E.   Anat. Rec. $\underline{127}$:302 (1957).

(14) Legrand, J., Kriegel, A., and Jost, A.   Arch. Anat. Microsc. Morph. Exp. $\underline{50}$:507 (1961).

(15) Hamburgh, M., and Burkart, J.   in Hormones in Development (Eds. M. Hamburgh and E.J.W. Barrington).   Appleton-Century-Crofts, New York, in press (1971).

(16) Balázs, R., Brooksbank, B.W.L., Davison, A. N., Richter, D., and Wilson, D. A.   J. Physiol., Lond., $\underline{201}$:28P (1969).

(17) Hamburgh, M.   Develop. Biol. $\underline{13}$:15 (1966).

(18) Myant, N. B. and Cole, L. A.   J. Neurochem. $\underline{13}$:1299 (1966).

(19) Pasquini, J. M., Kaplún, B., Garcia Argiz, C. A., and Gómez, C. J.   Brain Res. $\underline{6}$:621 (1967).

(20) Geel, S. E., and Timiras, P. S.   Endocrinology $\underline{80}$:1069 (1967).

(21) Fazekas, J. F., Graves, F. B., and Alman, R. W.   Endocrinology $\underline{48}$:169 (1951).

(22) Ghittoni, N. E., and Gómez, C. J.   Life Sci. $\underline{3}$:979 (1964).

(23) Gelber, S., Campbell, P. L., Deibler, G. E., and Sokoloff, L. J. Neurochem. $\underline{11}$:221 (1964).

(24) Schneck, L., Ford, D. H., and Rhines, R.   Acta Neurol. Scand. $\underline{40}$:285 (1964).

(25) Sokoloff, L., Kaufman, S., Campbell, P. L., Francis, C. M., and Gelboin, H. V.   J. Biol. Chem. $\underline{238}$:1432 (1963).

(26) Klee, C. B., and Sokoloff, L.   J. Neurochem. $\underline{11}$:709 (1964).

(27) Hamburgh, M., and Flexner, L. B.   J. Neurochem. $\underline{1}$:279 (1957).

(28) García Argiz, C. A., Pasquini, J. M., Kaplún, B., and Gómez, C. J.   Brain Res. $\underline{6}$:635 (1967).

(29) Balázs, R., Kovacs, S., Teichgräber, P., Cocks, W.A., and Eayrs, J. T.   J. Neurochem. $\underline{15}$:1335 (1968).

(30) Myant, N.B. in Hormones in Development (Eds. M. Hamburgh
      and E. J. W. Barrington). Appleton-Century-Crofts, New
      York, in press (1971).

(31) Balázs, R., Cocks, W. A., Eayrs, J. T., and Kovacs, S. in
      Hormones in Development (Eds. M. Hamburgh and E. J. W.
      Barrington). Appleton-Century-Crofts, New York, in press
      (1971).

(32) Geel, S. E., and Timiras, P. S. in Hormones in Development
      (Eds. M. Hamburgh and E. J. W. Barrington). Appleton-
      Century-Crofts, New York, in press (1971).

(33) Zamenhof, S., Van Marthens, E., and Bursztyn, H. in
      Hormones in Development (Eds. M. Hamburgh and E. J. W.
      Barrington). Appleton-Century-Crofts, New York, in press
      (1971).

(34) Gitlin, D., Kumate, J., and Morales, C. J. clin. Endocr.
      25:1599 (1965).

(35) Schapiro, S. Gen. Comp. Endocr. 10:214 (1968).

(36) Harris, G. W. Endocrinology 75:627 (1964).

(37) Barraclough, C. A. in Neuroendocrinology, Vol. II (Eds.
      L. Martini and W. F. Ganong), pp. 61-99. Academic Press,
      New York (1967).

(38) Harris, G. W. Phil. Trans. R. Soc. B 259:165 (1970).

(39) Goy, R. W. Phil. Trans R. Soc. B. 259:149 (1970).

GENERAL FEATURES OF THE SYNAPTIC ORGANIZATION IN THE CENTRAL NERVOUS SYSTEM

Constantino SOTELO

Laboratoire de Biologie Animale, Faculté des
Sciences (Paris 5) and Laboratoire d'Histologie
Normale et Pathologique du Système Nerveux de
L'I.N.S.E.R.M., Hôpital Port-Royal (Paris 14)

Nervous tissue is composed of two types of cells, neurons and neuroglia cells, derived from the primitive neuroepithelium. Based on its epithelial origin and on its morphological types of connections, the nervous tissue has been compared to an epithelium (1), where the cells are packed close together, leaving a very small extracellular space (150 to 200 A wide) between each other, and isolated from the blood vessels by a basal membrane. As will be detailed further on, the junctional complexes described by Farquhar and Palade (2) between epithelial cells are present, with additional characteristics, in the nervous tissue.

Nerve cells are differentiated for communication and transfer of coded information. The specialized structures capable of receiving and decoding such information are located in their surface membrane. The history of nerve cells is, in a way, parallel to the concept of the cell membrane. As the neuron theory was developed, the idea that these cells have a limiting membrane became well established. When Ramon y Cajal (3) wrote his last work, supporting the neuron theory, the first point discussed was the existence or non-existence of a cell limiting membrane. If morphologically the plasma-membrane was not visualized, and Ramon y Cajal only offer some indirect arguments, it was necessary, for acceptance of the individuality of nerve cells, that each of these units be separated from its environment by a limiting membrane. From a functional point of view, the plasma-membrane of the neuron must have, in some sites, different properties than the rest of the membrane, which allow the interaction between two nerve

cells in zones of close juxtaposition. Sherrington (4) gave
the name of SYNAPSES to these zones of nerve articulation.

Even if nerve cells have some of the characteristics of
epithelial cells, they differ from an epithelium because they
are differentiated to a specific function. Ramon y Cajal (5)
already understood that the primary difference between neurons and other cells is that neurons are polarized in order to
acomplish their function. His early "law of dynamical polarization" clearly distinguished two poles in a neuron : a receptive pole, and a conducting pole. Thus, the receptive pole
of a neuron was constituted by the dendritic field and the
perikaryon, whereas, the conducting pole corresponded to the
axon. According to this law, the nerve impulses pass from the
dendritic portion of the nerve cell to the axon in a one-way
conducting mechanism. Even if the one-way conducting mechanism has been denied by physiologists, and it has been proved
that antidromic conduction may occur, the idea of the neuronal
polarization has been confirmed, and the key for such polarization is provided by the valve-like behaviour of synapses,
that allow transmission in only one direction, i.e., from the
presynaptic to the postsynaptic neuron.

The environment of each individual neuron is mostly constituted by the number, the type (excitatory or inhibitory) and
the spacial distribution of the synaptic articulations attached to its receptive surface; in general terms, the environment is different and specific for each neuron. In consequence, each neuron forms a unique entity, and the whole of these
entities, communicating with each other by means of their
specific connections and distributing their information to the
periphery, make up the nervous system.

The physiological work developed during the past 20 years
has clarified the mechanisms of synaptic transmission. The
old discrepancy between the supporters of the electrical nature versus the chemical nature of synaptic transmission has
been closed with the triumph of the latter. Even if, by far,
the chemically transmitting synapses are those most commonly
found in the central nervous system of vertebrates, the possibility of an electrical coupling in the nervous system must be
kept in mind. In 1959, Furshpan and Potter (6) working on the
giant motor synapse of the crayfish, have demonstrated the
existence of such coupling between two nervous elements. It
is very tempting to refer to the electrical coupling, by analogy with the chemical transmission, as an electrical synapse.
One may wonder whether the concept of synapse is not too res-
trictive and should be used only to indicate junctions where a
chemical transmission takes place. For this reason, the morphological features of the sites of electrical coupling will

be described under the name of electrotonic junctions in this
review.

The present review will be confined to the description of
the morphology characterizing the different kinds of neuronal
articulations, and to the discussion of some questions arised
from the attempt to correlate structure with function.

I - CHEMICAL SYNAPSES

Since the earliest electron microscope studies (7,8), it
has been demonstrated that the chemical synapse consists of
a presynaptic element, a narrow extracellular space or synaptic cleft and a postsynaptic element. Recent reviews dealing
with the ultrastructure of these three elements may be found
in Gray and Guillery (9) and in Peters, Palay and Webster (10)

PLASMA-MEMBRANES AND JUNCTIONAL COMPLEXES AT THE SYNAPTIC
INTERFACE.

Nerve cells intercommunicate at the synaptic zones. These zones have the morphological features of junctions between
epithelial cells. Farquhar and Palade (2) described three
types of junctions in epithelia : zonula adhaerens, macula
adhaerens and zonula occludens. The three types of junctions
are present in the vertebrate central nervous system. The
macula adhaerens has only been found between glial cells in
fish (11). This junctional zone is not represented in the
mammalian central nervous system.

Between epithelial cells the zonula adhaerens seem to
have only the functional role of adhesion, and they consist
of zones where the apposed plasma membranes are parallel to
each other and underlined by a symmetrically disposed dense
portion of cytoplasm with a filamentous aspect. The extracellular space, at these junctions, is wider than at the non-
junctional regions and exhibits a dense intermediate plate.
In the nervous system at the synaptic interface these junctions may be present, in which case, they are generally small
in extension. For this reason, Palay (1) called them the
punctum adhaerens or attachment plates. As has been said
above, the special feature that characterizes the nervous junctions is that they are polarized in order to transfer the
coded information in only one direction (the valve-like mechanism of the synaptic junctions). To mark the polarization of
the synapses, the representative junctional complex of the
chemical synapses corresponds to a slight modification of the
punctum adhaerens, where the junction becomes asymmetrical.
This specialized junction, associated with the clumped presynaptic vesicles, has been called by Palay (12) the "synap-

tic complex". Even if there is not direct evidence to prove
that the synaptic complex corresponds to the site where the
chemical transmission takes place, there are numbers of assumptions indicating that the specialized zones of the neuronal
membrane where the neurotransmitter is released -in the presynaptic side- and where this substance exerts its effects -in
the postsynaptic side, or the receptor- may have their morphological correlations on the synaptic complex. Consequently,
Couteaux (13) has given the denomination of "active zone"to
such synaptic complex, and this name will be used in the present paper.

A) The "Active Zones"

As defined above, "active zones" are modified attachment
plates, at least in some instances, and have their origin in
typical puncta adhaerentia. In electron microscope studies
on some regions of immature nervous system (14,15) it has been
proposed that the first type of contact between nervous elements exhibits the features of attachment plates; later on,
when the junctions become functional, it is modified presenting the aspect of "active zones".

The "active zones" are formed by three different elements : the presynaptic membrane, its attached cytoplasmic
differentiation and associated vesicles; the synaptic cleft;
and the postsynaptic membrane and its subsynaptic speciali-
zations. The physico-chemical properties of the pre- and
post- synaptic membranes must differ from those of the non-
synaptic membranes; unfortunately, the study of these membranes, even with the highest electron microscope resolution
does not allow us to differentiate synaptic from non-synaptic
regions; for this reason, only the paramembranous differentiations of the "active zones" will be analysed.

The presynaptic grid : Gray (16) has described in the
spinal cord of the cat, when the tissue was stained with an

Fig. 1. Subfornical organ of the cat. Tangential section
through a presynaptic area illustrating the presynaptic vesicular grid. Bismuth-iodide block impregnation. X 90,000.
Micrograph by Dr. Pfenninger et al. (19).
Fig. 2. Subfornical organ of the cat. Presynaptic vesicular
grid. Bismuth-iodide block impregnation. X 90,000. Micrograph
by Dr. Akert et al. (21).
Fig. 3. Cerebellar cortex of the cat. Type I synapse between a
mossy fiber and a granule cell dendrite. X 67,000.
Fig. 4. Cerebellar cortex of the frog. Type II synapses
(arrows) between a stellate cell axon and a Purkinje cell
dendrite. X 46,000.

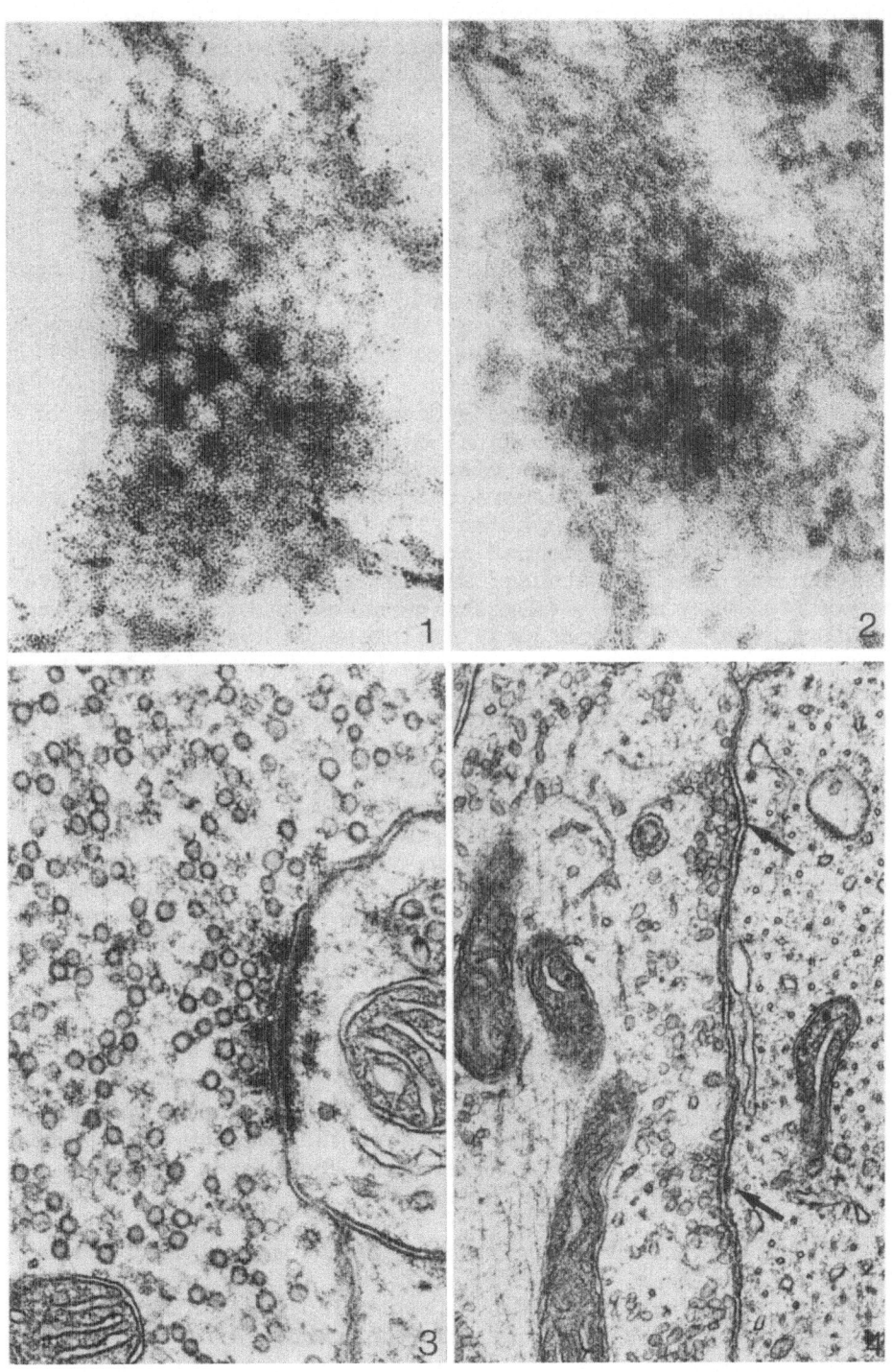

ethanol solution of phosphotungstic acid (PTA) in block
before embedding, dense cytoplasmic particles regularly spaced about 1,000 A apart and attached to the presynaptic membrane; the extension of these dense projections marks the
limits of the "active zone". In tangential sections through
presynaptic areas the dense projections appear as rounded
particles about 500 A in diameter and arranged in a hexagonal
pattern. The presynaptic dense projections seem to exist in
all the observed central synapses, with the exception of a
special synaptic contact found in the retina between the cones
and bipolar cells and the rods and bipolar cells; in these
instances, it has been suggested that the presynaptic ribbon
may share some of the functions of the dense projections (9).

The presynaptic dense projections are frequently seen in
material prepared with the usual staining methods (Fig. 3),
but it is obvious that more elective procedures, as is the
Gray's use of PTA, give a more complete staining of these projections. Bloom and Aghajanian (17) using glutaraldehyde
fixed nervous tissue, stained in block with the ethanolic-PTA
solution and plastic embedded without post-osmification, have
impregnated very selectively the paramembranous and the extracellular material of the junctional zones. Regarding the
morphology of the presynaptic dense projections, Aghajanian
and Bloom (18) demonstrated that they are often interconnected
by a thin line of the same material, suggesting that they are
the peaks of a plate-like structure lining the plasmalemma.
An exhaustive study of the presynaptic dense projections has
been made by Akert and his colleagues (19,20 and 21) by the
use of a new impregnation method, the bismuth-iodide block
impregnation. With this technique, they were able to describe
the three dimensional arrangement of the dense projections.
They form a lattice with the dense peaks hexagonally arranged
and interconnected with filamentous cross-bridges. The hollow
structures of the network have roughly the dimensions of a
synaptic vesicle. The whole assemblage, the lattice and the
intercalated synaptic vesicles, is called the "presynaptic
vesicular grid". According to the dimensions of the free
spaces in the grid, the authors claim that it is possible to
observe two types of "presynaptic vesicular grid" (Figs. 1
and 2), one with clear spaces allowing the passage of vesicles
measuring up to 500 A in diameter, and the second one with
narrower spaces for smaller vesicles.

If the electrically excitable membrane corresponding to
the site of release of neurotransmitters is characterized by
the attached "presynaptic grid", and if only at this specific
place synaptic vesicles enter into an immediate contact with
the excitable presynaptic plasmalemma, the grid may play an

important role in guiding and supporting the vesicles at the
release sites.

According to the thickness of the paramembranous material,
its disposition in the pre- and post- synaptic side and the
width of the synaptic cleft, Gray (16) working on the cerebral
cortex, described two types of synaptic contacts. In type I
synapses (Fig.3) the synaptic cleft is about 300 A wide and
there is a very prominent postsynaptic differentiation. In
type II synapses (Fig.4), the synaptic cleft is narrower,
and the postsynaptic differentiation is much less pronounced.
Gray (22,23) extended this classification to the synapses
present in the cerebellar cortex. For him, type I synapses
were always observed in synaptic contacts established on dendritic spines, and type II in contacts established on the
perikarya of neurons, whereas, synapses on dendritic trunks
may be of both types. This dichotomy of synaptic contacts
cannot be extended to all the central synapses. As already
Gray and Guillery (9) have pointed out, this classification
is not workable in non cortical structures, and intermediate
forms can be observed in the brain stem and spinal cord. In
a recent study on the synaptology of the cat cerebral cortex,
Colonnier (24) drew the conclusion that type I and type II
represent the extremes of a gradual morphological change and
all the intermediate forms can be observed. Colonnier pre-
fers to classify the synapses in asymmetrical and symmetrical;
but his nomenclature is unappropiated since, as has been
stated before, the polarization of the synapse implicates the
asymmetrical morphology of the synaptic complex, and even,
refering only to the paramembranous material, there is an
asymmetry, because the "presynaptic vesicular grid" differs
always from the postsynaptic differentiation.

It may be concluded that all the intermediate forms
between type I and type II synapses can be encountered in
the central nervous system, and the two synapses must be considered as "part of a continuum of differentiation" (25).

The synaptic cleft : this space, 200 to 300 A wide, depending on the type of synapse, is occupied by structures of variable morphology. In general, the synaptic cleft is occupied by an intermediate plate of filamentous aspect; sometimes it may exhibit a striated aspect, with the striae lying
perpendicular to the apposed synaptic membranes. After
histochemical studies (26,17) it has been demonstrated that
this region contains a relatively high concentration of proteinaceous material, of a probably glycoproteic nature (27).
Gray and Whittaker (28) after homogenization of brain in
isosmotic media, have observed that the fraction rich in
acetylcholine- containing particles consists of pinched off

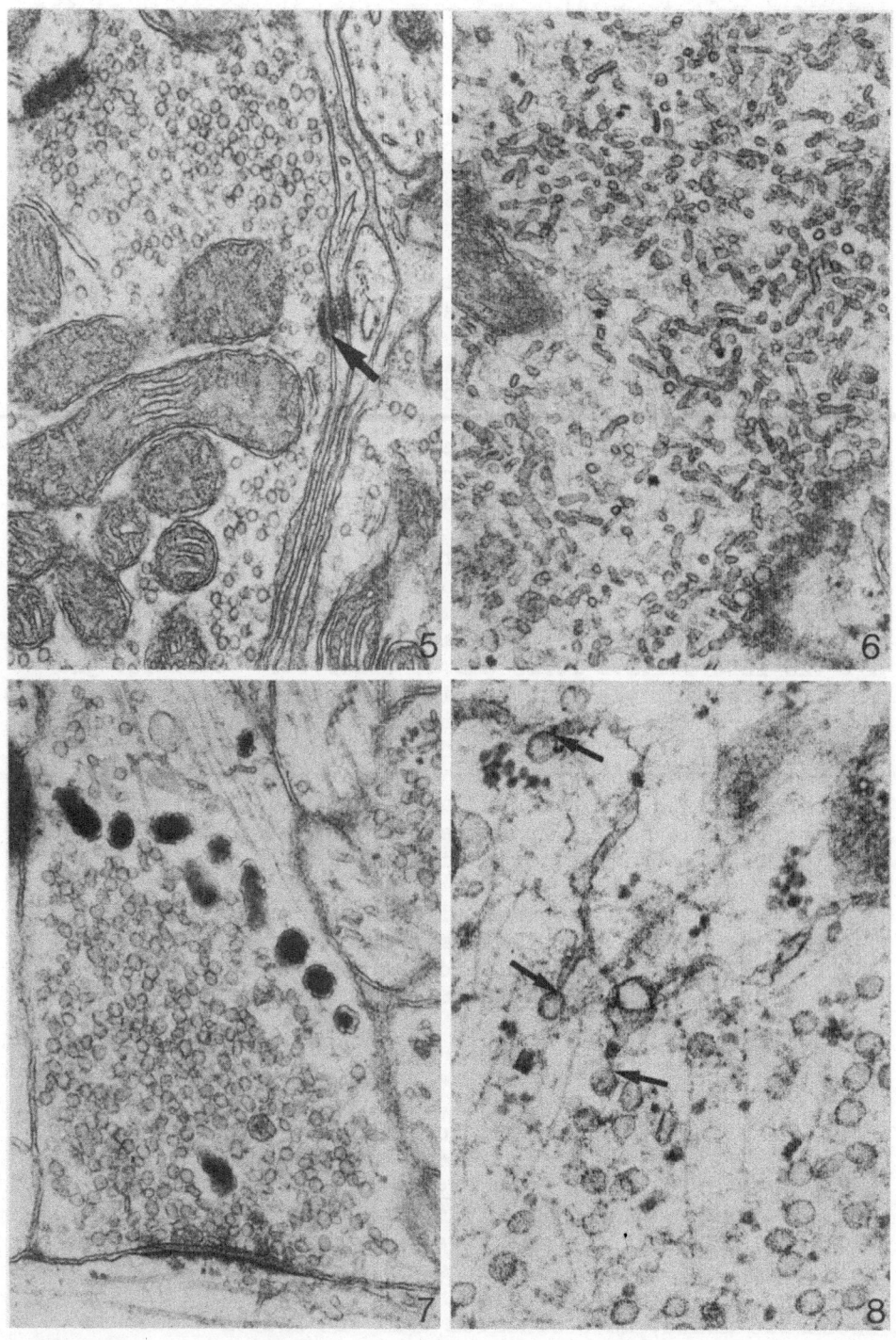

presynaptic elements, the synaptosomes. They often have
attached, at the region of the "active zone", part of the
postsynaptic membrane (Figs. 9 and 10) indicating that this
zone has kept its mechanical property for adhesion. It
has been supposed that the material present in the synaptic
cleft may be responsible for this strong adhesive property
of the synaptic membranes. The chemical composition of the
synaptic cleft may indicate that it is not a free diffusion
space for the rapid passage of small molecules. Over a long
period of time, large molecules as big as peroxidase can
diffuse through it (29). It may be speculated, that the molecular configuration of the extracellular material present
in the cleft acts as prefigurative channels to allow the
passage of the small molecules of neurotransmitters at high
speed.

   Postsynaptic differentiation : The surface of the neurons
is conceived as an heterogenous membrane in which some zones
have been specialized in order to receive information. These
receptive zones act in a specific way according to the diver-
se neurotransmitters, and their specificity may be related
to difference in the molecular pattern of the membrane itself
or of the proteinaceous material associated with it. This
hypothesis has been formulated as the "mosaic concept" of the
neuronal membrane organization. From a morphological point
of view, the nature of the receptors escapes analysis with
the present capabilities of electron microscopy. Recent
neurochemical studies (30) tend to prove that the receptors
involved in synaptic transmission are located in the postsynaptic membranes and the only structural feature that
allows their recognition is the cytoplasmic differentiation
associated with the postsynaptic plasmalemma. This cytoplasmic differentiation consists of a layer of dense filamentous
material attached to the postsynaptic plasmalemma. As already mentioned, variations in the thickness of this material

---

Fig. 5. Cerebellar cortex of the cat. Attachment plaque or
"punctum adhaerens" (arrow) between a mossy fiber and a thin
astroglial process. X 42,000.
Fig. 6. Spinal cord of the frog. Axon terminal filled with tubular profiles corresponding to synaptic vesicles after pro-
longate aldehyde fixation. X 45,000.
Fig. 7. Spinal cord of the frog. Axon terminal containing
rounded synaptic vesicles and large granular vesicles of
different sizes and shapes. X 45,000.
Fig. 8. Spinal cord of the frog. Rounded vesicles sometimes
related to tubular profiles of agranular endoplasmic reticulum (arrows) are seen in the presynaptic area. X 68,000.

provide one of the features that distinguishes the two types
of synaptic contacts described by Gray (22). During the ultrastructural study of the cerebellum of the pleurodeles (31)
a new variety of postsynaptic cytoplasmic differentiation
has been found. It consists of a spherical cytoplasmic dense
zone, measuring 1,500 to 3,000 A in diameter and immediately
apposed to the postsynaptic plasmalemma (Fig. 13). The dense
zone is not homogenous and is formed of small vesicles of
about 100 A in diameter, given to the postsynaptic differentiation an alveolate configuration.

B) Nonsynaptic Junctions

As has been pointed out above, besides the "active zones"
other junctional complexes can be found between nervous elements, even at the synaptic interface. Such contacts correspond to "puncta adhaerentia"; they are easily differentiated
from "active zones" since there is no "presynaptic vesicular
grid" and for this reason, the junction is symmetrical. The
"puncta adhaerentia" seem to have only a mechanical role of
adhesion. They have been first described in the cerebellar
glomeruli (32,23) between neighbouring granule cell dendrites.
Peters et al. (10) have reviewed the instances where attachment plates have been described : between dendrites, between
a dendrite and an axon, between dendrite and neuronal perikaryon, between adjacent perikarya, and between axon terminal
and initial segment of axon. Occasionally, attachment plates
have also been observed between neuronal and glial elements;
in the granular layer of the cat cerebellum, attachment plates
exist between a mossy fiber and a thin process of astroglial
cytoplasm (Fig. 5).

PRESYNAPTIC ORGANELLES

A large variety of organelles have been described in the
presynaptic element (9 and 10). Only three presynaptic organelles will be considered in the present review : the synaptic vesicles, the mitochondria and the smooth-surfaced
endoplasmic reticulum.

A) Synaptic Vesicles

These constitute the most distinctive morphological
feature of the presynaptic element. Without considering here
the neurosecretory vesicles, since they correspond to an entire different chapter of the neurobiology, the synaptic vesicles, according to the absence or the presence of a dense
material inside them, can be classified in agranular and granular vesicles.

__Agranular vesicles__ : are the most frequently found type of vesicle. They have a spherical shape with a diameter measuring between 400 to 500 A and they are delimited by a three-layered unit membrane of about 60 A thick. After the discovery of miniature end plate potentials (33) it was suggested that the quantal unit of acetylcholine postulated by Del Castillo and Katz (34) was packed in one single synaptic vesicle. Till now, there is no direct evidence of the presence of neurotransmitters inside the vesicles, since the preconized histochemical method for visualizing the acetylcholine (35) has not the required specificity. Nevertheless, the vesicular theory is strongly supported by biochemical arguments obtained by studies of fractions from brain tissue homogenates. The purification of a vesicular fraction (36, 37) associated with a high content of acetylcholine indicates the correlation between the agranular vesicles and the neurotransmitter. Recently, Israel and Gautron (38) working with pure cholinergic material, the nerve electroplaque junction of the electric organ of Torpedo marmorata, have obtained a vesicular fraction containing 85 % of the acetylcholine activity layered on the gradient. This fraction contains 150 to 300 times more acetylcholine than the equivalent fraction prepared from guinea pig cerebral cortex. This result proves with certainty the presence of acetylcholine in some vesicles with clear centers.

According to the vesicular theory, it is assumed that the vesicles move towards the strategic sites of the axon terminal membrane, where they are concentrated, and it seems that the "presynaptic vesicular grid" may play an important role in conducting the vesicles to these sites of the membrane. How does the discharge of neurotransmitters take place ? Two different theories try to explain this extrusion : i) the synaptic vesicles, after fusing with the presynaptic plasmalemma, open into the synaptic cleft and discharge their contents into the cleft. This exocytotic mechanism has been proposed by De Robertis (39), but images suggesting the existence of such mechanism are rarely observed with the usual electron microscopic methods. ii) The synaptic vesicles accumulated at the "active zone" and in close contact with the presynaptic plasmalemma release their contents into the surrounding axoplasm; it crosses the membrane by molecular diffusion. The current data seem to support the diffusion mechanism. It may be argued that if images of exocytosis are rarely seen, it could be due to a slight swelling of the axon terminals occurring during the fixation, effacing the small pits of the membrane, and giving it the appearance of a smooth surface. The study of the presynaptic membrane with new methods, such as the freeze-etching, may help to resolve this problem.

The vesicular theory, formulated from the convergence of three different approaches, electrophysiological, biochemical and morphological, seems to agree with all the present knowledge on synaptic transmission, though other possibilities may be considered. Marchbanks (40), working with isolated axon terminals (synaptosomes) and synaptic vesicles from cerebral cortex, has differentiated two pools of acetylcholine in the synaptosomes; one cytoplasmic pool, which has been called "labile bound" and is freely releasable by osmotic shock, and the other - vesicular pool - or the "stable bound" acetylcholine, which remains after the osmotic shock of synaptosomes. Both compartments are separated by a permeability barrier to acetylcholine. Acetylcholine synthesis takes first place in the extravesicular compartment, and experiments of cortical stimulation have led to the conclusion that the last synthetized acetylcholine, that is to say the one of the extravesicular pool, is the first to be released during the stimulation. According to these results, it may be speculated that the vesicular acetylcholine is only a reserve pool and it is the cytoplasmic acetylcholine which is released during the synaptic transmission. The accomplishment of quanta may be dependent on a mechanism other than vesicular storage, for instance, a membranous mechanism related to the axolemma.

However, recent experiments have been carried out in the motor end plate to bring substantial evidence and relate the synaptic vesicles directly with the quantal release of acetylcholine. Hubbard and Kwanbunbumpen (41), after bathing the rat diaphragm in a milieu with high potassium concentration, in order to induce a high rate of spontaneous firing of the motor end plate with an increased quantal release of neurotransmitter, have observed a reduction in the number of

---

Fig. 9. Spinal cord of the frog. Synaptosomal fraction fixed in aldehydes. The synaptosomes are filled with rounded synaptic vesicles, same large granular vesicles, glycogen grains and mitochondria. The arrows point to the postsynaptic membrane. X 40,000.
Fig. 10. Same material as in Figure 9. The synaptosome contains a heterogenous population of vesicles, where flatten vesicles are seen. X 85,000.
Fig. 11. Spinal cord of the frog. Freeze-etched preparation of an axon terminal. This specimen was fixed with aldehydes. The terminal contains some spherical profiles of synaptic vesicles and larger profiles representing large granular vesicles. X 40,000.
Fig. 12. Same material as in Figure 11. The axon terminal only contains spherical profiles of vesicles of different sizes. X 50,000.

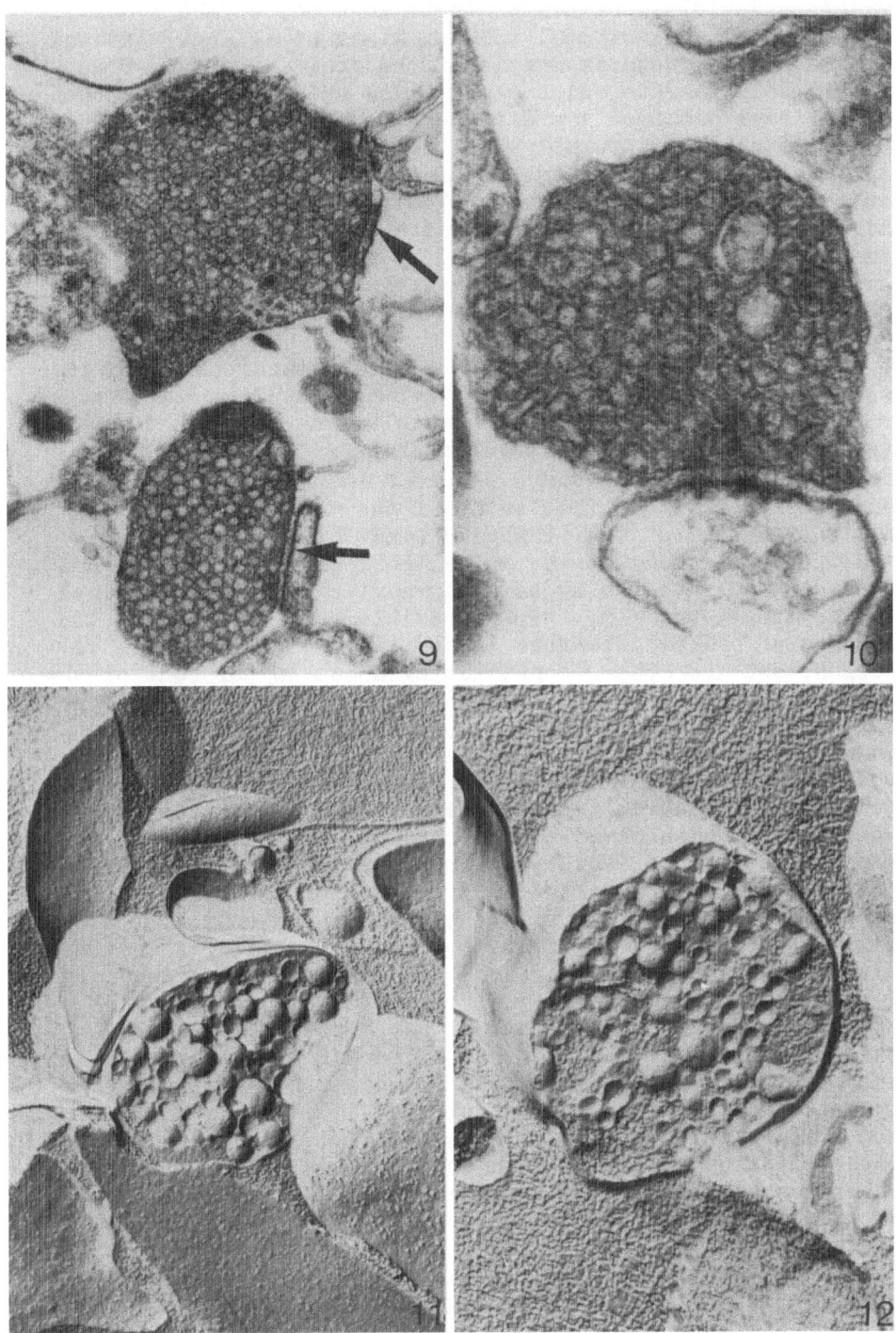

synaptic vesicles, mainly of those adjacent to the synaptic
cleft opposite junctional folds. Clark et al. (42) bathing
the cutaneous pectoris muscle of the frog in calcium free
Ringer solution, to which black widow spider venom has been
added, have obtained a great increase of the miniature end
plate potentials, probably due to a reaction of the venom with
the axolemma of the motor end plate, thereby inducing the
release of neurotransmitter. Electron microscopic observations of such preparations reveal a nearly complete disappea-
ring of the synaptic vesicles. The authors claim that these
experiments support the idea that the synaptic vesicle and
the quantum of chemical transmitter are equivalent.

     After the introduction, in the last years, of aldehydes
as primary fixatives of the central nervous tissue a new
variety of agranular synaptic vesicles has been described
(43,44). They have an elongated shape with a major axis
measuring about 500 A and a minor of 200 A. It seems well
established that the flattening of the vesicles is due to
a direct effect of the aldehydic fixation. During the ultrastructural study of the spinal cord of the frog (45) and
according only to the types of agranular vesicles contained
in the axon terminals, it was possible to differentiate two
classes of synaptic boutons : one filled with spherical vesicles, and the other, constituting about 30 per cent of the
terminals, filled with small tubular profiles (Fig. 6), much
more elongated than the flatten vesicles described in the
mammalian spinal cord (46). The presence of axon terminals
containing tubular profiles was also dependent on a prolonged primary aldehyde fixation, since, when the spinal cord
was perfused with osmic acid alone, or with a double perfusion, first with aldehydes followed by osmic acid, all synaptic boutons were filled only with spherical vesicles.
The spinal cord of the frog is an optimal material for tes-
ting the effect of fixatives upon synaptic vesicles and for
studying whether the elongated vesicles exist "in vivo".
A synaptosomal fraction (kindly prepared by Dr. M. Israel,
Service de Microscopie Electronique, Hôpital de la Salpê-
trière, Paris) was used to test different fixatives; potassium permanganate or osmic acid fixations have only provided
synaptosomes with spherical vesicles. However, a prolonga-
ted primary aldehyde fixation has allowed us to recognize,
besides the synaptosomes with spherical vesicles (Fig. 9),
a second category with heterogenous vesicular population,
where elongated vesicles were present (Fig. 10). To study
if in an unfixed material the tubular vesicle profiles exist,
the freeze-etching technique was used, since this technique
combined with electron microscopy provides the possibility

of studying the surface of cellular profiles and organelles
in the unfixed frozen state. From all the replica examined
with the electron microscope only axon terminals containing
spherical vesicles were observed, indicating that the elongated vesicles are created during the fixation and routine preparation of the tissue. More difficult to understand was the
fact that, in spinal cords fixed by aldehyde perfusion and
prepared with the freeze-etching technique, it was impossible
to detect axon terminals containing elongated vesicles. The
only terminals identified were filled with spherical vesicles
(Figs. 11 and 12). It can be concluded that the aldehydes
produce a sensibilization of certain category of vesicles
only, and the flattening occurs during the subsequent treatment of the tissue (post-osmification, dehydratation or
embedding).

   After the paper of Lenn and Reese (47) particular attention has been paid to the size of the agranular vesicles.
These authors, as well as Larramendi et al. (48), claim that
it is possible to distinguish two categories of axon terminals : in one the synaptic vesicles are larger than in the
other. For Larramendi et al. the size of the vesicles is
more important than their shape. The possible functional
significance of these two populations of agranular vesicles
will be discussed later.

   Granular vesicles : if by granular vesicles is understood
vesicles with internal electron-dense material, according to
their size and shape and without considering the differences
in the density of the content, a great variety of morphological kinds of granular vesicles can be found in the central
nervous tissue. The most frequent vesicles have a rounded
shape, but elongated large granular vesicles are also present.
For instance, they are very frequent in the neuropil of the
spinal cord of the frog (Fig. 7). The range of diameters for
the rounded granular vesicles varies from 400 to 4,000 A; the
largest granular vesicles have been described in neurosecretory neurons (49) and also in fish ependymal cells (50).

   It seems quite clear that a simplified classification of
granular vesicles in two distinct categories, small granular
vesicles (SGV) with a diameter of 400 to 600 A, and large
granular vesicles (LGV) with a diameter of 800 to 4,000 A,
is inoperative. In the peripheral nervous system, SGV are
directly related to the intravesicular pool of endogenous
norepinephrine (51). Recently, Hökfelt (52, 53) using 3%
potassium permanganate as a fixative has been able to demonstrate the same kind of SGV in the central nervous system.
Good evidence has been provided by Hökfelt to consider the
SGV in the central nervous system as storage sites for catecholamines. These results resolve the earlier existing

discrepancy between the localization of mono-amines in axon
terminals of the peripheral and central nervous system. The
problem is much more complex regarding the LGV. Considering
that vesicles of similar morphology can be found in the neurosecretory neurons (54) as well as in glia cells (50), it
is obvious that chemically as well as functionally there
must exist very different types of LGV, and that their presence in neurons and in axon terminals is not sufficient to
consider them as related to the mono-aminergic neurotransmis-
sion, even if in some instances, LGV have been demonstrated
to be one of the storage organelles for central mono-amines
(55).

It has been suggested that in central mono-aminergic
neurons, some monoamines would be stored in LGV of the Golgi
region, and by an active and rapid axonal flow they would be
carried down to the terminals. In a recent autoradiographic
study of the localization of tritiated norepinephrine in the
substantia nigra and the area postrema of the rat (56) no
correlation has been found between the sites of norepinephrine uptake and storage and the LGV. In the same study, a
high acid phosphate activity of the LGV associated with the
Golgi region has been proved, suggesting that these organelles may be considered, in some instances, as primary lyso-
somes (Fig. 17). On the other hand, labelled axon terminals have been found in the neuropil of the area postrema;
in most of them LGV were absent (Fig. 18).

B) <u>Mitochondria and Smooth-Surfaced Endoplasmic Reticulum</u>

Besides the synaptic vesicles, the mitochondria and the
tubular or vesicular profiles of smooth endoplasmic reticulum are the most outstanding organelles of the presynaptic
element, even if they seem to play no direct role in synaptic
transmission. It is generally accepted that mitochondria
are concerned with the production of energy required for the
different metabolic processes associated with the transmis-

Fig. 13. Cerebellar cortex of pleurodeles. The arrow points
to a peculiar postsynaptic differentiation. X 63,000.
Fig. 14. Cerebellar cortex of the cat. Typical spine apparatus are present in the molecular layer. X 28,000.
Fig. 15. Cerebellar cortex of the cat. Crest synapse in the
molecular layer. X 50,000.
Fig. 16. Sympathetic ganglion of the frog seven days after the
section of the preganglionic fibers. At the ganglion cell surface the arrow points to a postsynaptic differentiation and
subsynaptic formation covered by neuroglial cytoplasm.
X 48,000.

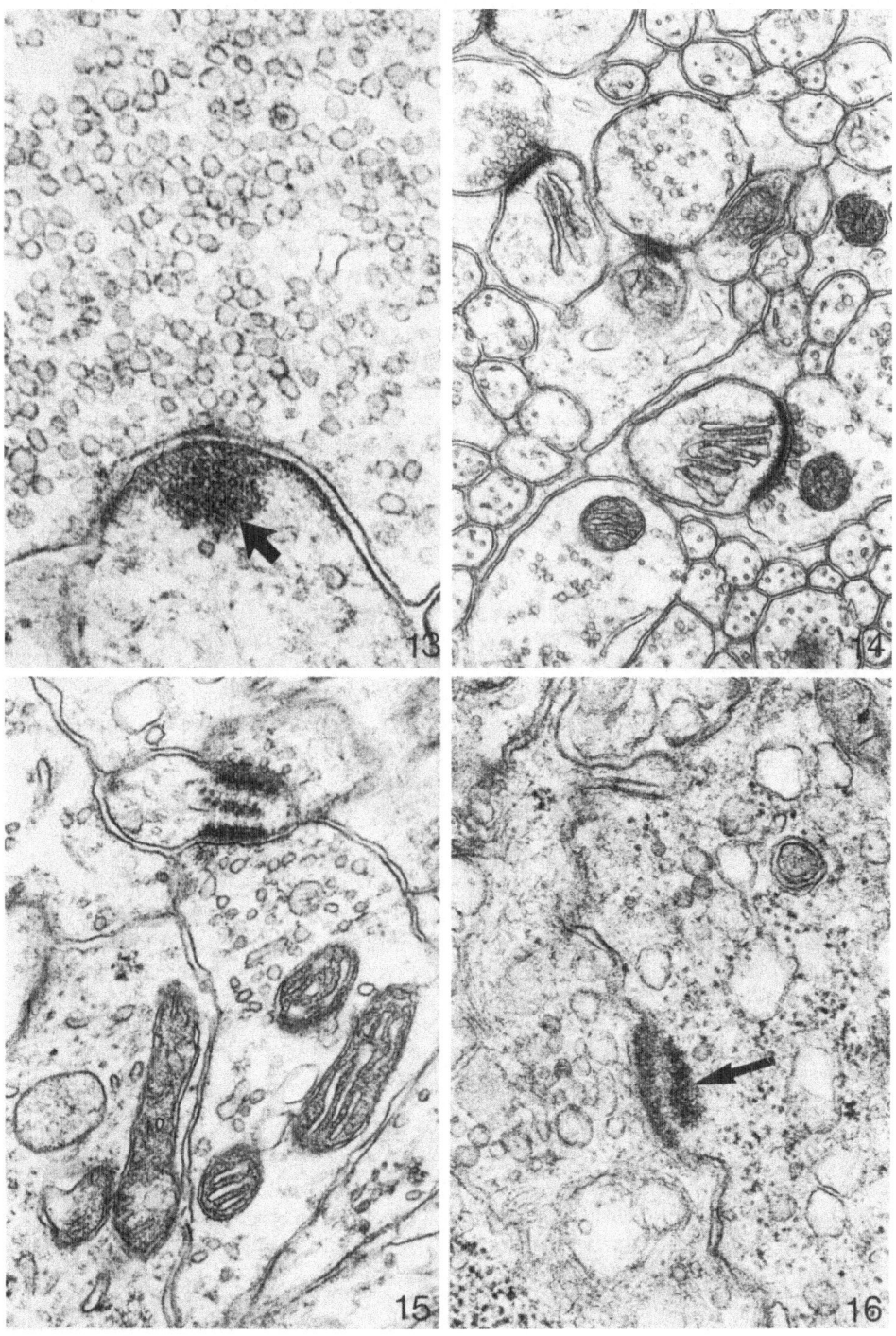

sion. Recently, it has been speculated that the mitochondrial RNA is responsible for the protein synthesis taking
place in the axon terminals (57).

Morphological evidence has been reported in an attempt
to designate the smooth endoplasmic reticulum as the source
of synaptic vesicles (12), but the important problem of the
vesiculogenesis cannot be resolved with only morphological
information. It would be of great interest to find a specific marker of the endoplasmic reticulum membrane, and to
investigate if such a marker exists in the vesicular membrane.
From a morphological point of view, clear evidence of transi-
tional stages between tubular profiles of smooth endoplasmic
reticulum and the synaptic vesicles (Fig. 8) has been
repeatedly reported. For the moment, and till new chemical
evidence may be found against the endoplasmic reticulum origin of the vesicles, it may be speculated that the most important function of the tubular and vesicular profiles of
agranular endoplasmic reticulum, present in most of the axon
terminals, is to produce the membrane of the synaptic vesicles.

Sotelo and Palay (58) have suggested that the stages of
the physiological degeneration of axon terminals in the cycle of degeneration and regeneration or axonal remodeling
are characterized by a tremendous increase of smooth endoplasmic reticulum membranes; thereby, this organelle may play
an important role in some degenerative processes.

SUBSYNAPTIC ORGANELLES

In the subsynaptic cytoplasm some specific organelles
have been described directly related to the synaptic contacts, indicating the possible participation to the synaptic
phenomenon of cytoplasmic regions located at some distance
apart from the postsynaptic plasmalemma. From all the structures so far described as specialized subsynaptic organelles
only two seem to retain a clear relationship with the synaptic junctions. They are the spine apparatus and the subsynaptic formation.

The spine apparatus was described by Gray (16) in dendritic spines of the occipital cortex of the rat. The apparatus consists of two or more flattened or distended sacs
of smooth-surfaced endoplasmic reticulum separated from each
other by zones of dense material. So far, the spine apparatus has been observed in the spines of all studied mammalian
cerebral cortical regions, including the hippocampus and
also, in some instances, in spines of the ventro-lateral geniculate nucleus of the monkey and of the dorsal horn of the

cat and rat spinal cord (for references see 9). Similar organelles have been described in dendrites, but in an extra-spinous position (59,60). In these instances, the outer membrane of the peripheral sacs was associated with attached ribosomes indicating that the sacs belong to the endoplasmic reticulum. Palay et al. (61) and Peters et al. (62) have published electron micrographs of initial segments of pyramidal cell axons of the rat, showing cisternal organelles similar to the spine apparatus. These observations indicate that structures resembling the spine apparatus may also have an extra-dendritic localization.

Herndon (63) has described "lamellar bodies" dispersed throughout the cytoplasm and dendritic tree of the Purkinje cells. The "lamellar bodies" have some resemblance to the spine apparatus since they derive from the endoplasmic reticulum, and consist of stacks of dilated cisterns, some of them in continuity with cisterns of granular endoplasmic reticulum, and separated from each other by dense particles, that give the appearence of a discontinous dense plate. Herndon considers the "lamellar bodies" as an arrangement of the endoplasmic reticulum induced during the fixation. Schultz and Karlsson (64), in a comparative study of the effects of different fixatives on the spine apparatus, drew the conclusion that the occurrence of this organelle can be quantitatively related to the type of fixation.

The usual component of the dendritic spines is the smooth-surfaced endoplasmic reticulum, which can be present under a vesicular or a cisternal form. In the spines of the Purkinje dendrites of the cerebellum, only tubes of endoplasmic reticulum have been described but the spine apparatus seems to be absent (for references see 65). During a current study of the ultrastructure of the cerebellar cortex of the cat, I have observed, in one animal, a great amount of spine apparatus (Fig. 14) in the spines of the terciary branches of the Purkinje dendrites; five other cats were also studied, but the spine apparatus was present in none of them. The conclusion which may be drawn is that the presence of this organelle is, in a way, dependent of the fixation and by this fact can be assimilated to the "lamellar bodies". It is relevant to indicate that if the typical aspect of the spine apparatus may be due to an aggregation of cisternal profiles of smooth-surfaced endoplasmic reticulum present in the spines, and that from a quantitative point of view the spine apparatus is a peculiarity of pyramidal neurons of the mammalian cerebral cortex, the fact that different fixatives reproduce the same image at the same location indicates that this organelle, even if its existence "in vivo" may be questionable, may express some functional features of the pyra-

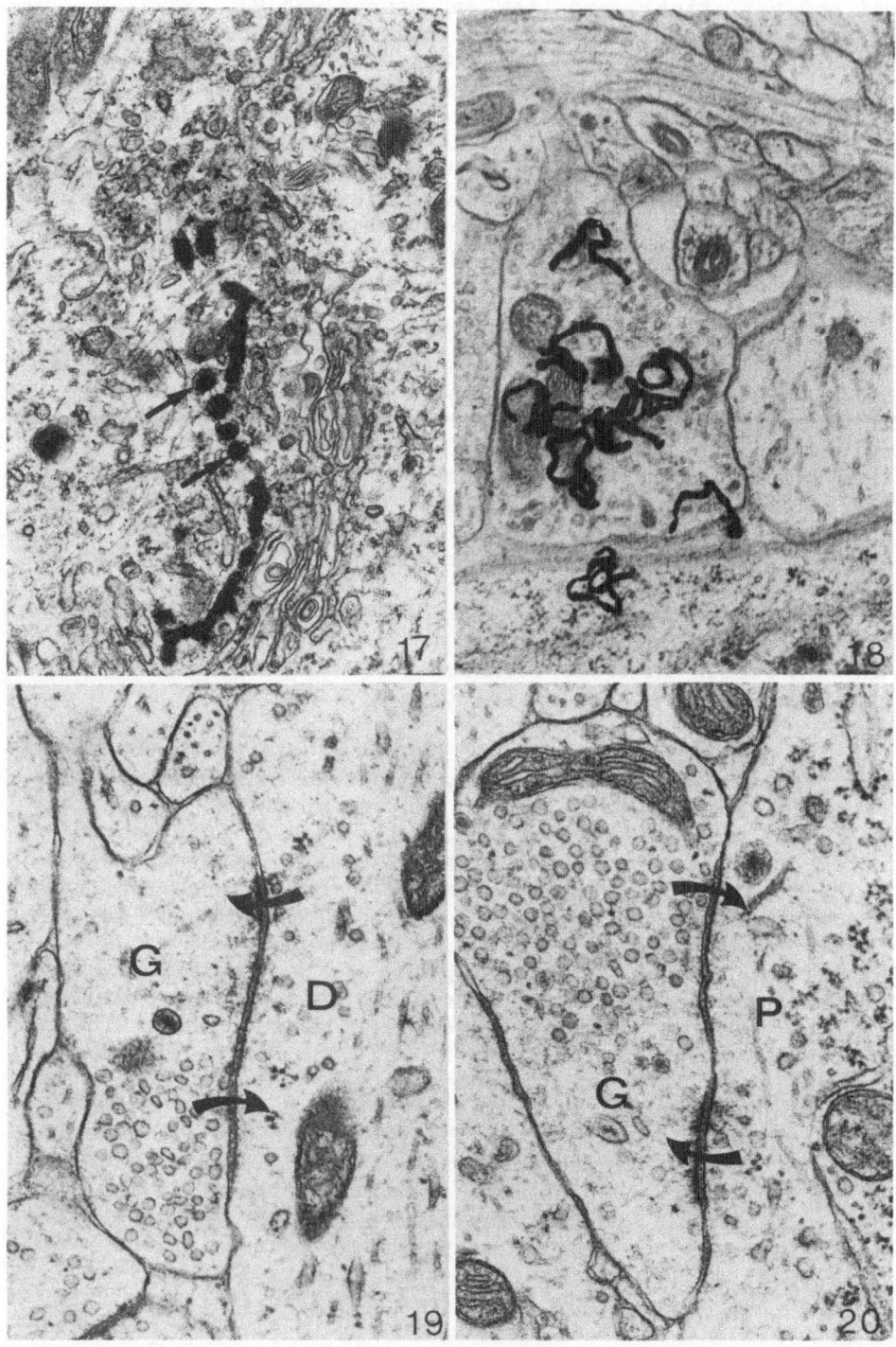

midal spines, probably related to the postsynaptic events.

The subsynaptic formation, first described by Taxi (66) in the sympathetic ganglia of the frog, consists of a dense plaque of material, 200 to 300 A thick, located 250 to 600 A below the postsynaptic differentiation and always shorter than it. This dense plaque resembles the postsynaptic differentiation and like it, is not visible when the material has been fixed in potassium permanganate (66). Occasionally, the subsynaptic formation of the frog sympathetic ganglia exhibits a discontinous aspect (67).

In the mammalian central nervous system a discontinous subsynaptic formation, consisting of a row of dense particles symmetrically spaced, has been described by Gray (68). As the different regions of the central nervous system have been studied with the electron microscope a more widespread distribution of this organelle has been reported (for references see 69). A special localization of the subsynaptic formation has been described by Milhaud and Pappas (70) in the interpeduncular and habenular nuclei and by Akert et al. (69) in the subfornical organ. It consists of a neck of spine or a narrow process of a dendrite sandwiched by twin presynaptic terminals. A single row of a discontinous subsynaptic formation is located in the central axis of the narrow postsynaptic cytoplasm and it is shared by the two presynaptic terminals. Akert et al. (69) have denominated the whole assemblage, the "crest synapse". Crest synapses have also been observed in different regions of the central nervous system. They are frequent in the substantia gelatinosa of Rolando (71) and they are occasionally found in the superior colliculus of the rat (72) and in the molecular layer of the cat cerebellum (Fig. 15) (71).

Fig. 17. Area postrema of the rat. Acid phosphatase activity in the innermost saccule of the Golgi apparatus and its associated large granular vesicles (arrows). X 39,000.
Fig. 18. Area postrema of a rat injected with 3 H-norepinephrine. An axon terminal devoid of large granular vesicles is well labelled. X 30,000.
Fig. 19. Olfactory bulb of the rat. Reciprocal synapse between a granule cell dendritic gemmule (G) and a mitral cell dendrite (D). The arrows indicate the direction of the synaptic transmission. X 48,000.
Fig. 20. Olfactory bulb of the rat. Reciprocal synapse between a granule cell dendritic gemmule (G) and the perikaryon of a mitral cell (P). The arrows indicate the direction of the synaptic transmission. X 48,000.

The function of the subsynaptic formation is still unknown. It seems clear that there is not a direct relationship between the subsynaptic formation and a specific type of
innervation (69,73). The study of the long term effect of
denervation on this organelle (67) has demonstrated that the
subsynaptic formation is not intimately related to the functional integrety of the synapse. In the frog sympathetic
ganglia, seven to twelve days after the section of the preganglionic fibers, almost all the axon terminals have disappeared, the neuronal surface is enterely covered by a sheath
of one or more layers of glial cytoplasm, but the subsynaptic
formation remains unchanged (Fig. 16). In the central nervous system a similar persistence of the subsynaptic formation has been reported by Lund (72).

VARIETIES OF NEURONAL SYNAPSES

The recent development of our knowledge on synaptic transmission makes it clear that the number of varieties of neuronal synapses exceeds the 11 types of axonal synaptic contacts
proposed by Ramon y Cajal (3). In most of the neuronal synapses encountered in electron microscopic investigations the
presynaptic element is an axon and the postsynaptic one either a neuronal perikaryon or a dendrite. Two large categories of axonal contacts can be described. In one of them,
the axonal enlargement constituting the presynaptic element
is located at the tip of the axon ("boutons terminaux"), and
in the other it is located along the length of the axon
("boutons en passant"). However, other varieties of neuronal
synapses have also been encountered with the electron microscope, in which the presynaptic element can be a dendrite or
even a neuronal perikaryon, or in which the postsynaptic
element can be an axon.

A) Axo-Axonal Synapses

The identification of the axonal nature of the postsynaptic element is only possible in two different situations :

Fig. 21. Spinal cord of the rat. Different types of axon terminals in synaptic contact with the initial segment of a mo-
toneuron axon. X 33,000.
Fig. 22. Cerebellar cortex of the cat. Initial segment of a
Purkinje cell axon. The arrow points to the synaptic contact
between a basket fiber and this initial segment. X 40,000.
Fig. 23. Spinal cord of the frog. Axo-axonal synapse. The
arrows point to the "active zones". The presynaptic axon
contains elongated vesicles. X 43,000.

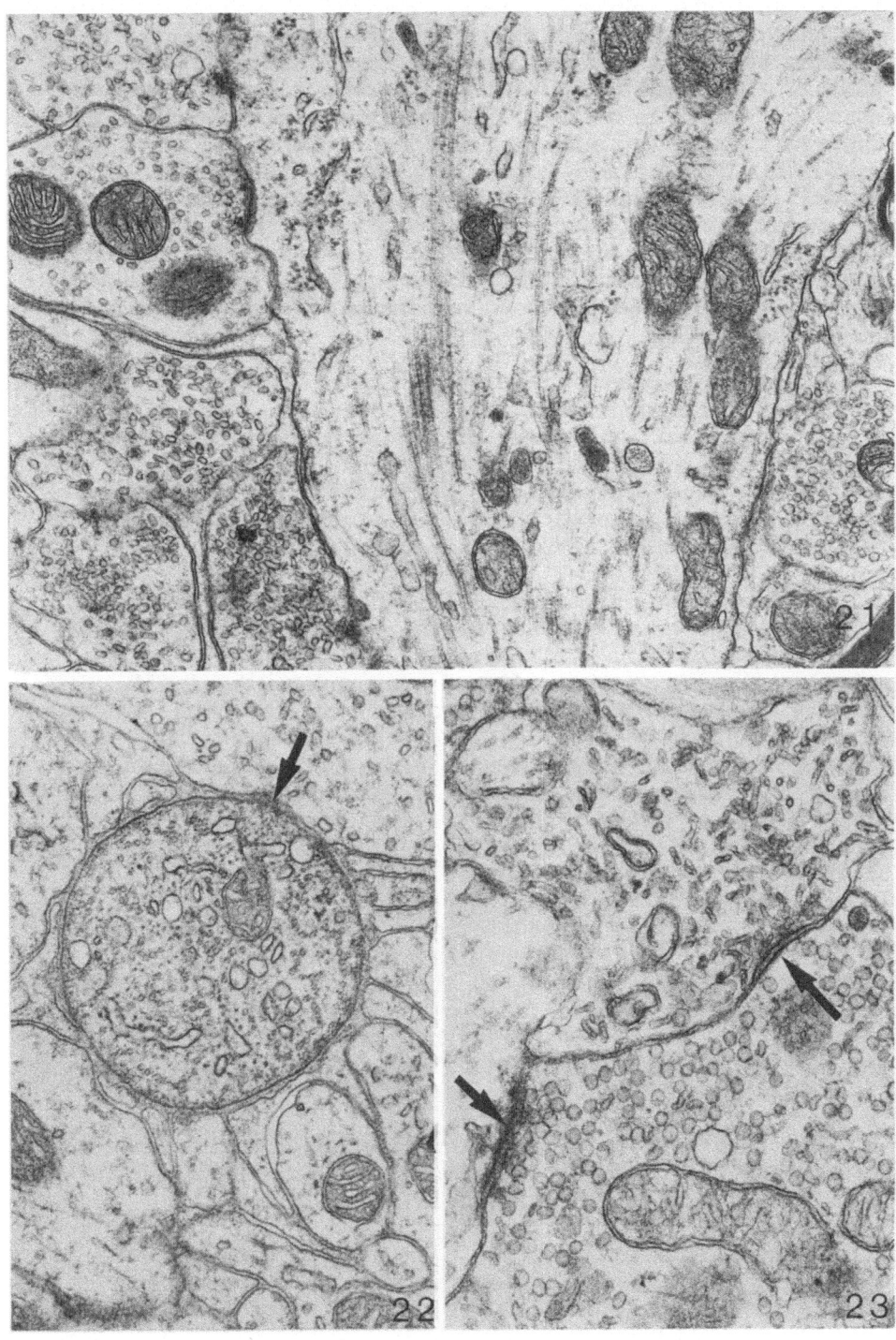

first, when the neuronal profile contains the typical aggregations of synaptic vesicles and mitochondria characterizing an axon terminal; second, when the profile exhibits the morphological features of an axon hillock or of an initial segment (61), that is to say, the bundles of microtubules, the undercoating of the axolemma and the clusters of polyribosomes. This kind of axo-axonal synapse has been described in several types of neurons (for a review, see reference 61). The presynaptic element contains the usual organelles, and according to the shape of its vesicles, two kind of axons are in synaptic contact with the surface of the hillock or of the initial segment; axon terminals with rounded vesicles and with elongated vesicles (Fig. 21).

When the postsynaptic element is the initial segment of an axon, the fine granular material undercoating the axolemma was interrupted at the sites of direct apposition between the pre- and postsynaptic axons; only at the location of the "active zone" there was an aggregation of dense cytoplasmic material beneath the axonal plasmamembrane to form the postsynaptic differentiation (Fig. 22).

In the review of Peters, Palay and Webster (10) a different type of axo-axonal relationship with synaptic function is considered. This type is only encountered in two situations : one is the axon cap surrounding the initial segment of the Mauthner cell axon (74) and the other is the final portion of the brush of the basket fibers around the initial segment of the Purkinje cell axon of the cerebellum (75,76). In both cases, a whorl of axons surrounds the initial segment of an axon and there is not synaptic contacts between them (10), even if physiological evidence of an inhibitory action of synaptic nature has been reported (77,78). In the cat cerebellum synaptic contacts between basket fibers and the initial segment of the Purkinje cell axon can be found (Fig. 22), but they are not numerous. On the contrary, synaptic contacts on the surface of the Mauthner cell initial segment are frequently observed (79). Therefore, even if this kind of synapses exhibits specific morphological features, the criteria for identification of synaptic contacts are also present.

The other variety of axo-axonal synapses corresponds to those where the postsynaptic element is also an axon terminal. They have been identified in several regions of the central nervous system, specially in the spinal cord (80). In the spinal cord of the frog (45), the presynaptic axon of these serial synapses contains, in all the cases observed in material perfused with aldehydes, the elongated or tubular vesicles described above. Figure 23 illustrates one of these

synapses, the probable direction of the impulse transmission
is given by the cluster of elongated vesicles associated to
the presynaptic membrane.

B) Dendro-Dendritic, Dendro-Somatic and Somato-Dendritic Synapses

Ramon y Cajal (81) has described in the central nervous
system of vertebrates the existence of cells which have all
the features of neurons, except that their axon is lacking.
These cells were located in the retina, in the olfactory bulb
and probably in other regions of the brain as, for instance,
in the ventral cochlear nucleus. Ramon y Cajal (81) has
suggested that one or several of the processes of such
amacrine-like cells, even if they do not have the features
of an axon, may have its functional role. Consequently, in
an indirect way, and only on special occasions, has Ramon y
Cajal admitted the existence of dendro-dendritic and dendro-
somatic synapses. Electron microscope observations of the
retina (82,83) and the olfactory bulb (84,85,86,87) have
succeeded in confirming the hypothesis of Ramon y Cajal.
Profiles exhibiting the typical features of dendrites, where
rough endoplasmic reticulum or large number of polyribosomal
clusters, and in addition clusters of synaptic vesicles are
present, have been found in a presynaptic position. The
vesicles were associated to a junctional complex wearing the
features of an "active zone". In the external plexiform
layer of the olfactory bulb, the gemmules of the radial dendrites of the grains establish synaptic contacts with the
surface of the mitral secondary dendrites. In some instances,
these synaptic contacts differ from more typical contacts in
that at the same synaptic interface two separate "active
zones" immediately adjacent to each other but with opposite
polarities coexist. These reciprocal synapses are illustrated in Figure 19. The "active zone" in which the direction
of the transmission if from the granule cell dendritic gemmule to the mitral cell dendrite is constituted by a cluster
of heterogenous vesicular populations, where elongated vesicles are evident, and the presynaptic vesicular grid as well
as the postsynaptic differentiation are practically absent.
However, the opposite "active zone", polarized from the mitral dendrite to the gemmule, has only spherical vesicles
and a junctional complex of the type I of Gray.

In recent electron microscopic investigations of different regions of the nervous system (88,72,89,90) dendritic
profiles assuming the role of presynaptic elements have been
identified. The morphological evidence that dendrites give
rise to synapses elsewhere than in the retina and the olfactory bulb, has not till now been supported by physiological

evidence, but it seems reasonable to consider the possibility of neuronal interaction mediated by dendro-dendritic or
dendro-somatic synapses. It may be speculated, as Ramon y
Cajal (81) has already done, that neurons lacking their axons
(amacrine-like neurons) are more widespread in the nervous
system than they were thought to be, and that these kinds of
associative neurons have a specific function. It is clear
that more work is needed in order to demonstrate the existence or not of a neuronal type of cell lacking its axon
in the different regions of the vertebrate central nervous
system where the dendritic synapses have been found. The
possibility, that the neurons which give rise to dendritic
synapses have their typical axons, remains open, but in this
instance, it would be more difficult to understand the reason for such synapses.

Using the morphological criteria to identify chemical
synapses, somato-dendritic synapses have also been observed
in the central nervous system. They are a frequent feature
in the olfactory bulb, where they can form reciprocal synapses between the perikarya of mitral cells and the gemmules
of granule cells (Fig. 20). Sétalo and Székely (91) have
described somato-dendritic synaptic contacts between the
stellate neurons and dendrites probably arising from pyramidal or pear-shaped neurons in the optic tectum of the frog.

CORRELATION BETWEEN STRUCTURE AND FUNCTION

The effect of the chemical transmission on the postsynaptic membrane may give rise to the depolarization or the hy-
perpolarization of such membrane. Physiological studies in
some molluscan neurons (92) have demonstrated that inhibition can be a functional property of the postsynaptic membrane. In other instances, the inhibition seems to be a
function of the presynaptic element; for example, when the
vesicles contain a chemical transmitter such as GABA. On
these bases, cytologists have attempted to correlate several
morphological features of the presynaptic terminals and of
the "active zones" with different functions, and tried to
establish morphological criteria to characterize excitatory
and inhibitory axon terminals.

Essentially two criteria have been used to distinguish
excitatory from inhibitory axon terminals, and both were
first established in cortical areas of the brain. The first,
suggested by Eccles (93) is based in the morphology of the
"active zones". The type I synapses, observed mainly at axo-
dendritic synapses, are considered to be excitatory; while
the type II, which are mostly axo-somatic synapses, are considered inhibitory. The second criterion is based on varia-

tions in the size and the shape of synaptic vesicles (44,46, 47,48). Boutons with an heterogenous population of vesicles, most of them elongated, and with an average diameter of about 400 A, are defined as inhibitory; whereas, boutons with larger rounded vesicles are excitatory. In the cerebral cortex, Colonnier (24) has generalized that terminals with rounded vesicles form synaptic contacts exclusively with "active zones" of the type I, while, terminals with the mixture of rounded and elongated vesicles establish only synapses of the type II. In the spinal cord of the frog (45) this generalization seems to be operative (Fig. 24).

Neither of the two criteria has been completely proved. The main argument against the hypothesis suggested by Eccles (93) that the type I synapses might be excitatory and the type II inhibitory, was brought by Palay (1), for whom the synaptic junctions between the climbing fibers and the Purkinje cell primary dendrites, known as powerfully excitatory (94), are always of the type II. This argument seems to be no longer valid since the climbing fibers identified in the frog (95), in the mouse (96) and in the cat (Fig. 25) cerebellar cortices always form synaptic junctions of the type I. More valuable arguments have recently been raised against this hypothesis. The fact that in cortical and most noncortical regions of the central nervous system, the morphology of the synaptic terminals, as has been mentioned above, does not exhibit such stereotype features, and many intermediate forms of "active zones" can be observed. In such cases, it is difficult to relate a morphological classification to a functional pattern. Another important argument is the existence of axon terminals which have two different synaptic interfaces, one of them exhibiting the features of type I synapses and the other of type II (1,97). If the Dale's principle is valid, it is difficult to accept, that in these instances, the morphology of the "active zones" is directly related to their function. Mugnaini (97) advances the hypothesis that during the development of synaptic connections some local factors particular to the postsynaptic surface engaged in the contact are responsible for the morphological features that will attain the "active zone". The morphological appearance of type I or type II synapses is related to the induction of the postsynaptic receptive surface. For instance, for Mugnaini, spines always form type I contacts independently of the nature of the presynaptic element. Here again it is difficult to construct a theory that includes all patterns. For instance, in the spinal cord (45) it is frequent to find synaptic contacts on long dendritic spines with the features of type II synapses (Fig. 27). Furthermore, in the same material analysed by Mugnaini (97),

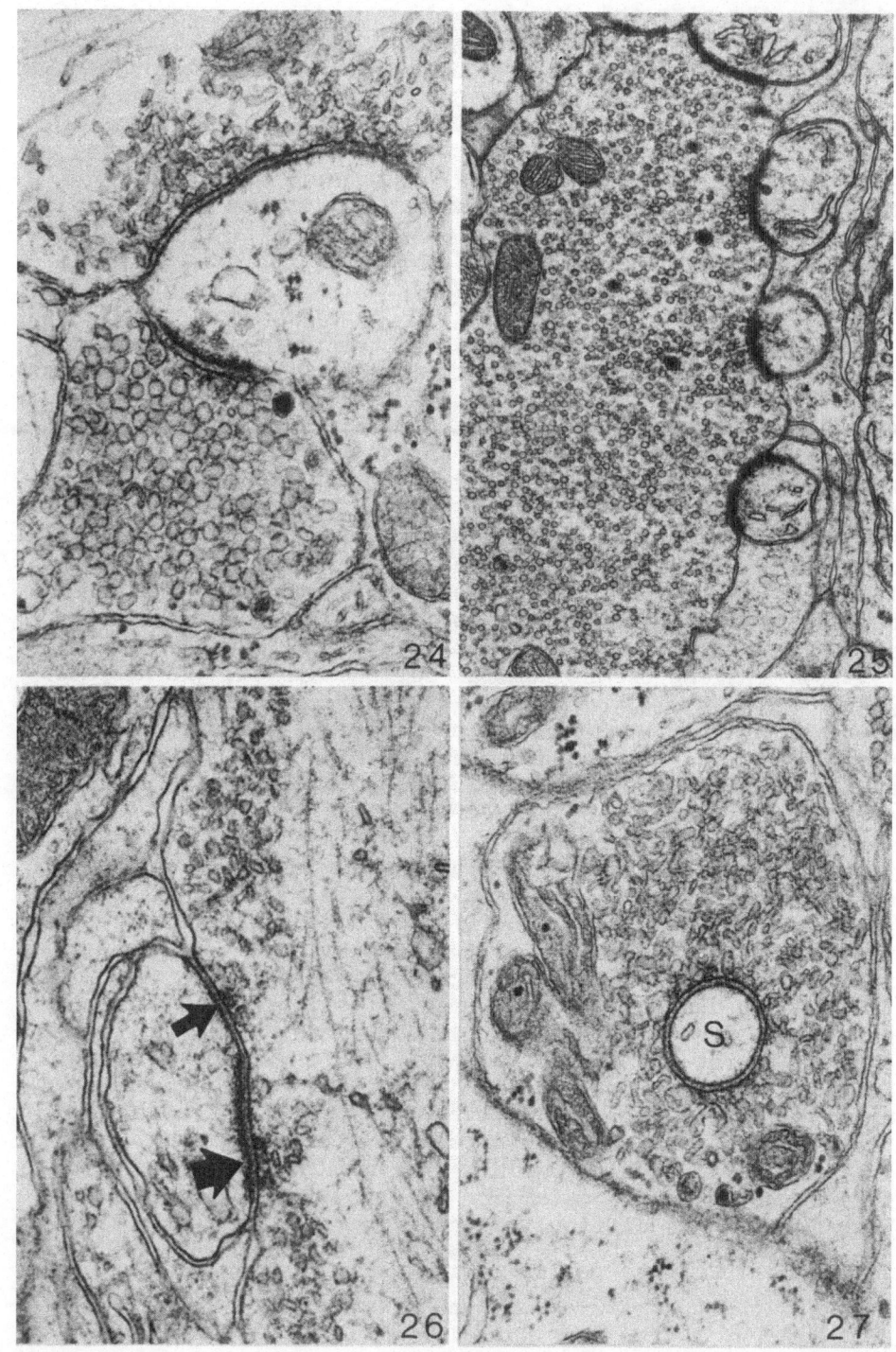

the cerebellar cortex of the cat, large axonal profiles belonging to basket fibers (Fig. 27) synapsing on a small dendrite or a large spine can be observed. At the synaptic interface, type I and II "active zones" can be found side by side. These observations are in contradiction to Mugnaini's generalization.

Strong circumstantial evidence is accumulating to suggest that axon terminals with small elongated synaptic vesicles may be related to inhibition. In the cerebellar cortex, where the identification of synaptic boutons is morphologically possible, and the electrophysiology is well known (94), it has been found that presumed inhibitory axon terminals contain elongated vesicles, while the rounded vesicles are present in excitatory terminals (98,96,95). In the developing spinal cord of the monkey, Bodian (99) has been unable to find axon terminals with elongated vesicles before the stage at which long intersegmental and crossed reflexes appear. At this stage one would expect to find the development of inhibitory synaptic inputs.

Experimental studies have revealed that in the cuneate nucleus, degenerating axon terminals, belonging to fibers of the pyramidal tract, with excitatory function, contain elongated vesicles (100). These results prevent, at the present stage of investigations, the generalization of Uchizono's hypothesis. The corollary of this hypothesis is that the transmitter substances are related to the synaptic vesicles in such a way that variations in vesicular morphology must reflect differences in their chemical content, indicating a difference in function of the axon terminal. The excitatory

---

Fig. 24. Spinal cord of the frog. Two axon terminals synapsing on the surface of a small dendrite. The axon terminal containing rounded vesicles has an "active zone" corresponding to the type I, and the terminal filled with elongated vesicles establishes a type II synapse. X 52,000.
Fig. 25. Cerebellar cortex of the cat. Climbing fiber in the molecular layer. The "active zones" between the climbing fiber and the spines of the Purkinje cell main dendrite are of the type I. X 27,000.
Fig. 26. Cerebellar cortex of the cat. Basket fiber in the molecular layer. The synaptic interface between the basket terminal and a small dendrite exhibits two "active zones", one of the type II (small arrow), and the other of the type I (big arrow). X 54,000.
Fig. 27. Spinal cord of the frog. Type II synapse between an axon terminal filled with elongated vesicles and a dendritic spine (S). X 48,000.

transmitter substances would be packed in vesicles which have
a rounded shape and maintain this shape after the routine
technical procedure for their observation on the electron microscope; while the inhibitory transmitters would be packed
probably in rounded vesicles, but during the aldehyde fixation and subsequent treatment, they would become elongated.
To accept this hypothesis it would be necessary to prove that
the inhibitory transmitters are present in elongated vesicles,
and the excitatory ones in rounded vesicles, and that the
postsynaptic effects for each neurotransmitter are always of
the same nature.

One of the possible ways to approach the study of such
correlation is offered by cytochemistry; for example, using
radioactive neurotransmitters and studying with high resolution autoradiography the uptake and storage of such exogenous
transmitters in the different axon terminals, it would be
possible to associate a specific transmitter substance with
the morphology of the synaptic vesicles contained in the
terminal, and to obtain valuable information about the possible functional role of such axon terminal. This technical
approach is illustrated in Figure 18. It belongs to a section of the area postrema of the rat, 30 min. after intraventricular injection of 3 H-norepinephrine. The silver grains
overlying the axon terminal indicate that it belongs to an
axon of a catecholamine- containing neuron.

II - ELECTROTONIC JUNCTIONS

In the last few years it has been clearly demonstrated
that the electrotonic coupling has its morphological correlation in a special junction of the "zonula occludens"
variety, where two plasmamembranes are in close apposition
constituting a junctional complex of 150 to 170 A thick.
The evidence to be considered, that namely, lowered membrane
resistivity occurs at this junctional region, has mostly
been provided by Bennett, Pappas and their coworkers (101,

Fig. 28. Spinal cord of the frog. Electrotonic junction between two dendritic profiles in the ventral horn. The "gap
junction" has an extension of about 1 micron (arrows) and its
inner surfaces are coated by a cytoplasmic fibro-granular
material. X 80,000.
Fig. 29. Spinal cord of the frog. Mixed synapse in the intermediate gray matter. The synaptic interface between the axon
terminal and the dendrite exhibits two different types of
junctional zone :"active zone" (S) and "gap junction" (arrow).
X 100,000.

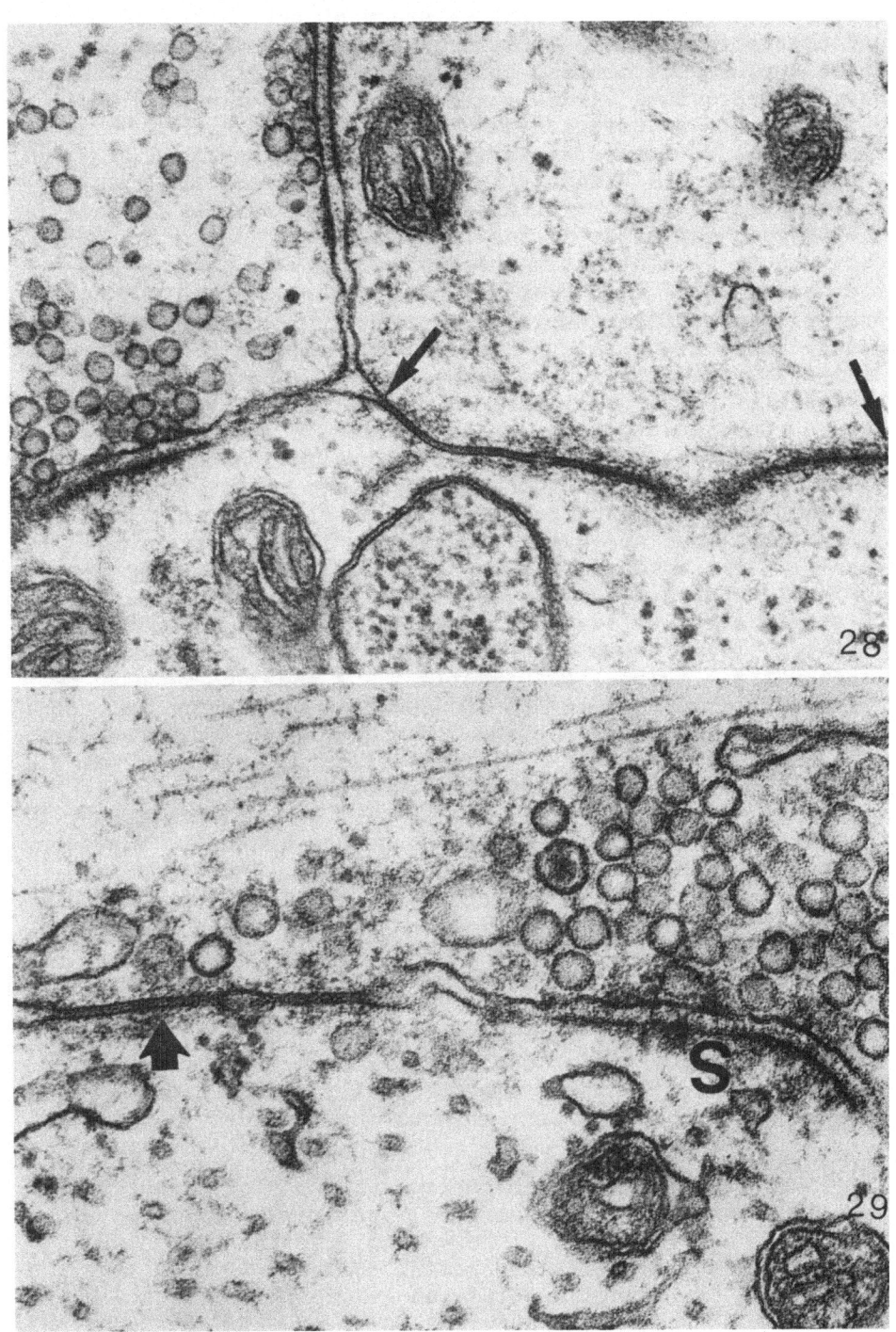

102,103,104).   In a correlative study of the morphological
and electrophysiological features of a number of neuronal
junctions, in the central nervous system of different fishes,
they have provided valuable arguments to sustain that the
electrotonic transmission takes place at the regions of membrane fusion.   A more direct proof has been provided by
Pappas, Asada and Bennett (105) who, submitting the giant motor synapse of the crayfish to various experimental conditions, such as, altering the ionic milieu, bathing the preparation in a solution in which propionate or acetate has
been substituted for chloride, have observed a substantial
increase in coupling resistance with failure to propagate
spikes between segments.   Electron microscopic observations
of these preparations have shown, in the "septae", a loss
of areas of close apposition of the axolemmae, with interpo-
sition of Schwann cell processes between the former sites
of tight contact.   When a normal salt solution is again applied to the preparation the electrotonic coupling reappears
and with it, the junctional regions of close apposition of
plasma membranes are again visualized.   From such dynamic
results it is evident that the low resistance pathway to the
ionic flux between the two cells has its morphological identity in the specialized junctions of the "zonula occludens"
variety.   It is reasonable to assume that if in morphological
studies these specialized junctions are observed between two
neuronal elements it may be deduced that an electrotonic coupling may occur at this place.   The correlation between structure and function seems better established for electrical
transmission than for chemical transmission.

Under the technical conditions used by Bennett, Pappas,
and coworkers (101 to 104) the electrotonic junctions are
described as a five-layered structure similar to the "zonula
occludens" observed in other epithelia (2).   With new electron microscopic techniques, such as the use of aqueous
uranyl acetate solution to stain small blocks of tissue prior
to the dehydratation and embedding (106) or the use of lanthanum as an electron microscopic tracer (107), it has been
observed that the outer leaflets of the apposed membranes in
the electrotonic junctions are not really fused, because
between the apposed plasmamembranes there is actually a nar-

Fig. 30. Lateral vestibular nucleus of the rat. Mixed synapse
between a large axon terminal and a perikaryon of a Deiters
cell. (S) "active zone"; (arrow) "gap junction". X 55,000.
Fig. 31. Lateral vestibular nucleus of the rat. High magnification of a "gap junction" between an axon terminal and a
neuronal cell body. The inner surfaces of the "gap junction"
are coated by a cytoplasmic dense material. X 127,000.

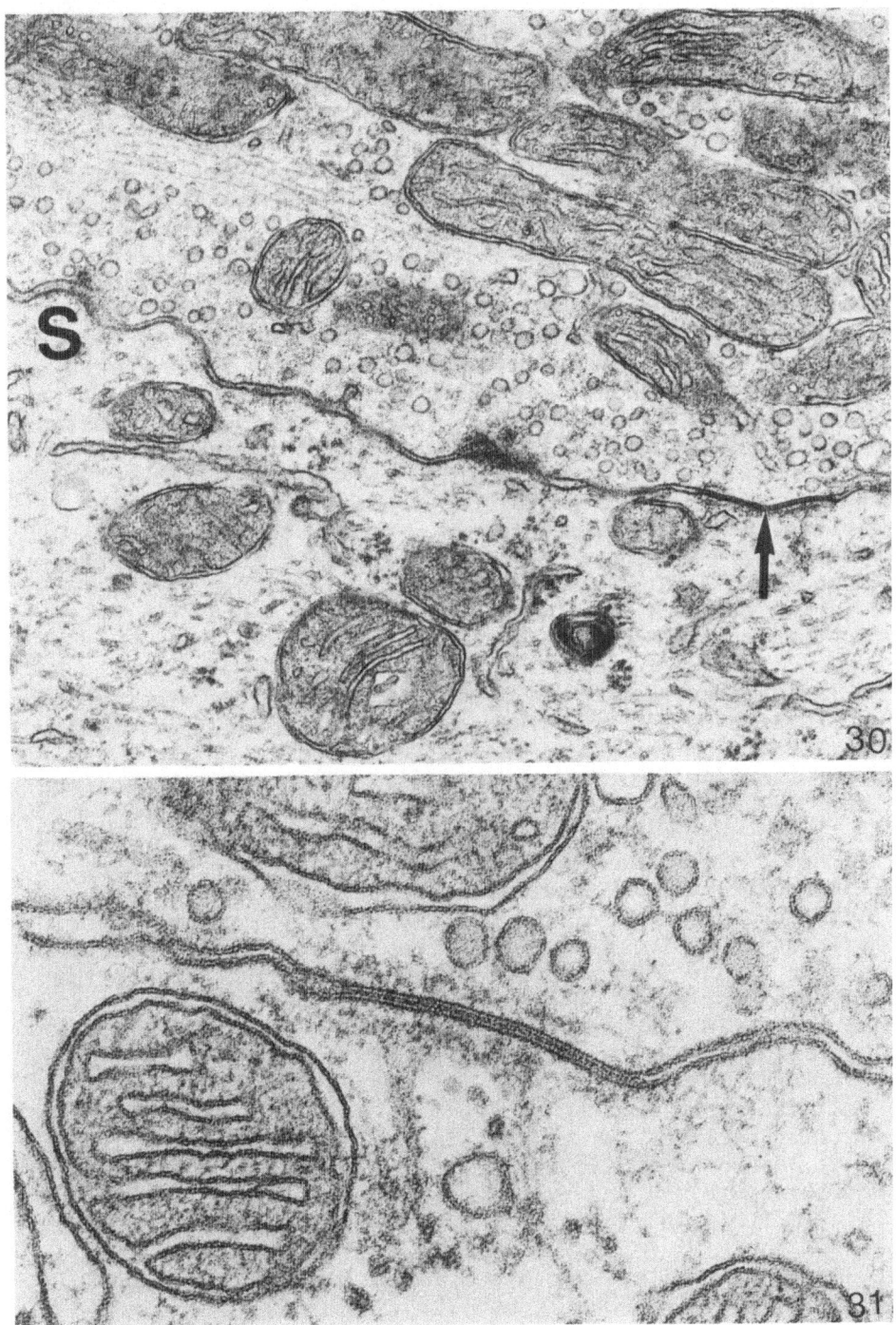

row gap 15 to 20 A wide (108,73); consequently, the electrotonic junctions have a seven-layered rather than a five-
layered structure. These close appositions are similar to
the "gap junctions" described by Revel and Karnovsky (107)
between heart muscle cells or liver cells.

The first description of a "gap junction" was by Robertson (109) in the electrotonic junctions between the club
endings and the lateral dendrite of the Mauthner cells of
the goldfish after permanganate and osmic acid fixation. In
this instance, the synaptic discs have an internal beading
which repeats itself at 90 A periods. From a frontal view
of the junction, a honeycomb pattern is seen, composed of a
system of lines and dots in hexagonal array. These images
are due to an intermediate lamina of prims or subunits closely packed in hexagonal pattern. The center-to-center distance between the subunits is 90-85 A and their diameter is
about 70-75 A. For Robertson (109) this mosaic pattern represents a structural protein on the outer surface of the
membrane. Using the lanthanum as a tracer the "gap junctions"
between heart muscle cells (107) or between astroglial processes (108) exhibit a similar hexagonal pattern. The minute gap is permeable to lanthanum, indicating its continuity
with the extracellular space.

In the central nervous system of vertebrates electrotonic
junctions have been discovered in fishes, amphibians, birds
and recently in mammals. The different parts of the nerve
cells which can come together to form electrotonic junctions
are :
1) Axo-dendritic and axo-somatic contacts are the most
frequently described. They exist in spinal and medullary
electromotor nuclei of mormyrid fish (102), in the giant
electromotor neurons of the catfish (103), in the medullary
electromotor nuclei of gymnotid fish (104), in the lateral
dendrite of the Mauthner cells of the goldfish (74), in the
cerebellar cortex of the mormyrid fish (110), in the spinal
cord (Fig. 29) and cerebellar cortex of the frog (111,95,
112), in the tangential nucleus of the chick (113), and in
the lateral vestibular nucleus of the rat (73).
2) Axo-axonal contacts have been described in the supra-
medullary cells of several species of teleosts (101).
3) Dendro-dendritic electrotonic junctions have been observed in mormyrid and gymnotic medullary electromotor nuclei (102,104) and recently between motoneurons in the spinal
cord of the frog (Fig. 28) (112). In the last instance,
electrophysiological evidence of electrotonic coupling of
dendrites has also been found (114).

4) Dendro-somatic contacts are rare; till now they have been described only in mormyrid electromotor nuclei (102).

5) Somato-somatic contacts are frequent in electric fish (102). This type of contact has also been described in the mammalian central nervous system between neural perikarya of the mouse mesencephalic fifth nucleus (115).

As a consequence of this classification, electrotonic junctions are quite widespread in the central nervous system of vertebrates, indicating that the electrotonic transmission may have an important functional role in the physiology of the brain. It is of interest to distinguish the electrotonic junctions from the "labile appositions" (108) since the last one may only represent artifacts of fixation. The fact that the electrotonic junctions are quite restricted in extent and in location and that the cytoplasm subjacent to the contact zone contains a dense layer of fluffy material (Figs. 28,29,30 and 31), that does not accompany non-junctional regions of the plasmalemma, provides good arguments against the possibility that the "gap junctions" are the result of fixation artifact.

Electrophysiological evidence of a dual mechanism of synaptic transmission, electrical and chemical, has only been obtained in the ciliary ganglion of the chick by Martin and Pilar (116). Electron microscope studies of this ganglion have shown the presence, at the same synaptic interface, of "active zones" and "gap junctions" (117,108). This kind of synaptic interface where different junctional regions are combined, has been called "mixed synapse" (10). The "gap junctions" are rare in the ciliary ganglion, and the surface covered by the "active zones" is much larger, making uncertain their responsibility for the electrotonic coupling. In the central nervous system of vertebrates "mixed synapses" have been found in electric fishes (102,103,104,110), in the frog (95,112) (Fig. 29), and in the rat (73) (Fig. 30). Bennett et al. (104) were unable to find electrophysiological evidence for chemical transmission in the gymnotid relay nuclei, where "mixed synapses" have been observed. These results presume that in some instances junctional complexes similar to those described here as "active zones", and commonly related to the site of the chemical transmission, may be devoid of functional meaning. This conclusion, from a morphological point of view, is untenable. For this reason, in a recent paper published in collaboration with Dr. Palay (73), we have advanced a series of functional speculations to remark that the combined junctions of the "mixed synapses" underlie a more subtle and refined kind of neuronal interaction than simply electrotonic coupling. More physiological

work is needed to understand the complicate role, suggested
by their morphology, of the "mixed synapses".

REFERENCES

1. Palay, S.L. in The Neurosciences (Eds. G.C. Quarton, T.
   Melnechuk, F.O. Schmitt), p. 24, Rockefeller Univ. Press.
   (1967)

2. Farquhar, M.G., and Palade, G.E. J. Cell Biol. 17:375
   (1963)

3. Ramon y Cajal, S. Trab. Lab. Invest. biol. Univ. Madr.
   29:1 (1934)

4. Sherrington, L.S. in A Text-Book of Physiology, vol. 3
   (Ed. M. Foster) Macmillan, (1897)

5. Ramon y Cajal, S. Rev. Trim. Microgr. 1:1 (1897)

6. Furshpan, E.J., and Potter, D.D. J. Physiol. 145:289
   (1959)

7. Palade, G.E., and Palay, S.L. Anat. Rec. 118:335 (1954)

8. De Robertis, E., and Bennett, H.S. Federation Proc. 13:
   35 (1954)

9. Gray, E.G., and Guillery, R.W. Intern. Rev. Cytol. 19:
   111 (1966)

10. Peters, A., Palay, S.L. and Webster, H. de F. The Fine
    Structure of the Nervous System - The Cells and their
    Processes. Hoeber/Harper and Row (1970)

11. Nakajima, Y., Pappas, G.D., and Bennett, M.V.L. Am. J.
    Anat. 116:471 (1965).

12. Palay, S.L. Exptl. Cell Res. Suppl. 5:275 (1958)

13. Couteaux, R. Actualités Neurophysiol. 3e série:145 (1961)

14. Pick, J., Gerdin, C., and Delemos, C. Z. Zellforsch. 62:
    402 (1964)

15. Glees, P., and Sheppard, B.L. Z. Zellforsch. 62:356(1964)

16. Gray, E.G. J. Anat. 93:420 (1959)

17. Bloom, F.E., and Aghajanian, G.K. J. Ultrastruct. Res. 22:361 (1968)

18. Aghajanian, G.K., and Bloom, F.E. Brain Res. 6:716 (1967)

19. Pfenninger, K., Sandri, C., Akert, K., and Eugster, C.H. Brain Res. 12:10 (1969)

20. Akert, K., Moor, H., Pfenninger, K., and Sandri, C. Prog. Brain Res. 31:223 (1969)

21. Akert, K., Moor, H., and Pfenninger, K. in First Int. Symp. on Cell Biol. and Cytopharmacol. Venice 1969, Raven Press (in press)

22. Gray, E.G. in Electron Microscopy in Anatomy (Eds J.D. Boyd, F.R. Johnson, and J.E. Lever) p. 54, Arnold,(1961)

23. Gray, E.G. J. Anat. 95:346 (1961)

24. Colonnier, M. Brain Res. 9:268 (1968)

25. Robertson, J.D. Neurosciences Res. Prog. Bull. 3(No.4): 17(1965)

26. Pappas, G.D., and Purpura, D.P. Nature 210:1391 (1966)

27. Rambourg, A., and Leblond, C.P. J. Cell Biol. 32:27 (1967)

28. Gray, E.G., and Whittaker, V.P. J. Anat. 96:79 (1962)

29. Brightman, M.W. Prog. Brain Res. 29:19 (1967)

30. De Robertis, E. in Cellular Dynamics of the Neuron (Ed. S.H. Barondes) p. 191, Academic Press (1969)

31. Sotelo, C. unpublished observations.

32. Palay, S.L. in Cytology of the Nervous Tissue; Proceedings of the Anatomical Society of Great Britain and Ireland, p. 82, Taylor and Francis, (1961)

33. Fatt, P., and Katz, B. J. Physiol. 117:109 (1952)

34. Del Castillo, J., and Katz, B. Prog. Biophys. Biochem. Chem. 6:121 (1956)

35. Akert, K., and Sandri, C. Brain Res. 7:286 (1968)

36. De Robertis, E., Rodriguez de Lores Arnaiz, G., Salgani-
    coff, L., Pellegrino de Iraldi, A., and Zieher, L.M.
    J. Neurochem. 10:225 (1963)

37. Whittaker, V.P., Michaelson, I.A., and Kirkland, R.J.A.
    Biochem. Pharmacol. 12:300 (1963)

38. Israël, M., and Gautron, J. in Cellular Dynamics of the
    Neuron (Ed. S.H. Barondes) p. 137, Academic Press
    (1969)

39. De Robertis, E. Intern. Rev. Cytol. 8:61 (1959)

40. Marchbanks, R.M. in Cellular Dynamics of the Neuron
    (Ed. S.H. Barondes) p. 115, Academic Press (1969)

41. Hubbard, J.I., and Kwanbunbumpen, S. J. Physiol. 194:
    407 (1968)

42. Clark, A.W., Mauro, A., Longenecker, H.E., and Hurlbut,
    W. Nature 225:703 (1970)

43. Walberg, F. J. Ultrastruct. Res. 12:237 (1965)

44. Uchizono, K. Nature 207:642 (1965)

45. Sotelo, C., and Taxi, J. unpublished observations.

46. Bodian, D. Science 151:1093 (1966)

47. Lenn, N.J., and Reese, T.S. Am. J. Anat. 118:375 (1966)

48. Larramendi, L.M.H., Fickenscher, L., and Lemkey-Johnston,
    N. Science 156:967 (1967)

49. Bodian, D. Bull. Johns Hopkins Hosp. 118:282 (1966)

50. Kruger, L., and Maxwell, D.S. Am. J. Anat. 119:479
    (1966).

51. Van Orden, L.S., III, Bloom, F.E., Barrnett, R.J., and
    Giarman, N.J. J. Pharmacol. Exptl. Therap. 154:185
    (1966)

52. Hökfelt, T. Z. Zellforsch. 79:110 (1967)

53. Hökfelt, T. Z. Zellforsch. 91:1 (1968)

54. Palay, S.L. Anat. Record 138:417 (1960)

55. Richards, J.G., and Tranzer, J.P. Brain Res.17:463 (1970)

56. Sotelo, C. in preparation.

57. Droz, B., and Barondes, S.H. Science 165:1131 (1969)

58. Sotelo, C., and Palay, S.L. Lab. Invest. (in press).

59. Westrum, L.E., and Blackstad, T.W. J. Comp. Neurol. 119:281 (1962)

60. Gray, E.G., and Guillery, R.W. J. Anat. 97:389 (1963)

61. Palay, S.L., Sotelo, C., Peters, A., and Orkand, P.M. J. Cell Biol. 38:193 (1968)

62. Peters, A., Proskauer, C.C., and Kaiserman-Abramof, I.R. J. Cell Biol. 39:604 (1968)

63. Herndon, R.M. J. Cell Biol. 20:338 (1964)

64. Schultz, R.L., and Karlsson, U. J. Ultrastruct, Res. 14:268 (1966)

65. Scheibel, M.E., and Scheibel, A.B. in Communications in Behavioral Biology, Part A, 1:231 (1968)

66. Taxi, J. Compt. Rend. Acad. Sci. 252:174 (1961)

67. Sotelo, C. Exptl. Brain Res. 6:294 (1968)

68. Gray, E.G. J. Anat. 97:101 (1963)

69. Akert, K., Pfenninger, K., and Sandri, C. Z. Zellforsch. 81:537 (1967)

70. Milhaud, M., and Pappas, G.D. J. Cell Biol. 30:437 (1966)

71. Sotelo, C. unpublished observations.

72. Lund, R.D. J. Comp. Neurol. 135:179 (1969)

73. Sotelo, C., and Palay, S.L. Brain Res. 18:93 (1970)

74. Robertson, J.D., Bodenheimer, T.S., and Stage, D.E. J. Cell Biol. 19:159 (1963)

75. Palay, S.L. in Brain Function, Vol. 2, RNA and Brain
    Function; Memory and Learning (Ed. M.A.B. Brazier)
    p. 69, Univ. of Calif. Press (1964)

76. Hamori, J., and Szentagothai, J. Acta Biol. Hung. 15:
    465 (1965)

77. Furukawa, T., and Furshpan, E.J. J. Neurophysiol. 26:
    140 (1963)

78. Eccles, J.C., Llinas, R., and Sasaki, K. Exptl. Brain
    Res. 1:1 (1966)

79. Caravita, S. personal communication.

80. Gray, E.G. Nature 193:82 (1962)

81. Ramon y Cajal, S. Histologie du Système Nerveux de l'Hom-
    me et des Vertébrés. Vols. I and II. Reprinted by C.S.I.
    C. (Madrid) (1952)

82. Dowling, J.E., and Boycott, B.B. Proc. Roy. Soc. Série B
    166:80 (1966)

83. Raviola, G., and Raviola, E. Am. J. Anat. 120:403 (1967)

84. Hirata, Y. Arch. Histol. Jap. 24:293 (1964)

85. Andres, K.H. Z. Zellforsch. 65:530 (1965)

86. Rall, W., Shepherd, G.M., Reese, T.S., and Brightman,
    M.W. Exptl. Neurol. 14:44 (1966)

87. Price, J.L. Brain Res. 7:483 (1968)

88. Hirata, Y. Arch. Histol. Jap. 27:373 (1966)

89. Guillery, R.W. Z. Zellforsch. 96:1 (1969)

90. Ralston, H.J. III, and Herman, M.M. Brain Res. 14:77
    (1969).

91. Sétalo, G., and Székely, G. Exptl. Brain Res. 4:237
    (1967)

92. Tauc, L., and Gerschenfeld, H.M. Nature 192:366 (1961)

93. Eccles, J.C. The physiology of Synapses, Springer
    Verlag (1964)

94. Eccles, J.C., Ito, M., and Szentàgothai, J. The Cere-
    bellum as a Neuronal Machine, Springer Verlag (1967)

95. Sotelo, C. in Neurobiology of Cerebellar Evolution and
    Development, (Ed. R. Llinas), p. 327, Am. Med. Associa-
    tion (1969)

96. Larramendi, L.M.H., and Victor, T. Brain Res. 5:15
    (1967)

97. Mugnaini, E. Brain Res. 17:169 (1970)

98. Uchizono, K. Exptl. Brain Res. 4:97 (1967)

99. Bodian, D. J.Comp.Neurol. 133:113 (1968)

100. Walberg, F. Exptl. Brain Res. 2:107 (1966)

101. Bennett, M.V.L., Nakajima, Y., and Pappas, G.D.
     J. Neurophysiol. 30:161 (1967)

102. Bennett, M.V.L., Pappas, G.D., Aljure, E., and
     Nakajima, Y. J. Neurophysiol. 30:180 (1967)

103. Bennett, M.V.L., Nakajima, Y., and Pappas, G.D.
     J. Neurophysiol. 30:209 (1967)

104. Bennett, M.V.L., Pappas, G.D., Giménez, M., and
     Nakajima, Y. J. Neurophysiol. 30:236 (1967)

105. Pappas, G.D., Asada, Y., and Bennett, M.V.L. Anat. Record
     157:297 (1967)

106. Karnovsky, M.J. J. Cell Biol. 35:213 (1967)

107. Revel, J.P., and Karnovsky, M.J. J. Cell Biol. 33:67
     (1967)

108. Brightman, M.W., and Reese, T.S. J. Cell Biol. 40:648
     (1969)

109. Robertson, J.D. J. Cell Biol. 19:201 (1963)

110. Kaiserman-Abramof, I.R., and Palay, S.L. in Neurobiolo-
     gy of Cerebellar Evolution and Development (Ed. R. Llinas)
     p. 171, Am. Med. Association (1969)

111. Charlton, B.T., and Gray, E.G. J. Cell Sci. 1:67 (1966)

112. Sotelo, C., and Taxi, J. Brain Res. $\underline{17}$:137 (1970)

113. Hinojosa, R., and Robertson, J.D. J. Cell Biol. $\underline{34}$:421 (1967)

114. Grinnell, A.D. J. Physiol. $\underline{182}$:612 (1966)

115. Hinrichsen, C.F.L., and Larramendi, L.M.H. Brain Res. $\underline{7}$:296 (1968)

116. Martin, A.R., and Pilar, G. J. Physiol. $\underline{168}$:443 (1963)

117. Takahashi, K., and Hama, K. Z. Zellforsch. $\underline{67}$:174 (1965)

SOME BIOCHEMICAL ASPECTS OF THE DEVELOPMENT OF AVIAN OPTIC CENTRES AND THE EFFECTS OF DEAFFERENTATION

Pier Carlo Marchisio[*]

Department of Human Anatomy, University of Turin
C.so M. D'Azeglio 52, 10126 Turin (Italy)

The optic pathway of birds, connecting the retina to the tectum mesencephali, has often provided a suitable model for studying the pattern of connection between two developing nervous centres. Also for studying the mutual influences between two populations of neuroblasts, the complex optic pathway-optic centres lends itself as a good study model because its structure is rather well known and its anatomical position allows easy accessibility to experimental manipulations during embryonic life. In particular, since the chick optic pathway is completely crossed and provides the largest supply of fibres to the tectum, the removal of one retina deprives the contralateral tectum of most afferent fibres but leaves the ipsilateral one quite unaltered for control purpose.

When the eye is removed at the stage of optic cup (i.e., between the 3rd and the 4th day of incubation in the chick embryo), the retina fails to form and the neuroblasts of the contralateral tectum will never be contacted by the terminals of the optic fibres; the tectum will thus develop being deprived of its physiological connection with the retinal ganglion cells.

---

[*] From the Institute of Histology and General Embryology of the University of Turin.

The whole work done   in   our laboratory during
several years was aimed at investigating the maturation
of a nervous centre   in the absence of most transneuronal
influence.   In particular, this experimental approach
had the final goal of identifying the influence of one
neuron upon the development of a second neuron; the latter
being the normal target of the axonal terminals of the
former. Such an experimental situation may be fulfilled
by the   system   formed   by retinal ganglion cells and
tectal neurons.

The initial experiment consisted in preventing one
neuron   to contact   the other neuron by mechanically
removing one optic cup (1,2). Very recently, a more
refined type of experiment has been developed by which
changes may be introduced in the metabolism of the
retinal ganglion cell while leaving the retino—tectal
connection anatomically unaltered;   the effects of
metabolic changes in one developing neuronal population
may be thus analyzed also for their influence upon the
maturation of the other one. Preliminary evidence has
been obtained indicating that rather deep alterations
may be induced in the tectum after blocking protein
synthesis or axonal flow in the retinal ganglion neuron
(3).

The work of Filogamo (1) provided early information
about the effects of monolateral extirpation of chick
embryo optic cup upon the development of tectum opticum
structure. He excised the optic cup by means of sharpened
needles between the 48th and the 52nd hour of incubation;
then, he followed the maturation of the contralateral
tectal layers by light microscopy. No apparent changes
were found until the 12th day of incubation if one
excludes the absence of retinal fibres: the thickness
of the cellular layers and the cell number were not
changed when compared to those of ipsilateral control
tecta. After the 12th day, on the contrary, the thickness
of the deafferented tectal layers decreased significantly.
This finding was accounted for by the progressive under
development of the nerve cell bodies; at hatching, a
large proportion of the perikarya of the deafferented
side had disappeared.

FIG. 1

Effects of optic cup removal upon amino acid incorporation into protein of chick   embryo cerebral hemispheres and optic tecta.

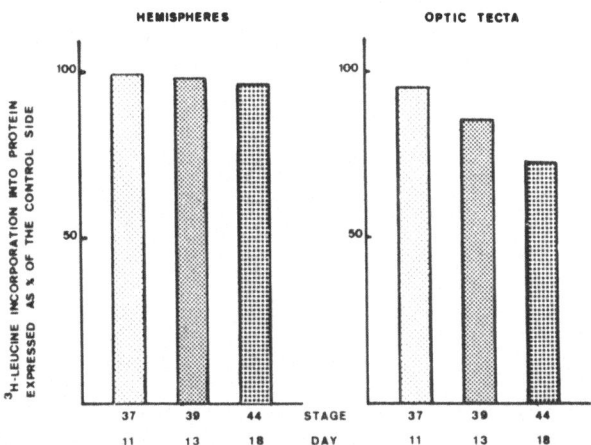

^{3}H-Leucine (10 µCi) was injected into the amniotic cavity 4hr before dissecting chick embryo brains. Protein was precipitated from sucrose homogenates with 10% TCA containing 10mM "cold" leucine, washed several times with 5% TCA and ether-ethanol (5:1), dissolved in Toluene (Packard), and counted by liquid scintillation spectrometry. From Marchisio and Gremo (3).

     The above results were interpreted as the maturation of the tectal neurons were independent of afferent fibres until the 12th day but that further maturation were in some way controlled by the onset of retino-tectal connection. Kollros (4) reached similar conclusions after analogous experiments on amphibian larvae; also Levi-Montalcini (5) had obtained comparable results working on the acoustic centres of the chick embryo.

     We have recently studied amino acid incorporation into protein in the tecta of optic cup deprived chick embryos (3). It has been found that amino acid incorporation is clearly depressed in deafferented tecta while it is not in cerebral hemispheres; the depression of protein synthesis becomes more marked in embryos approaching to hatching (Fig. 1).

FIG. 2

Changes in choline acetyltransferase activity of chick
embryo optic lobes after early removal of the right
optic cup.

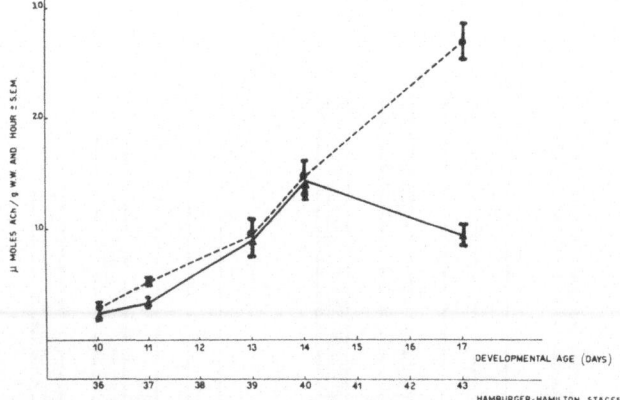

Choline acetyltransferase was assayed according to
McCaman and Hunt (9). Each value is the mean of three
optic lobes ± S.E.M.
Dashed line : control lobes; solid line : deafferented
lobes. From Marchisio (8).

On the basis of the results reported above, the
ontogenesis of the acetylcholine system was studied in
normal and deafferented tecta.

Earlier studies on the tectum opticum of frog
embryos (6) supported the idea that the appearance of
acetylcholinesterase activity within tectal layers could
be accounted for by the development of synaptic endings.
The latter interpretation could not be confirmed by
Filogamo and Strumia (7)and Filogamo (2) on the basis
of work done on chick embryos. On sections stained for
acetylcholinesterase activity by the histochemical
technique of Koelle, it was found that the enzyme is
already present in the earliest tectal neuroblasts which
are still actively migrating, not yet being part of any
established structure. The enzyme makes then its appearance
in all the cellular layers and spreads to the fibrous
layers only when dendrites grow out of neuroblasts
(about the 12th day of development when retinal fibres
have arrived to the tectum).

When the tectum develops in the absence of retinal fibres, the distribution of acetylcholinesterase occurs as in control tecta until the 12th day; then it disappears gradually from most tectal neuroblasts as a result of the generalized underdevelopment of tectal neurons which had been previously described by simpler histological methods (1).

The information obtained by Filogamo (2) about the ontogenesis of acetylcholinesterase distribution in tectal layers has been recently supported by a biochemical investigation (8). A quantitative study of choline acetyl-transferase activity was carried out in chick embryo optic lobes after early monolateral removal of the optic cup. It has been found (Fig. 2) that also the enzyme responsible for acetylcholine synthesis increases in activity in deafferented tecta until the 14th day, quite similarly as it does in controls; afterwards; it falls rapidly to less than 30% of the control value.

From these studies it could be established that the early appearance of the acetylcholine system in the tectum is not dependent on interneuronal connections but is an intrinsic property of tectal neuroblasts as it is of most neuroblasts (10). However, the final maturation and even the maintenance of the acetylcholine system in typical cholinergic neurons like those of the tectum opticum is strictly controlled by retinal fibres which are not cholinergic(11). This observation obviously implies that the lack of an "input" provided by retinal fibres to tectal neuroblasts hinders the whole mechanism of neuronal differentiation to begin with such an evidential phenomenon like the inhibition of dendritic outgrowth both "in vivo" (2) and "in vitro" (12).

When afferent fibres from retina reach the tectum they spread over it in a highly ordered and specific fashion which involves recognition of the tectal targets (13). When recognition has occurred both "synapse-forming" surfaces, namely the membranes of the retinal fibre tips and those of tectal neuroblasts, could, at least theoretically, exchange information. The information does not likely pass from one neuron to the other through mature synapses but may be transferred, probably coded

TABLE 1

Effects of Colcemid upon the axoplasmic transport of
protein to the optic tectum of 13-day chick embryos.

	% incorp. of ^{3}H-leu in the right retina	dis./min/ g prot. in the left tectum	$\dfrac{L-R}{R}$
Control	93.2	329,300	+ 0.31
Colcemid	94.8	60,000	− 0.08

Five μl of Colcemid Ciba (.1 mg/ml in saline) were
injected into the right eyeball of the embryo; the control
received the same volume of saline. After 6 hr, 5 μCi of
^{3}H-leucine were injected into the same eye. The retina
and the tectum were dissected from the embryo after
further 6 hr and their protein-bound radioactivity
determined (cf. legend to Fig.1). Percent incorporation
of labelled leucine was calculated according to Sjöstrand
and Karlsson (21). The ratio L-R/R is a "normalized
parameter of transport" (22).

in macromolecules, either through pynocitosis or through
temporary low-resistance junctions which occur and may
play a specific role in embryonic differentiation (14,
15, 16, 17, 18).

Could however macromolecules, acting as effectors
of gene regulation according to the operon hypothesis
(19), be transported along the optic nerve of embryos
and interact with the gene expression of tectal neuroblasts? Any answer to the latter question is still a
matter of speculation; to the former question, on the
contrary, we have recently contributed with some
confirmatory evidence, by showing axoplasmic transport
of protein along the optic pathway of chick embryos (20).

When labelled leucine is injected into one eyeball
of 7-day chick embryos, labelled protein is preferentially
recovered in the contralateral tectum containing the fibres
of the retinal ganglion neurons which have previously
incorporated the amino acid. Transported protein-bound
radioactivity could first be measured on the 10th day
when most nerve fibres from retina have reached the

FIG. 3

Effects of Colcemid induced inhibition of axoplasmic
transport in the optic pathway of chick embryos upon the
amino acid incorporation in cerebral hemispheres and
optic tecta.

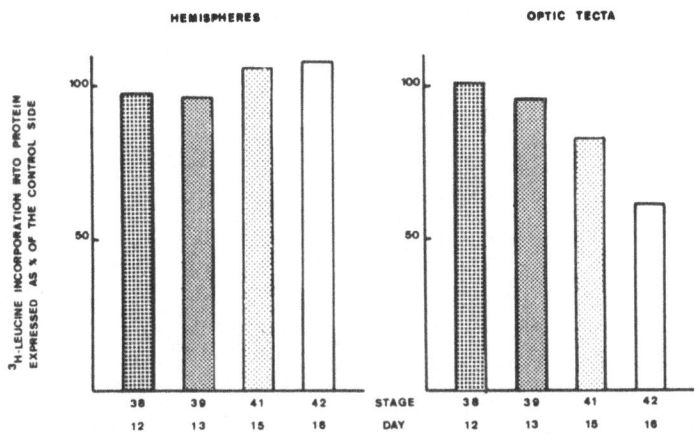

Colcemid solution (see legend to Table 1) was injected
into the right eyeball 24 hr before administering 3H-
leucine into the amniotic cavity of chick embryos.
See, for further details the legend to Fig.1. Each bar
is the mean of four determinations.

contralateral tectum. On the 13th day of incubation, when
the retino-tectal connection has been completed (13),
protein is transported to the tectum at a high rate (20).
Such an axoplasmic transport can be blocked when a
colchicine derivative (Colcemid, Ciba) is injected
intraocularly at a low dose, before giving labelled
leucine. (Table 1) (23).

The above results demonstrate that macromolecules
can be conveyed along the embryonic nerve fibres and
may reach their endings. Whether a macromolecule may
pass across the intercellular space and enter tectal
neuroblasts is not known since there is little
information about the process of synapse formation in
chick embryo tectum opticum and no evidence is available
on the permeability of the neuroblast plasma membrane.

However, in addition to the results reported above

FIG. 4

Effects of puromycin inhibition of protein synthesis in
chick embryo retina upon amino acid incorporation   in
cerebral hemispheres and optic tecta.

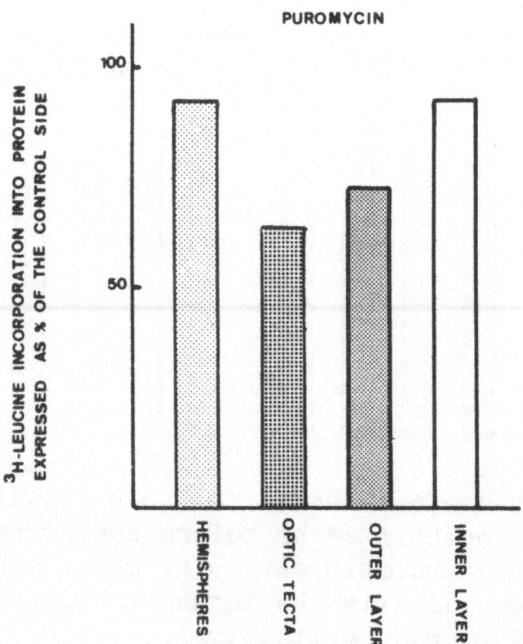

Puromycin (20μg) was injected into the right eyeball,
24 hr before administering ^{3}H-leucine (10 μCi) into the
amniotic cavity of chick embryos. Percent incorporation
of labelled leucine in the right retina was less than
10% of the control.See, for further details, the legend
to Fig.1. Each bar is the mean of four determinations.

about the transneuronal effects following optic cup
extirpation, some preliminary evidence is now available
indicating that the maturation of the tectum is influenced
by an unknown "factor" which seems actually synthesized
in the embryonic retina.In fact,when the axoplasmic flow
is blocked in one optic pathway,amino acid incorporation
into protein is depressed in the correspondent tectum
(Fig. 3) (3). It is noteworthy that the effect is
apparent only when the retino-tectal connection has been
completed and is more marked in the last days of
incubation.

Similarly, amino acid incorporation is depressed in tecta when protein synthesis has been previously blocked in the contralateral retina by puromycin. The effect is comparatively more marked in the outer layers of the tectum which contain higher density of retinal fibres than the deeper ones (Fig.4) (3).

It must at last be reminded that the protein synthesis depression obtained by blocking either the axonal transport or the protein synthesis in retina is in close agreement with that induced by optic cup extirpation. (Fig.1).

Though the latter part of this investigation is still at a very preliminary stage, its results may add support to the conclusions drawn from the experiments of optic cup extirpation (24). The whole evidence seems in favour of an intimate interaction between the neuronal components of the chick embryo optic pathway when very important events of the tectum morphogenesis occur. As a working hypothesis, we are inclined to think that the retino-tectal interaction is mediated by a specific "factor" produced in the embryonic retina and transported to the tectum via the axoplasmic flow.

Acknowledgement. These investigations have been supported by research grants to Prof.G.Filogamo from the Italian National Research Council (C.N.R.)

REFERENCES

(1) Filogamo, G. Riv. Biol. (Perugia) 42:73 (1950)

(2) Filogamo, G. Archs.Biol.(Liège)   71:159 (1960)

(3) Marchisio, P.C., and Gremo, F. To be published (1970)

(4) Kollros, J.J. Anat. Rec. 99:654 (1947)

(5) Levi-Montalcini, R.J. comp.Neurol.91:909 (1949)

(6) Boell,E.J., Greenfield, P., and Shen, S.C. J. exp. Zool. 149:415 (1955)

(7) Filogamo, G., and Strumia, E. Rend. Acc. Naz.Lincei (Roma) 25:119 (1958)

(8) Marchisio, P.C. J. Neurochem. 16:665 (1969)

(9) McCaman, R.E., and Hunt, J.M. J. Neurochem.12:253 (1965)

(10) Filogamo, G., and Marchisio, P.C. In Neurosciences Research, vol. 4, in press (Eds. S. Ehrenpreis and O.C. Solnitzky) (1970)

(11) Hebb, C.O. Quart J. exp. Physiol. 40:176 (1955)

(12) Olenev, S.N. Acta Anat. (Basel) 68:68 (1967)

(13) DeLong, G.R., and Coulombre, A.J. Exp. Neurol. 13:351 (1965)

(14) Loewenstein, W.R. Ann.N.Y. Acad. Sc.137:441 (1966)

(15) Potter, D.D., Furshpan, E.J., and Lennox, E.S. Proc.Nat. Acad. Sc. U.S. 55:328 (1966)

(16) Trelstad, R.L., Hay, E.D., and Revel, J. - P. Develop. Biol. 16:78 (1967)

(17) Sheridan, J.D. J. Cell Biol. 37:650 (1968)

(18) Pannese, E. J.Ultrastruct. Res. 21:233 (1968)

(19) Jacob, F., and Monod, J. In Cytodifferentiation and Macromolecular Synthesis (Ed.M. Locke) (1963)

(20) Marchisio, P.C. , and Sjöstrand, J. Brain Research (1970) In press.

(21) Sjöstrand, J., and Karlsson, J.-O. J. Neurochem. 16:833 (1969)

(22) McEwen, B.S., and Grafstein, B. J. Cell Biol. 38:494 (1968)

(23) Marchisio, P.C. Unpublished results (1970)

(24) Filogamo, G. In Symposium of the International Society for Cell Biology, Vol. 8, (Ed.S.H.Barondes) (1969)

THE NUCLEAR-RIBOSOMAL SYSTEM DURING NEURONAL

DIFFERENTIATION AND DEVELOPMENT

C. Radouco-Thomas, Gl. Nosal, and Simone Radouco-Thomas,
Department of Pharmacology, Faculty of Medicine, and
Toxicological Center, Laval University Medical Center,
Laval University, Quebec, Canada

Electron microscopic investigations on the maturing cytoplasmic organelles and on the growth of neuronal processes were formerly reported in several developing neurons (1,2,3,4,5,6,7). In contrast, the differentiation of the nuclear components in the maturing neuron was little investigated. To the best of the authors' knowledge, no systematic ultrastructural study on the nuclear maturation is as yet available.

In a comprehensive review on neuronogenesis, Nosal and Radouco-Thomas (8) have previously described the overall maturation of the nerve cell taking as a model the rat Purkinje cell (Pk). The present paper especially deals with the ultrastructural changes occurring in the nuclear-ribosomal system (NRS) of the immature neuron as compared to the adult nerve cell. The ultrastructural findings are related to the recent data obtained in the field of molecular biology.

MATERIALS AND METHODS

All investigations were performed on Sprague-Dawley albino rats. Fetuses were obtained by caesarean section from timed-pregnant females (16-21 gestation days). Young animals of both sexes were sacrificed at 3, 12, 24, 48, 72 and 120 hr, each day from the 6th to the 12th day and at 15, 18, 21 and 24 days after birth. For comparison, adult rats (90 days old) were used. At the specified age, examinations were done on the same litter and on different litters. A minimum of three animals were checked at each developmental stage.

Fixation with a phosphate buffered (0.1M pH 7.2) 2,5-3% glutaraldehyde was performed by immersion (fetuses) and by perfusion (post-

291

natal rats) according to the procedure previously described (9). Pieces of vermian cortex were postfixed in 1% osmic acid (2 hr), dehydrated and embedded in Epon (10). Thin sections mounted on un-coated copper grids were double-stained with 5% alcoholic uranyl acetate and lead citrate for 5-15 min. Electron micrographs were obtained with a Siemens Elmiskop I A and a Philips EM 300 operated at 60 KV with an objective aperture of 50μ. Direct magnification is indicated for each image.

RESULTS AND DISCUSSION

The described data and corresponding discussion related to the developing and adult neurons are presented under three headings: ultrastructural and molecular aspects and their relationships.

Ultrastructural aspects

During the maturation of the Pk cell, five developmental stages have been previously described (8): undifferentiated cell UC (ante-natal period), early neuroblast EN (neo-natal period), intermediate and late neuroblasts IN, LN (first two weeks) and young neuron (YN).

Nuclear components. In the interphasic nucleus of the adult neuron, we suggest tentatively to distinguish two types of components: (a) well-defined or rather compact structures and (b) an apparently less defined or disseminated material.

The first type is represented by some well organized chromatin structures and the nucleolus proper.

Chromatin -in the interphasic nuclei- has been conventionally identified as "heterochromatin" and "euchromatin". The terms of "condensed" and "dispersed" chromatin have also been proposed (11). On the basis of ultrastructural and macromolecular analysis, we suggest two levels of organization of chromatin which are illustrat-ed by two general models: the "multifiber" and the "monofiber" models (Fig. 11).

The "multifiber" model would correspond to the functionally non-active chromatin which appears as masses (chromocenters) and as a loose fibrillo-granular network, the chromonemata. The ratio between the condensed chromatin and the chromonemata is variable depending on the type of cells. The germ cells have homogenously dense nuclei consisting mainly of condensed chromatin. In some somatic cells such as plasmocytes, oligodendrocytes and even in small neurons, e.g., the cerebellar granule cells, the condensed chromatin is predominant and a "speckled" nucleus is observed. In contrast, in large neurons such as the Pk cell, the "multifiber" model is represented mainly by the reticulated chromonemata and the nucleus appears "vesicular" with

homogenously dispersed chromatin. However, small chromatin masses
(Figs. 1,2,7,10,23) may be observed along the inner nuclear membrane
(marginated chromatin), around and inside the nucleolus (peri- and
intranucleolar associated chromatin). One or two nucleolar satel-
lites, the presumed heterochromatin X chromosome, may also be found
in the female preparation (Fig. 19). Investigations on the interphasic nuclei either <u>in situ</u> or after isolation and disruption (12)
show that several levels of fibrillar organization could be detected: the protochromonema, the subchromonema, the chromonema, the
chromatides and finally the interphasic chromosomes. Figure 11 gives
a bidimensional schematic view of three orders of fibrillar magnitude: small (protochromonema), intermediate (subchromonema) and
large (chromonema) fibers. The chromatin basic element, most frequently observed in electron micrographs, is the subchromonema which
consists of fibrils of 100-150 Å (Figs. 1,5,7,10,19,23). A tridimensional model of the chromatin "multifiber" system is difficult to
work out on the basis of present ultrastructural data.

The "monofiber" model would represent the functionally active
chromatin. It is found in well defined structures such as the nucleolar organizer (NO) or disseminated in the nucleus at the periphery of both kinds of "multifiber" chromatin, namely the chromocenters and the chromonemata. The NO consists in light zones filled
with fibrillar material and located inside the pars fibrosa of the
nucleolus proper (Figs. 1,5,7,10,19,23). The NO has also been observed in the interphasic nucleus of other eucaryote cells (13-16).
The basic element is represented by the protochromonema consisting
of DNA double helix (about 25 Å) which forms, with the attached
proteins, the DNP fibrils of about 50 Å (Fig. 11).

The nucleolus proper in the Pk cell has been described in detail elsewhere (8). It appears generally as a network (nucleolonema)
containing vacuoles and intermingled granular and fibrillar components embedded in a protein matrix. The pars fibrosa is predominantly constituted by fibrils and the pars granulosa mostly contains
granules. A high contrast is visible between the two contiguous
fibrillar zones: the dense pars fibrosa and the light region of the
nucleolar organizer (Figs. 1,23).

The second type of nuclear components in adult neuron would be
represented by disseminated fibrillar and granular material which
includes the loose "multifiber" chromatin network (chromonemata),
some more or less defined granule-like elements (macrogranules and
interchromatin granules) and finally some fibrillar elements. The
macrogranules (mg) i.e., peri- and intrachromatin granules, have a
diameter of about 450 Å and often appear in continuity with trabecules. Both elements show a fibrillar structure consisting in coiled
fibrils of about 30 Å in diameter which also connect them to the
neighbouring "multifiber" chromatin (Figs. 1,2,13,23). The interchromatin granules (ig) appear to form a network composed of fibrillar

and granular material. The particles are of irregular shape and with an average diameter of 200–300 Å (Figs. 1,14,16,19,23). These various granule-like elements have been previously described in the adult (17) and in the developing neuron (8) and are similar to those reported in other interphasic nuclei of animal cells (18-21). The last component of the disseminated material would be the "monofiber" elements which are found at the periphery of the condensed chromatin (20) and of the loose network of the chromonemata (Fig. 23).

The developing neuronal nucleus which displays a gradually increasing complexity represents a useful model for the ultrastructural study of nuclear components. The "multifiber" condensed chromatin which constitutes the speckled nucleus of the UC (Figs. 4,5) exhibits a progressive dispersion during the early neuroblast stage (Figs.6-10) becoming homogeneously distributed as "multifiber" chromonemata vesicular nucleus) in the intermediate neuroblast (Fig. 19). The "monofiber" model is visualized as a nucleolar organizer, firstly as single or few light zones in the UC and EN (Figs. 5-7,10) and shows in the IN and LN several light regions, representing transversal or oblique sections of the nucleolar loop apparatus.

The nucleolus proper undergoes a series of ultrastructural changes, mainly at the onset of neuronal differentiation. In the UC stage (Fig. 5) and at a very early neuroblast period (Figs. 6,7), it appears as a compact mass, apparently only composed by fibrillar elements surrounding the light areas of the presumed NO (Figs. 5,7,10). Soon after, a "segregation-like" phenomena consisting in central fibrils and peripheral granules may transitorily be encountered. This particular repartition of the nucleolar components are found either as the so-called "macro-segregation" reviewed by Busch and Smetana (15) which is characterized by the distribution in three separate zones: the chromatin, the fibrillar and the granular elements; or as a presumably transitory state showing "nucleolar microbodies" (Fig. 10), similar to those described in other normal and drug-treated cells (15). In the intermediate neuroblast (Fig. 19), both components become intermingled to form the adult-type nucleolar network (cf. Figs. 1,23).

Disseminated fibrillar and granular material such as the macro-granules, the interchromatin granules and the "fibrillar elements" are regularly observed in all developmental stages (Figs. 4-10, 16, 19,23).

Furthermore, two additional compact structures have been observed in the differentiating neuronal nucleus. The first one appears to be a coarse, chromatin-like entity designated as "granular mass" (gm) which appears to be transitorily formed at the onset of neuronal differentiation (Figs.6,8). Secondly, the "spotted body"(sb) is a constant component of the maturing nucleus starting from the early neuroblast. The sb exhibits a fibrillar texture resembling the pars fibrosa of the nucleolus. It is postulated that the sb originates from the nucleolar body. This hypothesis may be supported by the sequence of events illustrated in Figs. 15-18, 23 and by similarity of the two components. Further evidence of its probable RNP nature is expected from the cytochemical investigations now under progress. A more detailed study of these two additional nuclear entities has been reported in a previous paper (8). The "spotted body" has also been found in several adult cells (20,22) but the sb is reported for the first time by us in a growing cell. Other nuclear inclusions described in some adult neurons (23,24) have not been observed in immature and mature Pk cells.

Nuclear Envelope. In the adult as well as in the developing neuron, the nuclear envelope (NE) exhibits a basic double-membraned structure (Figs. 1-3,5,7,9,12,20,23). The inner membrane appears somewhat denser than the outer one. The external membrane is irregularly studded by small polysomal clusters (Figs. 3,7,9,20,23). At the early onset of neuronal differentiation, the nuclear envelope forms "blebs" which, in turn, may become detached and released as vesicular elements in the surrounding cytoplasm (Figs. 5,23). The ultimate fate of these vesicles is as yet not elucidated. The blebbing mechanism as a source of membranous organelles has also been postulated during transitory stages of the oocyte differentiation (25,26). Later on, during the EN, IN and LN stages, continuity of this external membrane and presumed newly-formed granular endoplasmic reticulum (GER) cisternae is obvious (Figs. 20,23). Previous observations (8) and literature data confirm this point (6,7).

Pore-complexes are found at irregular intervals along the nuclear envelope. Although granular elements may be present on both sides, they only contain either a fine filamentous material or an annular material as shown in Figures 2,3,5,12,20 and 23. The diagram of Figure 12 illustrates the pore-complex in transverse section as nuclear pores (B) and in tangential sections as annuli (A). The latter are preferentially observed along the nuclear envelope invaginations (NEI) where their fine structure may show some resemblance to the adjacent ribosomal spirals and rosettes (Figs. 2,23). Following microscopical examination, the occurrence and repartition of pore-complexes do not seem to vary significantly during neuronal development.

More or less complex NEI have been detected for the first time in the EN (Figs. 6,7) and appeared preferentially located at the apical pole. They are predominantly filled by a profusion of polysomal clusters (Figs. 2,7,23); in addition, mitochondria in contact with GER profiles are often present at their emergence. This observation suggests that the NEI may be metabolically very active, representing a selective site for ribosomal assemblage. A more detailed study on their morphology and possible functional significance has been previously reported in adult Pk cell (8,9,27) and in the developing neuron (8).

Ribosomal system. The cytoplasmic organization of the adult Pk cell
has been previously described (8,28-30). The striking features are
illustrated in Figures 1 and 3 and outlined in the diagram (Fig. 23).

In the adult neuron, the ribosomal system displays a typical
pattern of distribution in the form of the so-called Nissl bodies
(NB) which are unevenly distributed as relatively small areas among
other organelles and polysomal areas. The constituting GER cisternae
interspersed by free ribosomal clusters, are oriented in parallel
arrays and display a characteristic arrangement as nuclear cap (NC)
in the apical perinuclear region (Figs. 3,23).

In the differentiating neuron, typical changes take place mainly
as regard to the development and distribution of organelles. An increase in number and variety of cytoplasmic components occurs firstly
at the apical pole, subsequently in the lateral perikaryon and finally in the basal region which matures relatively later. A transient
demarcation into two zones appears in the lateral cytoplasm of the
late neuroblast (Figs. 20,21). The central zone is crowded by numerous organelles, especially GER cisternae and accompanying mitochondria and therefore it resembles the more advanced apical region.
In contrast, the peripheral zone is almost exclusively occupied by a
large population of polysomes, very few ER profiles adjacent to the
plasmalemma and rare mitochondria; consequently, it appears somewhat
similar to the immature basal cytoplasm.

During neuronal development, the ribosomal organization follows
a variable course. In the earliest stages, free ribosomes (Rb)
clustered into rosettes and spirals (4-8 particles) are the unique
or the predominant component (Figs. 4,5,7,9). In the subsequent
neuroblast stages, membrane-bound ribosomes increase progressively
until the YN state. Rough-surfaced cisternæ are found at first in
the perinuclear apical cytoplasm of the EN (Fig. 9). Thereafter
the GER undergoes a continuous, progressive proliferation presumably
deriving, at least partly, from the outer nuclear membrane; later
it acquires more complex, branched profiles in the LN (Fig. 20).
It has been hypothesized that these cisterns are produced by an
intrinsic growth of the ER membranes (7,8). The ribosomal system
appears fully developed only in the YN where the previously dispersed
GER cisternae become arranged in an adult pattern as NB and NC
(Figs. 3,23). A close contact between ER profiles and mitochondria
is a constant feature in developing as well as in mature cytoplasm
(Figs. 3,20,23). This topographical relationship may reflect an
association functionally active.

Molecular aspects

The central dogma in molecular biology for the transmission of
genetic information in the interphasic cell is that, after the nuclear transcription process, the various RNAs migrate via the nuclear
pores into the cytoplasm where the ribosomal machinery carries out

the translation of the information necessary for protein synthesis. Figure 22 gives a schematic view of the available data concerning the nuclear RNA species and their relationships with cytoplasmic RNA species.

Ribosomal RNA. Important contributions to the molecular aspects involved in the biogenesis of the ribosomal RNA have been afforded by Birnstiel (31,32), Darnell (33), Mandel (34), Nomura (35), Perry (36,37). According to these data, the transcription occurs in the NO where the r-DNA acts as a template for the ribosomal RNA. Recently, Birnstiel (31,32) and Miller and Beatty (38) have presented biochemical and ultrastructural data which suggest that the NO is represented by a repetitive linear sequence of basic units: each unit consisting of a r-DNA transcription subunit and an inactive spacer DNA. A high degree of redundancy or amplification occurs in the NO.

The 18 S gene and 28 S gene, which are intermingled with each other along the DNA, are transcribed as an initial large 45 S RNA-precursor molecule. This nascent r-RNA is methylated (2'-0-methyl-ribose) and become associated with proteins very soon after or even concomitant with its synthesis. The complete 45 S r-RNA is released and broken down into two ribosomal particles which, after maturation, will become the 40 S small subunit (18 S RNA) and the 60 S large subunit (28 S RNA) of the ribosome.

During their intranuclear maturation, the two ribosomal RNAs associate with more proteins and make progressive conformational changes starting with the mono-dimensional (primary) and terminating with the tridimensional (tertiary) structure. However, whereas the 18 S molecule is rapidly released and exported to the cytoplasm (18 S cytoplasmic pool), the intranucleolar maturation of the 28 S molecule is of longer duration and involves the successive formation of several transitory RNA molecules of decreasing S values (41 S, 36 S, 32 S, 28 S RNAs). About 40% of the initial RNA, mainly non-methylated portions, is lost in the maturation process. The 5 S RNA which associates with the new 60 S molecule is not believed to be transcribed in the NO.

The new ribosomes enter the cytoplasm not as whole ribosomes but as subunits. Although the nuclear preribosomal 18 S RNA and 28 S RNA particles have approximately the same sedimentation behaviour as the mature units, they might be incomplete with respect to proteins. Anyhow, the final cytoplasmic stages of ribosomal maturation apparently succeed rather rapidly. The mature small 40 S subunit (18 S RNA) and large 60 S subunit (28 S RNA + 5 S RNA) combine to constitute the functional unit in protein synthesis (80 S ribosome), when they attach themselves onto a strand of m-RNA.

Other RNAs. Besides the r-RNA, the nucleus also provides templates for other RNAs to be transported into the cytoplasm: the t-RNA which

like the r-RNA has a role to play in the synthesis of every protein
and the m-RNA which accomplishes the actual dictation of amino acid
sequence. A nucleoplasmic and cytoplasmic heterogenous RNA has been
described but its role and function are still unknown.

Relationship between morphological and molecular data

In Figure 23, we shall try to interpret the ultrastructural
data obtained in the adult and developing neuron in the light of the
recent cytochemical (11,15,20) and biochemical findings (31-38).
Three kinds of structures will be considered, DNP- and RNP-containing structures as well as the disseminated presumed DNP-RNP material.

DNP-containing structures. As nuclear DNA, both the "monofiber"
(euchromatin) and "multifiber" (heterochromatin) types have been
systematically observed (Fig. 11). Only the first type which represents the functionally active DNA will be discussed. We suggest
that the r-DNA cistrons and thus the r-RNA transcription would be
located in the well defined "monofiber" system in the nucleolar
organizer (Figs. 1,22,23). The NO is evident in all stages of neuronogenesis and is in close contact with the dense fibrillar part of
the nucleolus proper (Figs. 5,7,10,19).

The DNA cistrons synthesizing the other RNAs could correspond
to the disseminated "monofiber" chromatin located at the periphery
of the "multifiber" system of both dense chromatin and chromonemata
(Figs. 22,23). Other studies (20,39,40) suggest also that the periphery of the condensed chromatin may play a role in transcription.

Concerning a possible extramitochondrial cytoplasmic DNA, no
data are available at the present time in the developing and adult
neuron. During neuronogenesis (8), nucleoplasmic vesicles may be
observed by evaginations of the nuclear envelope (Figs. 5,23).
Similar data have been reported during the development of other
cells (25,26,41). Investigations are now under progress in our
laboratory in an attempt to establish the structure, the function
and the fate of this nuclear migrating material in the developing
and possibly in the adult perikaryon. It could be speculated that
this material represents a local DNA template for a cytoplasmic RNA
synthesis,presumably located in the internal membranous system
(e.g., the granular endoplasmic reticulum).

RNP-containing structure. The two main compact corpuscular bodies:
the nucleolus proper and the "spotted body"would be RNP carrying material. The nucleolus, constituted mainly of pars fibrosa and pars
granulosa, is the essential site of maturation of r-RNA from nascent
45 S RNA till 40 S and 60 S RNP particles (Figs. 22,23). During the
earliest stages of neuronogenesis, the fibrillar part is the only
one present and is in contact with the NO (Figs. 5,7). The granules,
at first located at the periphery, become thereafter disseminated in
the nucleolus (Figs. 10,19). The "spotted body," either attached to

the nucleolus or free in the nucleoplasm, is always of fibrillar texture (Figs.15-18,23).No relationship between the synthesis of the various RNAs and the sb has as yet been established.

The disseminated DNP-RNP material corresponds to the already mentioned "multifiber" chromonemata and the "monofiber" chromatin as well as to the RNP fibrillar and granular elements distributed throughout the nucleoplasm (Fig. 23). Recent cytochemical findings (11,20,21) carried out in mammalian cells have shown that the macro-granules are RNP-containing structures and that the interchromatin granules would represent a special class of RNP with well-protected RNAs. It has been suggested that the fibrils located at the periphery of condensed chromatin would also be RNP-carrying elements and that they are the morphological expression of the RNA transcription. (20) As working hypothesis we suggest that -in the neuron- the disseminated DNP-RNP material would represent the extranucleolar site for transcription and maturation of various RNAs except for r-RNA (Figs. 22,23).

The fibrillar material found in the nuclear pores would be migrating RNP material (Figs. 2,3,5,22,23). As above-mentioned, the NEI could represent a selective site for the assembly of free polysomes (Figs. 2,6,7,23).

The ribosomal system (Figs. 3,9,20-23) constituted of free and attached ribosomes may have separate functions in the immature (8) and in the mature neuron (42). The free ribosomal component which appears at first (Figs. 4,5) may be devoted to protein synthesis related to the cell body development and maintenance whereas the membrane-bound ribosomes (Figs.20,21) which mostly proliferate at the time of growth of the neuronal processes may be specifically involved in the conduction and transmission of the nerve impulse.

This research was supported by Medical Research Council MT-1438.

REFERENCES

1. Wechsler, W. Z. Zellforsch. 74: 401 (1966)
2. Wechsler, W. Exp. Biol. Med. 1: 153 (1967)
3. Meller, K.,Eschner,J.,and Glees,P. Z.Zellforsch. 69: 189 (1966)
4. Karlsson, U. J. Ultrastruct. Res. 17: 158 (1967)
5. Kornguth, S.E., Anderson, J.W., and Scott, G. J. Comp. Neurol. 130: 1 (1967)
6. Caley, D.W., and Maxwell, D.S. J. Comp. Neurol. 133: 17 (1968)
7. Pannese, E. J. Comp. Neurol. 132: 331 (1968)
8. Nosal, Gilliane, and Radouco-Thomas, C. Proc. 1st Int. Symp. Cell Biology and Cytopharmacology, Venice 1969 (in press)
9. Nosal, Gilliane, Chouinard, L.A., and Radouco-Thomas, C. Can. J. Zool. 45: 17 (1967)
10. Luft, J.H. J. Biophys. Biochem. Cytol. 9: 409 (1961)

11. Bernhard, W., and Granboulan, Nicole in The Nucleus (Eds. A.J. Dalton and F. Haguenau) p. 81. New York, Academic Press (1968)
12. Ris, H. in The Interpretation of Ultrastructure (Ed. R.J.C. Harris) p. 69. New York, Academic Press (1962)
13. Lafontaine, J.G. in The Nucleus (Eds. A.J. Dalton, and F. Haguenau) p. 151. New York, Academic Press (1968)
14. Lord, A., and Lafontaine, J.G. J. Cell Biol. $\underline{40}$: 633 (1969)
15. Busch,H., and Smetana,K. The Nucleolus. N.Y. Acad. Press (1968)
16. Chouinard, L.A. J. Cell Sci. $\underline{6}$: 73 (1970)
17. Radouco-Thomas, C., Nosal, Gilliane, and Chouinard, L.A. Proc. Can. Fed. Biol. Soc. $\underline{10}$: 122 (1967)
18. Watson, M.L. J. Cell Biol. $\underline{13}$: 162 (1962)
19. Swift, H. Exptl Cell Res. Suppl. $\underline{9}$: 54 (1963)
20. Monneron,A., and Bernhard,W. J. Ultrastruct. Res. $\underline{27}$: 266 (1969)
21. Bernhard, W. Natl. Cancer Inst. Monographs $\underline{23}$: 13 (1966)
22. Hardin, J.H., Spicer, S.S., and Greene, W.B. Anat. Rec. $\underline{164}$: 403 (1969)
23. Siegesmund,K.A.,Dutta,C.R.,and Fox,C.A. J.Anat. $\underline{98}$: 93 (1969)
24. Sotelo, C., and Palay, S.L. J. Cell Biol. $\underline{36}$: 151 (1968)
25. Scharrer, B., and Wurzelman, S. Z. Zellforsch. $\underline{96}$: 325 (1969)
26. Baker, T.G., and Franchi, LL. Z. Zellforsch. $\underline{93}$: 45 (1969)
27. Nosal, Gilliane, Chouinard, L.A., and Radouco-Thomas, C. Z. Zellforsch. 1970 (in press)
28. Herndorn, R.M. J. Cell Biol. $\underline{18}$: 167 (1963)
29. Palay, S.L. in Brain Function II (Ed. M.A.B. Brazier) p. 69. Los Angeles, Univ. Calif. Press (1964)
30. Peters, A., Palay, S.L., and Webster, H.deF. The fine structure of the nervous system. New York, Hoeber (1970)
31. Birnstiel, M. Ann. Rev. Plank Physiol. $\underline{18}$: 25 (1967)
32. Birnstiel, M., and Chipchase, M. Science J. $\underline{6}$: 41 (1970)
33. Darnell, J.E. Bacteriol. Rev. $\underline{32}$: 262 (1968)
34. Mandel, P. Proc. 1st Int. Symp. Cell Biology and Cytopharmacology, Venice 1969 (in press)
35. Nomura, M. Scient. Am. $\underline{221}$, 4: 28 (1969)
36. Perry, R.P. Progr. Nucleic Acids Res. $\underline{6}$: 219 (1967)
37. Perry, R.P. in Handbook of Molecular Cytology (Ed. A. Lima-de-Faria) p. 620. New York, Elsevier (1969)
38. Miller, O.L., and Beatty, B.R. J. Cell Physiol. $\underline{74}$, Suppl. 1: 225 (1969)
39. Allfrey, V.G., Pogo, B.G.T., Pogo, A.O., Kleinsmith, L.J., and Mirsky, A.E. in Histones, Ciba Found. Study Group no. $\underline{24}$: 42. London, Churchill (1966)
40. Gall,J.G.,and Collan,H.G. Proc. Natl Acad. Sci. $\underline{48}$: 562 (1962)
41. Hadek, R., and Swift, H. J. Cell Biol. $\underline{13}$: 445 (1962)
42. Eschner, J., and Glees, P. Experientia, $\underline{19}$: 301 (1963)

EXPLANATION OF FIGURES

Figs. 1-3. Adult neuron: 90 p.n. day. (1) Nucleolus near the nuclear envelope (NE) exhibiting the presumed nucleolar organizer (→) dense fibrillar (f) and granular (g) regions; "multifiber" chromatin (Ch); macrogranules (⇒); mitochondria (M). 20.000X. (2) Nuclear envelope (NE) invagination tangentially cut with a polysomal content. Nuclear pore-complexes in transverse section with fibrillar material (see inset) and in tangential section, annuli (an) resembling adjacent ribosomal spirals and rosettes. Macrogranules (→). 12.000X. (3) Apical perinuclear cytoplasm with nuclear cap (NC). Nuclear pores with annular and fibrillar-like material inside (—) and with particulate components on both sides. 12.000X.

Figs. 4-5. Undifferentiated cell (UC): 20 days fetus. (4) Portion of three UC cells showing the typical clumping of heterochromatin (Ch) and the perinuclear rim of cytoplasm with clusters of free ribosomes (Rb). Restricted extracellular spaces. 3.000X. (5) Part of an UC: compact nucleolus with presumed nucleolar organizer (→) or r-DNA surrounded by dense fibrils (f); marginated chromatin (mCh); chromonemata (Cht); macrogranules (⇒). Note blebbing of the NE leading to nuclear-like inclusions (—) and lucid vesicles (b). Nuclear pore (←), cytoplasmic ribosomal clusters and vesicles (V); plasmalemma (P). 6.000X.

Figs. 6-8. Early neuroblast (EN): 3 hours after birth. (6) Overall view of the nucleus showing a more homogenous chromatin distribution, the presence of two maturing nucleoli (Nu), of an additional granular mass (gm) and of a nuclear envelope invagination (NEI). 6.000X. (7) Apical side. Compact nucleolus with NO and fibrils(f) separated by heterochromatin (Ch) from the NEI which contains numerous polysomes. 10.000X. (8) Detail of the granular mass (gm) with two "macrogranules" (→). 10.000X.

Figs. 9-10. More advanced EN: first neo-natal day. (9) Rough-surfaced cisternae (⇒) appearing in the perinuclear apical cytoplasm (AP). Note a pseudo-interruption of the NE and dense trabecules apparently crossing transversely the nucleo-cytoplasmic boundary (→). Extensive extracellular spaces (Ex); plasmalemma (P). 8.000X. (10) Detail of a more advanced nucleolus in the vicinity of a NEI. The NO (→) partly encircled by dense fibrils (f). Note the particular arrangement referred to as "nucleolar microbodies" transitorily found in the maturing nucleolus. Associated chromatin (ACh). 20.000X.

Fig. 11. Suggested models for the organization of chromatin in the interphasic nucleus of neurons and other mammalian cells. (A) "Monofiber" model found in the active derepressed site of chromatin (euchromatin) such as the nucleolar organizer and the perichromatin fibrillar regions. (B) "Multifiber" model found in the functionally inactive chromatin (heterochromatin) such as the condensed chromatin (chromocenters) and chromonemata.

Fig. 12. Schematic diagram of a pore-complex of the neuronal nuclear envelope (NE). (A) Tangential section showing the annuli. (B) Transverse section illustrating the external and internal membranes (Ex M,IM) as well as the intrapore presumed annular material.

Figs. 13-19. Nuclear components in developing and mature neurons. (13) Nucleoplasmic area containing several macrogranules, some of them being in continuity with trabecules (⟹). 12.000X. (14) Cluster of interchromatin granules apparently forming a fibrillo-granular network. 20.000X.   (15) Part of a spotted body (sb) attached to the nucleolus. 12.000X.   (16-17) Typical sb attached to the nucleolus, exhibiting a fibrillar texture. 15.000X and 12.000X.   (18) A sb free in the nucleoplasm. 15.000X. (19) Intermediate neuroblast: 5 p.n. day (female rat). Loosely packed nucleolus showing an ultrastructural organization resembling that of the adult state: NO (→); intermingled fibrils (f) and granules (g) forming the nucleolonema. Associated chromatin and nucleolar satellite (NS); macrogranules (⟹).   12.000X.

Figs. 20-21. Late neuroblast: 12 p.n. day. Lateral cytoplasm with a clear-cut demarcation in the distribution of the organelles. (20) The central zone (CZ) is crowded by rough-surfaced cisternae, some of them in close contact to mitochondria (→); long branched GER profiles more centrally located. Note continuity between the outer NE and rough-surfaced cisternae (⟹).   (21) In contrast, the peripheral zone (PZ) is mostly occupied by free polysomes, with only a few profiles in the vicinity of the plasma membrane (—). Plasmalemma evaginations (PE) and adult-type synapse with vesicles (Sy).   12.000X.

Fig. 22. Molecular aspects of the nuclear-ribosomal system in the eucaryote interphasic cell. Diagrammatic representation of the current concepts concerning the transcription, maturation and transport of the nucleolar and extranucleolar RNAs.

Fig. 23. Schematic diagram of the nuclear-ribosomal system in the mature and developing neuron. (1) DNP-containing structures. "Multifiber" chromatin (heterochromatin) represented by the reticulated chromonemata (Cht) as well as the condensed chromatin: marginated chromatin (mCh) nucleolus-associated chromatin (ACh) and nucleolar satellite (NS). "Monofiber" chromatin (euchromatin) as nucleolar organizer (NO). (2) RNP-containing structures. Nucleolus proper with pars fibrosa (PF) and pars granulosa (PG) and the "spotted body"(sb). (3) Disseminated DNP-RNP material (dfg): macrogranules (mg), interchromatin granules (ig) and perichromatin fibrils (f); (4) Nuclear envelope showing blebbing (b), invaginations (NEI),pore-complexes with annuli (an) and nuclear pores (np) filled with fibrillar (f) and annular material (am). (5) Ribosomal system with free and attached ribosomes forming the Nissl bodies (NB) and the nuclear cap (NC).

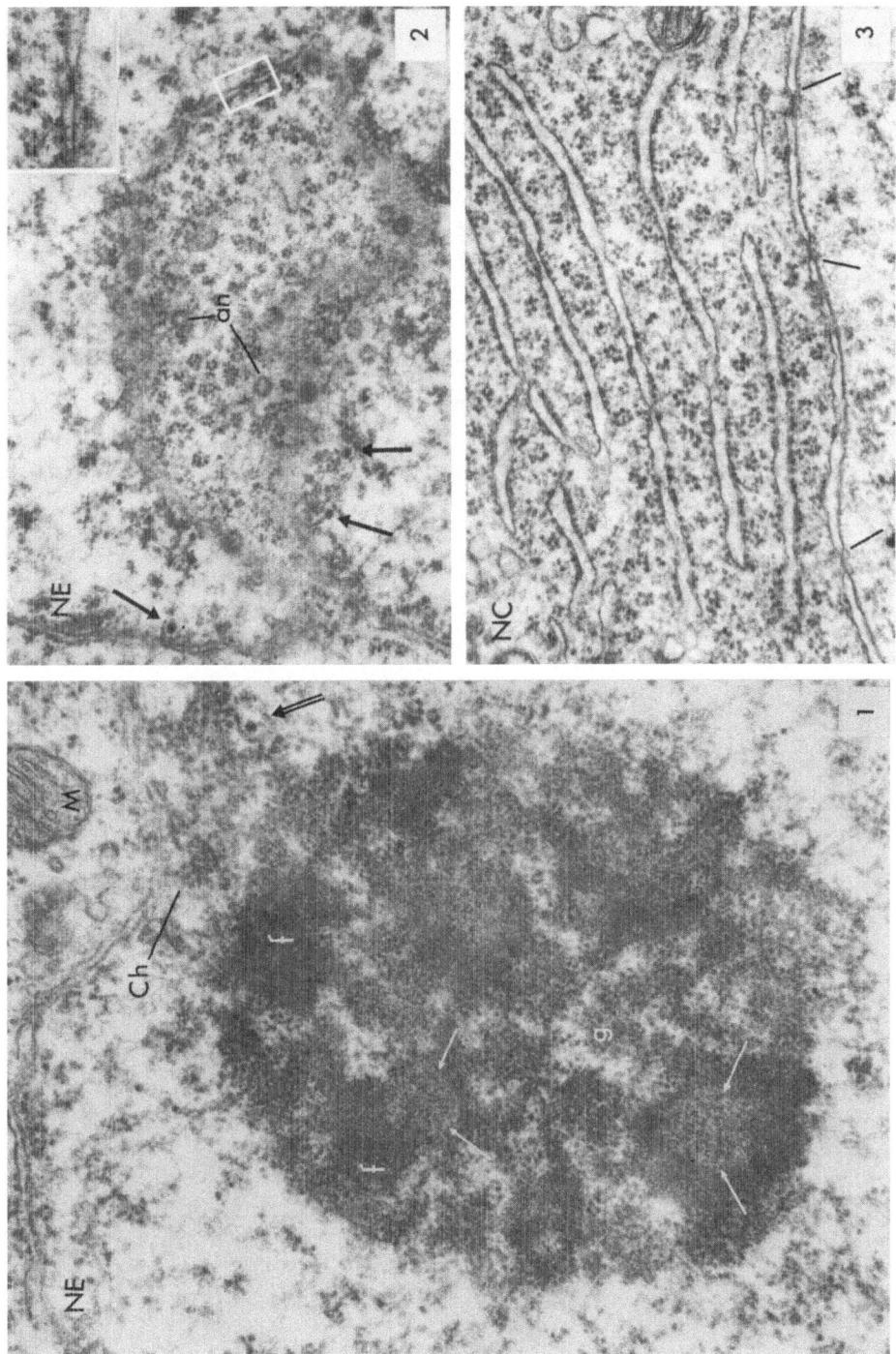

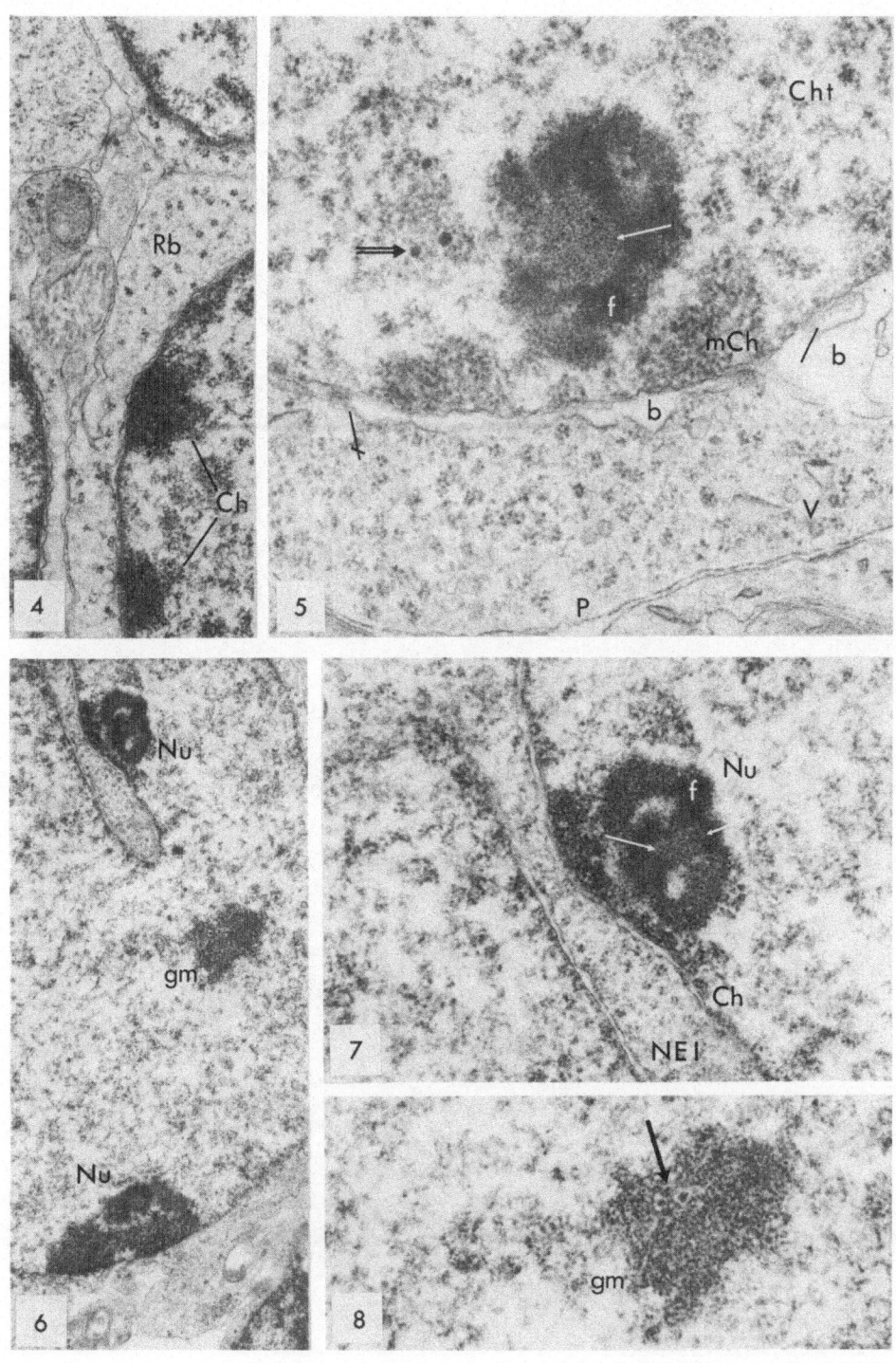

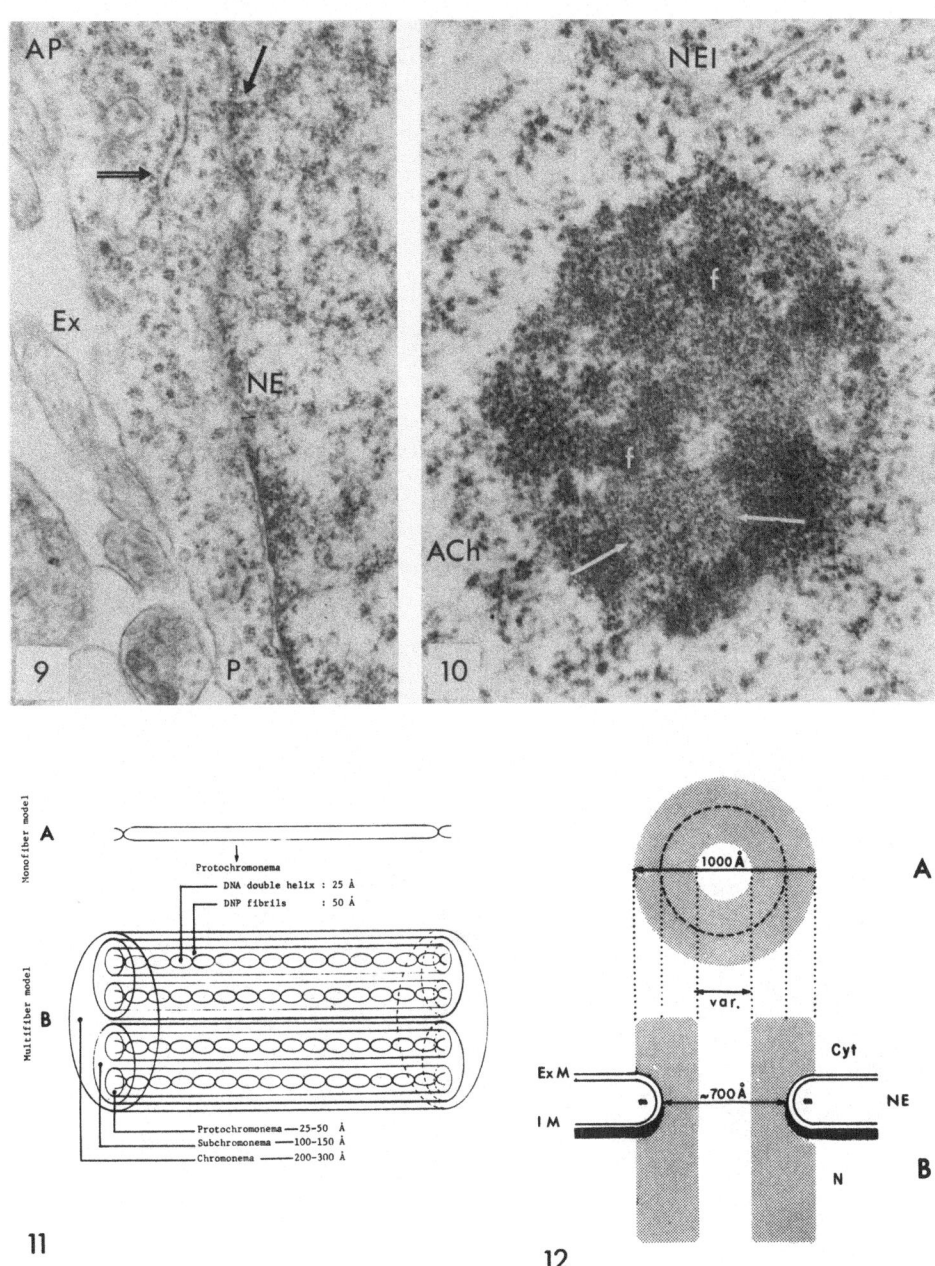

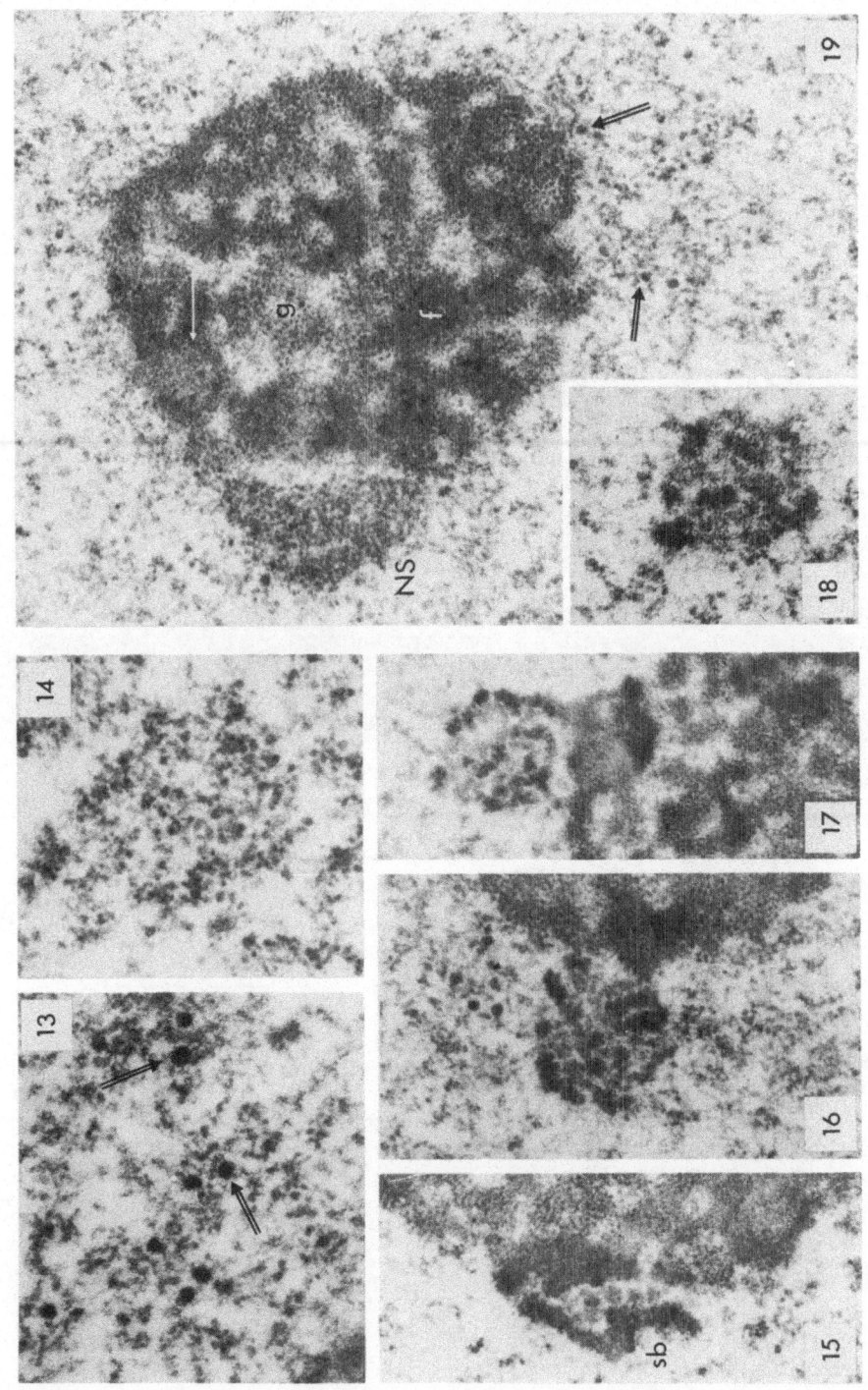

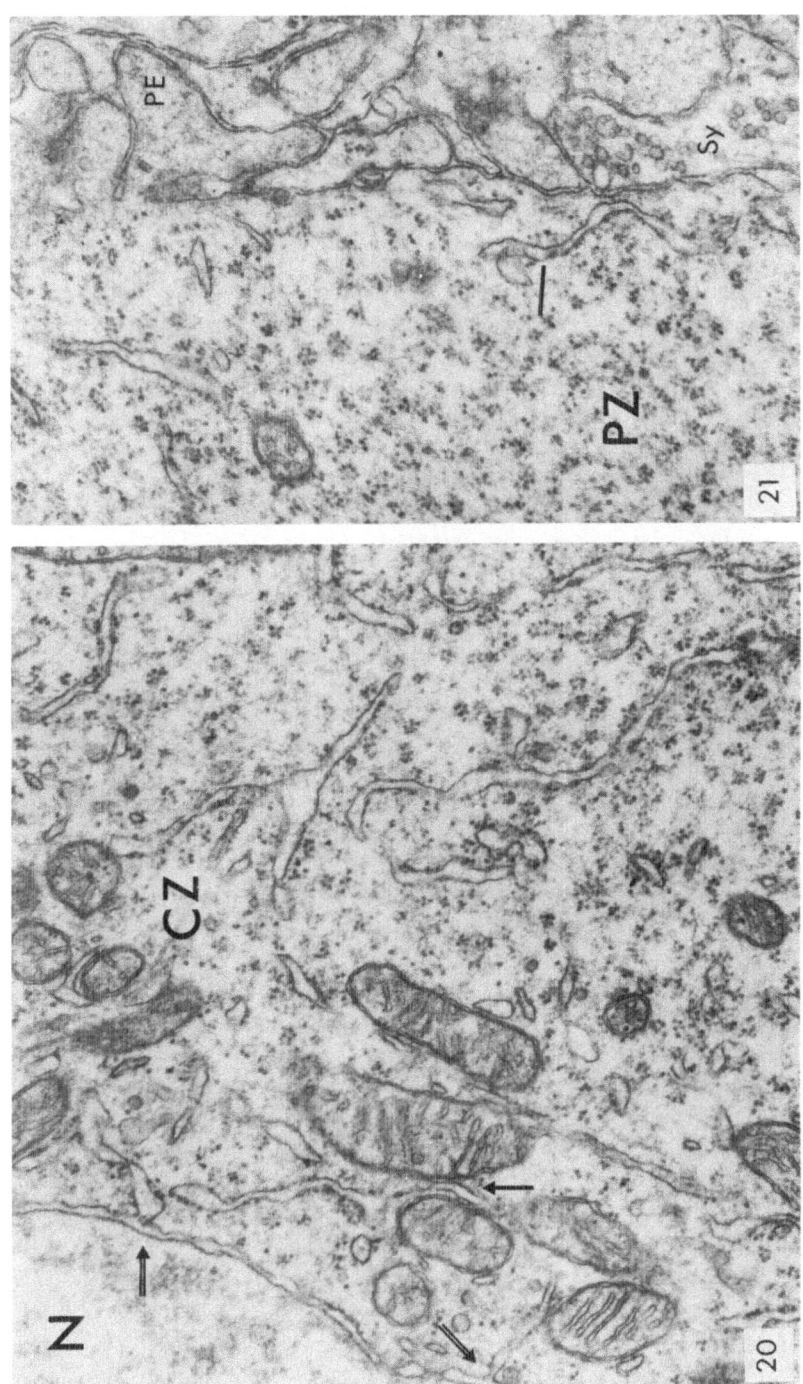

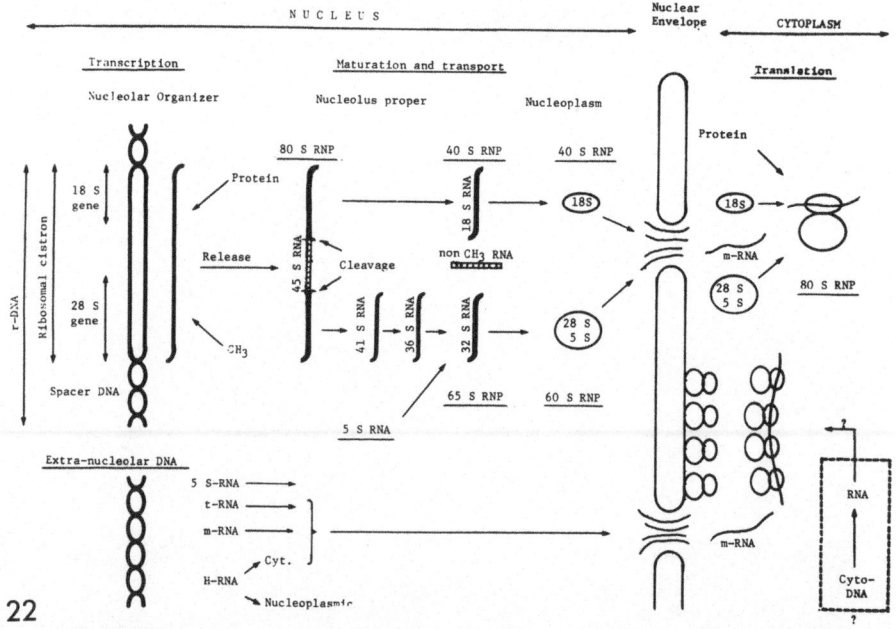

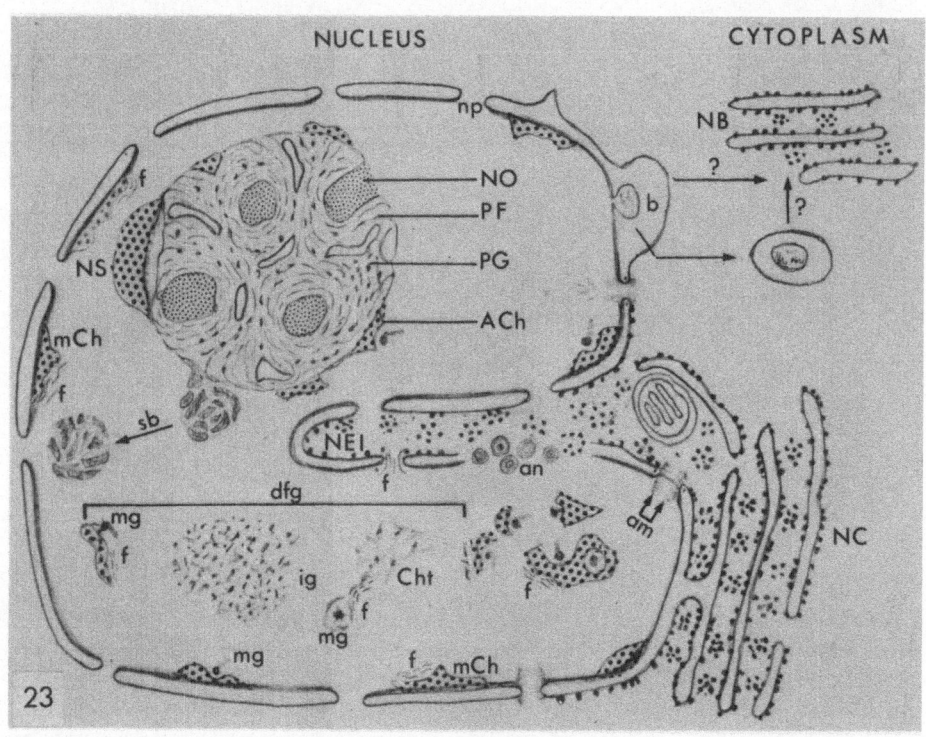

SECTION III

MEMBRANE FORMATION AND FUNCTION

STRUCTURE OF CELLULAR MEMBRANES AND REGULATION OF THEIR LIPID
COMPOSITION

George Rouser

City of Hope National Medical Center

Duarte, California 91010

ABSTRACT

Two lipid class substitution groups have been defined. Members
of a group can replace each other in membranes. Similar substitution
groups are found for animal cells, fungi, and bacteria, although a
lipid class that occurs in many organisms may be completely absent
from others. Myelin is a membrane that has the maximum amount of lipid
(fully packed) and essentially all of the protein appears to bind
polar lipid. Other membranes contain less lipid and more protein,
some of which does not appear to bind lipid. In membranes, lipid is
postulated to lie upon the protein to which it is attached by apolar
bonding of carbon chains as well as ionic and hydrogen bonds through
polar groups. Lipid molecules lie close to each other and interact
through their carbon chains (apolar bonding). In myelin, cholesterol
probably binds to acidic lipids. The maximum cholesterol level is
seen in myelin and the human red blood cell in which one-half the
molar amount of cholesterol equals the amount of acidic lipid (acidic
phospholipids, phosphatidyl ethanolamine, sulfatide, and ganglioside).
It is thus postulated that two molecules of cholesterol can bind to
one of acidic lipid. The cross section of the membrane is postulated
to consist of four layers of lipid and two of protein in the sequence
lipid-protein-lipid-lipid-protein-lipid. In myelin which is maximally
packed with lipid, the four layers are completely filled. In other
membranes with less lipid and lipid-binding protein, the lipid layers
are incompletely filled. In the human red blood cell, lipid appears
to cover only about one-third of the protein surface. Membrane permeability to many ionic substances appears to be related to the total
amount of lipid, those membranes having the most lipid being the
least permeable. It is suggested that the blood-brain barrier exists
because plasma membranes of cells of the nervous system contain more

311

lipid than those of other organs and are thus more slowly penetrated
by some ionic substances. The different lipid compositions of subcellular particulates of the same cell type and from different organs
is correlated with the relative proportions of the different types of
membranes. It is postulated that, during differentiation, levels of
each type of subcellular particulate are set and that thereafter
lipid composition changes follow a similar course in all organs. Some
changes in membrane lipids in abnormal states are discussed.

I. INTRODUCTION

Recent data (summarized in ref. 1) have cast doubt upon the
validity of the classical Davson-Danielli model of the cell membrane and different models have been proposed (reviewed in ref. 2).
In the new models, the classical model of a bilayer between two
layers of protein with lipid-lipid interaction through carbon chains
and lipid-protein interaction through polar groups is modified in
various ways. Recognition of the lipid class substitution groups (3),
new data on lipid removal from the red blood cell (23), and re-
examination of data from the literature led us to the development of
a new model for cell membrane structure. The development of the
model, the way it correlates with data from the literature, and a
postulate concerning the means by which membrane lipid composition
is determined are described in this report.

II. LIPID SUBSTITUTION GROUPS AND
LIPID-BINDING MEMBRANE PROTEINS

Polar lipids fall into two general substitution groups (3).
Members of each group can substitute for (replace) other members of
the same group and thus bind to the same membrane protein sites.
The same substitution groups are found for whole organs and subcellular particulates of vertebrates, invertebrates, and fungal
mycelia. The only major qualitative differences among animal species
are the presence of ceramide aminoethylphosphonate and ceramide
phosphorylethanolamine that replace sphingomyelin in some invertebrates (3,4).

Since all subcellular particulates have the same substitution
groups, it is possible to correlate lipid substitution groups with
the data for lipid-binding proteins of any subcellular particulate.
Myelin is an ideal membrane for this purpose and its protein composition has been determined (5). Myelin appears to contain acidic,
basic, and neutral proteins all of which probably bind lipid. The
tetramer model for lipid-binding membrane protein offers an explanation for the variable composition noted for myelin proteins and
for the existence of five subgroups for each lipid class substitution group (3). Myelin proteins from different parts of the
nervous system are different (6). It seems probable that lipid-

binding proteins from other organs may differ in composition and properties even though they appear to have the same lipid-binding specificity as judged by the uniformity of the substitution groups (3).

The substitution groups differ in the nature of the polar groups. This plus the fact that, whereas all membranes use the same substitution groups, the fatty acid composition of lipid classes of different membranes may differ widely, points to high specificity for polar groups and little or no specificity for particular carbon chains.

Although lipids appear to bind only to certain sites on proteins under normal conditions, lipid composition data for human erythrocytes (reviewed in ref. 7) indicate that, in some abnormal states, lipid may bind to protein that does not normally bind lipid. Most membranes contain less lipid than myelin and thus probably have less lipid-binding protein. Also, most membranes have a more complex protein pattern (8) than myelin, although only three different proteins were observed in microsomes from influenza-transformed cells (8).

III. LIPID CLASS SEQUENCES IN MEMBRANES

Data obtained by physical methods indicate that lipid-lipid interaction through carbon chains is extensive (12,13). It thus appears that lipid molecules must lie very close to each other. Since substitution group I makes up over 50% of the total, some group I molecules must lie next to each other. Diphosphatidyl glycerol is a double molecule and thus would replace two molecules of the other lipid classes. Since diphosphatidyl glycerol appears to be a component of substitution group II only, it would appear that group II molecules also occur next to each other. This suggests a general sequence of the type

$$I\ I\ I\ II\ II\ I\ I\ I\ I$$

Divalent ions (calcium, magnesium) appear to be important in stabilization of membrane structure (9,14). The acidic phospholipids of group II all form relatively poorly dissociated divalent ion salts. The salts of sulfatide dissociate freely. Since the phospholipids with amino groups (phosphatidyl ethanolamine, ceramide phosphorylethanolamine, and ceramide aminoethylphosphonate) are the only members of group I that form relatively stable divalent ion salts, placement of the lipids with free amino groups as in the sequence

$$I\ I_A\ I\ I\ II\ II\ I\ I\ I_A\ I$$

in which A = amino group lipids gives a structure with the lowest

level of phosphatidyl ethanolamine in which a molecule not capable
of forming a relatively stable divalent ion salt appears next to
one that does. The amino lipids make up 30-100% (in some bacteria)
of group I. Thus, even at the lowest level, the flanking by
molecules capable of forming stable divalent ion salts can be
achieved. The exceptions to this are human aorta and organs in
Niemann-Pick disease (see Section VIII). Alternatively, if each
lipid layer is two molecules thick (see below), the more strongly
bound lipids (acidic lipids including phosphatidyl ethanolamine)
could bind to protein and most of the choline lipid, cerebroside,
and cholesterol bind to the acidic lipids to form the second layer
of molecules.

IV. CHOLESTEROL BINDING IN MEMBRANES

Sterol (cholesterol) is present in animal cells, although it
is absent from most bacteria and may be present in membranes of
some fungi only when it is supplied in the medium (10). Cholesterol
is found in largest amount in plasma membranes and their derivatives
such as myelin. Since cholesterol does not appear to replace polar
lipids in membranes (3), it appears that cholesterol is bound to
polar lipids rather than directly to proteins in agreement with
the suggestion of Finean extended by Vandenheuvel (11). We postulate that two molecules of cholesterol can bind to one molecule of
some lipids. The lipid composition of myelin and red blood cells
indicates that cholesterol is normally bound only to acidic lipids
(acidic phospholipids including phosphatidyl ethanolamine as well
as sulfatide and ganglioside) because one-half the molar amount of
cholesterol is very close to the amount of acidic lipid (ratios
0.97 to 1.02). Some membranes contain less cholesterol and thus
appear not to bind the maximum amount possible. Conversely, in
some pathological states, red blood cell cholesterol is elevated
(7) and binding to other polar lipids is indicated. Binding of
different amounts of cholesterol may be an important means for
changing membrane permeability.

V. POSITION OF LIPID RELATIVE TO PROTEIN

Physical methods (nuclear magnetic resonance, infrared spectro-
photometry, and differential scanning calorimetry) indicate apolar
bonding in both lipid-lipid and lipid-protein interaction (12,13).
Also, a large mass of data exists that demonstrates divalent ions
(calcium,magnesium) to be important in membrane structure (9,14).
The ions appear to be associated with lipids as judged by the ease
of removal of lipid with complexing agents such as ethylenediaminetetraacetate. The mode of interaction of an acidic phospholipid can
be visualized from the side and from above as

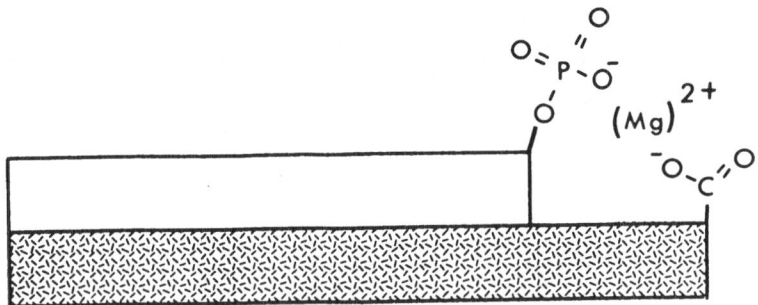

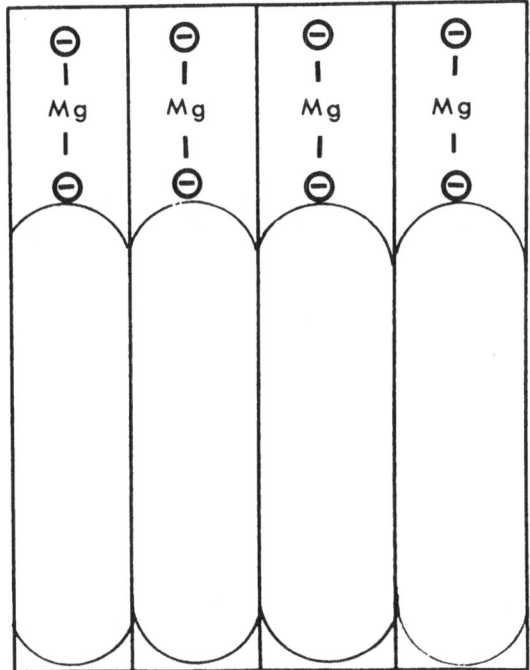

Electron microscopy has shown all membranes to give two dark staining lines. These lines are from protein since they are present after extraction of lipid (15,16). Thus, the presence of two protein layers seems established. The lipid/protein (wt/wt) ratio of myelin is about 3/1. Myelin contains about 23% (by wt) cholesterol that appears to bind to lipid rather than protein as noted above. Conversion to a vol/vol ratio with values of 0.9 and 1.3 for lipid and protein densities gives a polar lipid/protein ratio of 4.3/1 for myelin. Thus, myelin has enough lipid to make four tightly packed layers, each layer being two molecules thick.

A cross-section view of the fundamental unit of myelin can thus be visualized as

| Lipid |
| Lipid-binding protein |
| Lipid |
| Lipid |
| Lipid-binding protein |
| Lipid |

In the model, lipid coats both sides of each protein layer. Such a tightly packed structure is compatible with the poor permeability and insulator properties of myelin.

The human red cell ghost has a lipid/protein wt/wt ratio of 1.4 (23) and the cholesterol content is about 25% (by wt). Thus, there is enough lipid to fill only about one-third of the space in the myelin model. This is in agreement with the finding that red cells have only enough lipid to make 1.4 maximally packed lipid layers (17). The red cell model can be visualized as

	Lipid
	LBP
Protein	Lipid
(no bound lipid)	Lipid
	LBP
	Lipid

in which LBP is lipid binding protein.

There is much evidence that polar groups of both lipid and protein are available for interaction at the surface of the red cell membrane. Thus, the negative charge of the red cell membrane is lost by treatment of intact cells with neuraminidase which removes completely the sialic acid moiety of a glycoprotein at the cell surface (18). This protein is easily removed from the membrane without removal of lipid (19), as are some other membrane proteins (20). Free polar groups of proteins have been reacted with various chemical reagents (21). Phospholipids of the red cell ghost can be degraded by phospholipase C which splits polar groups from the lipids and leaves the carbon chains attached to the membrane (22). Glycolipids appear to be a part of the blood group system and thus at least some of their polar groups should be present at the surface of the cell.

Turner and Rouser (23) found that up to one-half of the lipid of red cell ghosts could be removed by washing with aqueous solutions containing ethylenediaminetetraacetate without loss of ghost protein. This plus the observation that up to 25-35% of the lipid of intact red cells can be lost by in vitro incubation (24,25) suggests, in agreement with the proposed model, that one-quarter of the lipid is located on the outer surface and one-quarter on the inner surface with the other one-half between the two layers of protein. In our model, the lipid is not distributed uniformly over the surface of red cell membranes. Murphy (26) reported data indicating nonuniform distribution. He introduced labeled cholesterol into red cells (by exchange) in vivo and by in vitro incubation without an increase in the amount of cholesterol. Autoradiography disclosed the label to be confined to the periphery of the cell. Some membranes have even lower lipid/protein ratios and thus probably have even less lipid-binding protein in their membranes than the red cell ghost.

Some data indicate that membranes which contain more lipid are less permeable to some polar substances than those with less lipid. Thus, myelin appears to be rather impermeable in vivo, and red cells in various pathological states where the lipid is increased are less permeable to ions than normal cells (24,25). Conversely, bacterial cells grown in a vitamin deficient medium have lower than normal lipid content and cannot concentrate amino acids, but when the lipid deficiency is repaired by incubation under conditions where the organisms do not grow, the permeability defect is no longer apparent (27).

Since a higher lipid content appears to be associated with decreased permeability, it seems possible that the blood-brain barrier, a term used to describe the slow entry of some ionic substances into brain, could arise from the high lipid content of plasma membranes in the nervous system. It is apparent that glia cell plasma membranes must be like myelin since myelin is an extension of the cell surface membrane. The same high lipid content seems probable for neurones as well. Also, the blood-brain barrier is established before myelination begins. It seems unlikely that brain capillaries are a part of the blood-brain barrier because they have the same lipid content and lipid class distribution as capillaries (glomeruli) from the kidney (29).

TABLE I

SIMILARITY OF LIPID ANALYSES OF DIFFERENT TYPES OF CELLS AND ORGANS[1]

	GROUP I		GROUP II					
	Rat heart	Rat brown fat	Bovine aorta	Bovine capil-laries	Human RBC	Human brain[4]	Human brain[5]	Lympho-cytes[6]
PC[2]	39.3	38.5	31.8	32.7	32.7	28.0	32.2	32.3
PE[3]	34.0	32.3	24.4	25.5	25.6	28.0	32.2	29.6
Sph	3.4	3.0	21.4	21.2	23.4	21.2	17.5	22.8
PS	3.3	2.7	10.2	10.2	11.6	12.2	12.3	11.0
PI	3.6	3.7	5.1	4.2	1.3	1.6	2.2	0.1
DPG	11.5	11.6	1.7	1.0	ND	0.09	0.09	0.06

	GROUP III					GROUP IV	
	Mouse liver	Tumor cells[7]	Neurospora crassa[8]	Bovine liver	Mouse skeletal muscle	Rat kidney	Bovine spleen
PC	46.9	45.3	44.2	52.1	51.8	35.4	37.1
PE	27.3	27.8	24.2	23.6	23.6	27.6	25.3
Sph	4.0	3.4	ND	4.4	3.5	12.8	14.9
PS	4.1	4.3	5.5	3.6	3.8	7.5	7.3
PI	8.2	7.8	9.8	8.8	6.5	6.0	4.3
DPG	4.9	3.5	5.3	2.7	4.9	6.6	6.0

[1]Values as percentage of the total polar lipid (phospholipid + glycolipid). Abbreviations: PC, phosphatidyl choline; PE, phosphatidyl ethanolamine; Sph, sphingomyelin; PS, phosphatidyl serine; PI, phosphatidyl inositol; DPG, diphosphatidyl glycerol. [2]Plus cerebroside for brain; [3]Plus sulfatide for brain; [4]Conceptual age 31 months; [5]Conceptual age 17 months; [6]Cultured from human blood; [7]Derived from human amnion and grown in tissue culture; [8]Mycelium grown in shaker culture; ND, not detected.

VI. REGULATION OF SUBCELLULAR PARTICULATE
LIPID CLASS COMPOSITION

Each different subcellular particulate of the same organ has a different and characteristic lipid class composition. Preparations of the same particulate isolated from different organs are found to differ in composition, although in some cases the differences are relatively small (4). The findings can be correlated with several observations. Organs differ in the total membrane mass present and the relative amounts of the different subcellular particulates.

As shown in Table I, lipid analyses of different cells and organs fall into characteristic groups. In Groups I, II, and III the predominant membranes appear to be mitochondrial, plasma (cell surface), and microsomal (endoplasmic reticulum) respectively, whereas in group IV there is no clearly predominant membrane. Cells capable of growth and reproduction fall into group III. The major differences noted among animal cells and fungal mycelia are the absence of sphingomyelin in some invertebrates and fungi and the presence of ceramide phosphorylethanolamine or ceramide aminoethyl-phosphonate. These differences do not appear to be specific for any particular species or general morphological classification group and all of the lipid classes are in the same substitution group.

Brain and heart are of special interest to compare. A characteristic feature of brain is the very high proportion of lipid contributed by plasma membranes with mitochondria contributing only a small amount. Both neurones and glial cells elaborate many cell surface processes. The reverse situation is seen in heart where mitochondria contribute much of the total lipid. There is little or no organ specificity in the regulation of the levels of phosphatidyl choline, sphingomyelin, or phosphatidyl ethanolamine. Organs differ in a characteristic manner, however, in phosphatidyl serine, phosphatidyl inositol, and diphosphatidyl glycerol content. Brain contains the highest percentage of phosphatidyl serine and the lowest percentage of diphosphatidyl glycerol found in organs. Heart is exactly the reverse with the highest percentage of diphosphatidyl glycerol and the smallest percentage of phosphatidyl serine. Other organs are intermediate between brain and heart (3,4). It is thus clear that the lipid patterns correlate with the presence of large mitochondrial or large plasma membrane masses.

The correlation for brain and heart can be checked readily for general applicability. The human red blood cell has only one membrane, the cell surface membrane, and thus the relative amounts of phosphatidyl serine and diphosphatidyl glycerol should be like brain. This is indeed the case. No diphosphatidyl glycerol is found in red blood cells that do not contain mitochondria, and the phosphatidyl serine content is high as in human brain. Also, the high percentage

of phosphatidyl serine in myelin lipids is in agreement with the
correlation. Myelin contains at most minute traces of diphosphatidyl
glycerol. Human blood lymphocytes fit the pattern also. These blood
cells have few mitochondria and a large cell surface. Their phosphatidyl serine regulation is like brain and no diphosphatidyl glycerol
at all was detectable in some preparations we examined (28). Endothelial cells (capillaries isolated from human and bovine brain and
human kidney) also have few mitochondria and a large surface area
and have a phosphatidyl serine content like brain (29). Diphosphatidyl glycerol is at a very low, but detectable, level.

At the other extreme from brain, and, like heart, brown fat is
found to have mitochondria contributing much of the total lipid and
we have obtained phospholipid analyses almost exactly like those for
heart (30). Other organs and cell types lie between brain and heart.
Brain differs from most organs in that the increase of plasma membrane mass is larger. The correlations of lipid class distribution
with the relative proportions of membranes of cells with grossly
different total membrane masses, e.g. adult brain total phospholipid
of 7.0 and human red cells of 0.4 mM/100 gm fresh tissue, points to
the relative amounts of the different membranes rather than total
membrane mass as being of primary importance.

Since the percentages of diphosphatidyl glycerol, phosphatidyl
serine, and phosphatidyl inositol are characteristic for different
cell types and their relative amounts correlate very well with the
relative proportions of different membranes, factors regulating the
amounts of these lipid classes may contribute to the control mechanism causing cells to differentiate along different lines.

VII. FACTORS DETERMINING THE FATTY ACID
COMPOSITION OF POLAR LIPIDS

Fatty acid composition and metabolism is different in different
organs. As noted for lipid class composition (Section V), brain
and heart provide an important contrast of extremes. Brain obtains
most of its energy by oxidation of carbohydrate, has a relatively
low mitochondrial membrane mass, and contains only a minor amount
of triglyceride, the storage form of lipid occurring in fat droplets
rather than membranes. In direct contrast, heart obtains over 90%
of its energy from fatty acid oxidation, a large proportion of the
total membrane mass is contributed by mitochondria, and a large
store of triglyceride is present. Also, the fatty acid compositions
of lipid classes of brain and heart are quite different.

Differences in fatty acid composition do not necessarily change
the lipid class composition of a membrane as shown by the data for
changes in whole brain and myelin from rats fed a diet deficient in
essential fatty acids, presented elsewhere in this volume. A broad

look at fatty acid composition data of lipids from many sources
suggests that fatty acid composition is the result of establishment
of a particular metabolic balance in different organs. Fatty acid
oxidation is an important factor. Saturated fatty acids are preferentially oxidized and thus heart will have more unsaturated fatty acids
than brain lipids. Since the biosynthetic enzymes for each lipid
class have somewhat different specificities for fatty acid composition, there is a limit to the extent to which different fatty acids
can enter into lipid classes of each organ. In brain and heart, biosynthesis of phosphatidyl choline and phosphatidyl ethanolamine
appear to take place largely by different mechanisms involving
different enzymes which can be expected to have different specificities for fatty acids.

The major fatty acid composition changes in whole human brain
glycerolphospholipids are (31) a decrease of 16:0 with increase of
18:0 and a decrease of polyunsaturated fatty acids balanced by an
increase of 18:1. In addition, in phosphatidyl ethanolamine, there
is a rise of plasmalogen. In brain sphingolipids, fatty acids of
16 to 20 carbons decrease in amount whereas 22 and 24 carbon acids
increase in amount. These changes can be explained in a general way
as follows. First, with rising lipid content, polyunsaturated fatty
acids become less available and more 18:0 is converted to 18:1 which
is incorporated into lipids to a greater extent. The lower 18:0
from conversion to 18:1 results in less 18:0 incorporation into
sphingolipids which instead use longer chain length fatty acids
derived from more acetyl Co A going directly to 22:0, 22:1, 24:0,
and 24:1. The drop in 18:0 and extensive utilization of 18:1 is
associated with production of more 16 and 18 carbon aldehydes from
sphingosine. Since ethanolamine and aldehydes are formed from
sphingosine, the reaction can lead primarily to formation of ethanolamine plasmalogen. In heart, in contrast, the total lipid level is
not as high, and polyunsaturated fatty acids are present· in greater
amount. There is thus less 18:1 and more 18:0. Heart contains
both ethanolamine and choline plasmalogens in large amount. Thus,
it appears that the route through sphingosine is used more extensively with formation of choline plasmalogen as well. This general
line of reasoning appears to hold for the composition of other
organs as well.

VIII. DISORDERS OF MEMBRANES AND MEMBRANE LIPIDS

There are a number of human diseases in which alterations of
membranes are apparent from abnormal excretion, absorption, etc.
Some of the conditions most relevant to neuropathology have been
listed and classified recently (31). Of special interest here are
those conditions in which an abnormal accumulation of lipid is
found, usually in the form of undegradable membrane fragments or
large membrane masses (whorls). The sphingolipidoses are particularly

informative disorders because the enzymatic defect is known for
some.   Thus, degradative enzyme deficiencies are known for Gaucher's
disease (cerebroside deposition), Niemann-Pick disease (sphingomyelin deposition), Tay-Sachs disease (ganglioside deposition),
and Fabry's disease (ceramide polyhexoside deposition).   An interesting feature of these conditions is that the affected lipid
classes are found to be increased only slightly (Tay-Sachs disease,
Fabry's disease) or not at all (Gaucher's disease, Niemann-Pick
disease) in erythrocytes and leukocytes, i.e. cells with short life
spans. Thus it appears that control mechanisms can prevent large
elevations of lipid classes for which there are deficiencies of
degradative enzymes. Large accumulations occur only in the cells
with long life spans (when recycling does not prevent a rise) and
in cells of the reticuloendothelial system that become loaded with
undegradable membrane fragments. The situation in Niemann-Pick
disease is of particular interest because an acidic glycerolphospholipid, lysobisphosphatidic acid, that is normally a minor component
in liver and spleen increases (to 14% of the total phospholipid)
(32). In the reticuloendothelial system there is accumulation of membrane masses containing sphingomyelin, lysobisphosphatidic acid, and
cholesterol. Graphic analysis discloses that the lipid content of
spleen in Niemann-Pick disease is like that of aorta (Figs. 1 and 2),
except that phosphatidyl serine is the major acidic lipid in aorta
and lysobisphosphatidic acid is the major acidic lipid in the abnormal spleen. Niemann-Pick brain lipid regulation is even more like
that of aorta because phosphatidyl serine in brain is adequate to
balance the need for acidic lipid.

     Although undegradable membrane components in phagocytic cells
are initially outside the cell membrane system, they can enter it
and distort the lipid pattern because the normal cellular system
appears to consist of a transport system for carrying lipids from
the site of synthesis to an appropriate binding site.

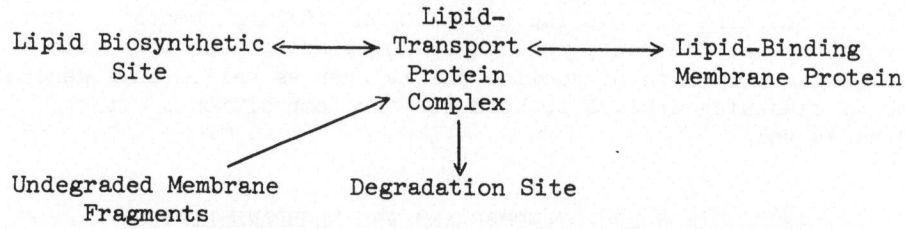

The close similarity of the findings for aorta and the abnormal
spleen suggest that human aorta also undergoes a similar process of
sphingomyelin accumulation. This conclusion is in keeping with the
enzyme changes that have been reported (33). The same general pattern is apparent for the lens of the human eye.

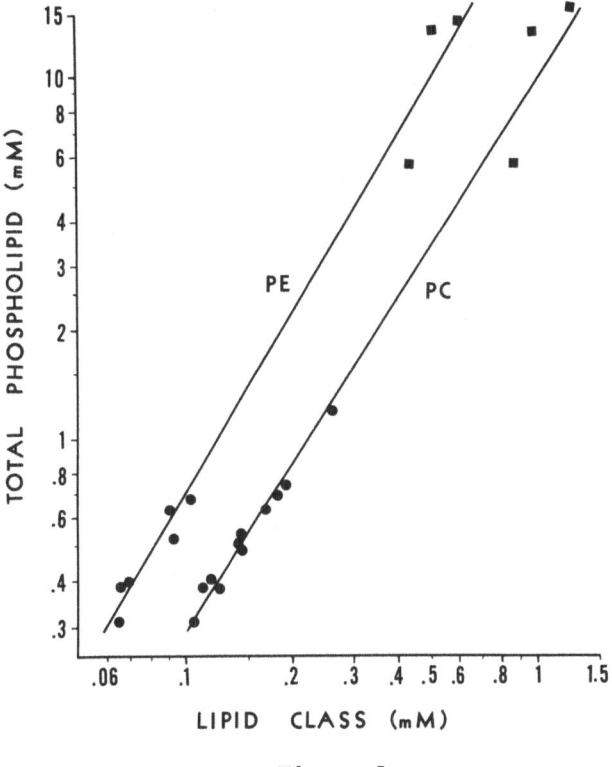

Figure 1

Plot of normal human aorta (solid circles) and
Niemann-Pick disease spleen (solid squares)
phosphatidyl ethanolamine (PE) and phosphatidyl
choline (PC) values against total phospholipid.
The same lipid class regulatory mechanism is
indicated when values fall on the same line.

The prediction according to the tetramer model (3) is that
during differentiation some cells may lose the ability to make
some tetramers. Thus, some patterns of differentiation may lead to
organ specific pathological changes. It can be predicted that
differentiation of human brain to give only the extremes of B[+] or
A[−] peptide chains would result in pathology because brain metabolism
is not able to provide the necessary amounts of the lipid classes
and we have not found any normal human brains to have subgroups I
or 5 of substitution group I. Since the protein of plasma lipoprotein is almost entirely the A[−] type, elimination of this type
should cause marked lowering of plasma lipids and abnormalities in
other parts of the body. The disorder fitting these expectations is
known as abetalipoproteinemia (acanthocytosis).

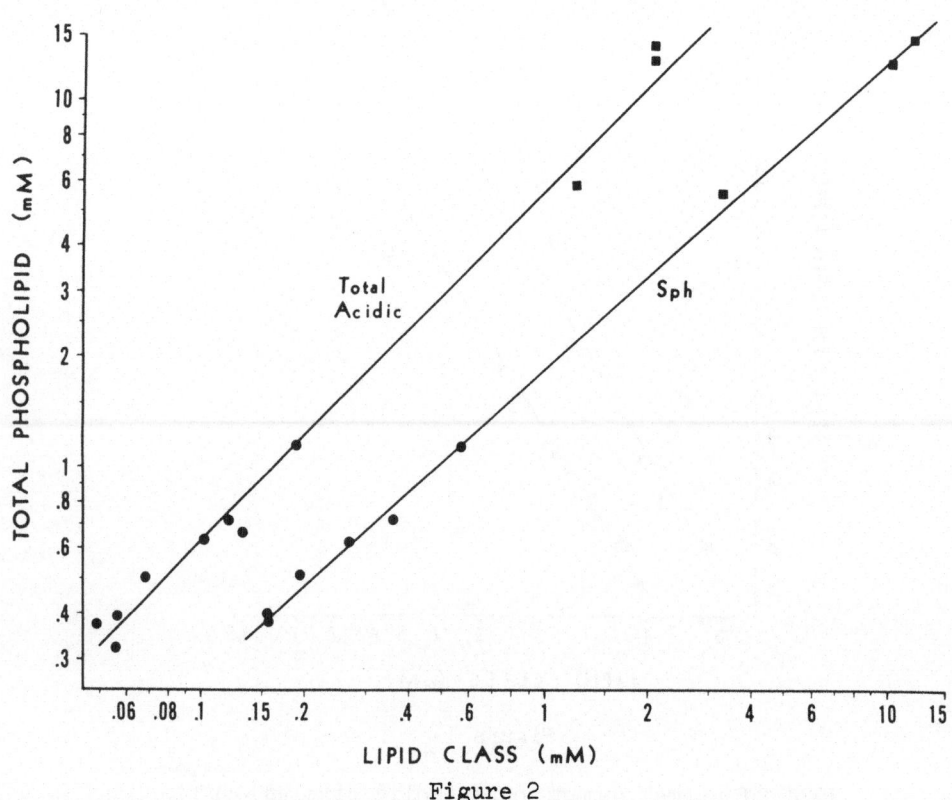

Figure 2

Plot of normal human aorta (solid circles) and
Niemann-Pick disease spleen (solid squares) total
acidic phospholipid and sphingomyelin (Sph) values
against total phospholipid. The same lipid class
regulatory mechanism is indicated when values fall
on the same line.

ACKNOWLEDGEMENTS

This work was supported in part by US. Public Health Service
Grants NS 01847 and NA 06237 from the National Institute of
Neurological Diseases and Stroke.

REFERENCES

(1) Korn, E. Science 153: 1495 (1966).

(2) Stoeckenius, W., and Engelman, D.M. J. Cell. Biol. 42: 613 (1969)

(3) Rouser, G., Yamamoto, A., and Kritchevsky, G. This Volume.

(4) Rouser, G., Nelson, G.J., Fleischer, S. and Simon, G. in
    Biological Membranes (Ed. D. Chapman), Academic Press (1968)
    p. 5-69.

(5) Eng, L.F., Chao, F.C., Gerstl, B., Pratt, D., and Tavaststjerna,
    M.G. Biochemistry 7: 4455 (1968).

(6) Mehl, E., and Wolfgram, F. J. Neurochem. 16: 1091 (1969).

(7) Cooper, R.A. Semin. Hematol. 7: 296 (1970).

(8) Kiehn, E., and Holland, J.J. Biochemistry 9: 1716 (1970).

(9) Op den Kamp, J.A.F., van Deenen, L.L.M., and Tomasi, V. in
    Structural and Functional Aspects of Lipoproteins in Living
    Systems (Eds. E. Tria and A.M. Scanu), Academic Press (1969)
    p. 227-325.

(10) Hendrix, J.W. Science 144: 1028 (1964).

(11) Vandenheuvel, F.A. J. Am. Oil Chemists' Soc. 40: 455 (1963).

(12) Chapman, D., and Salsbury, N.J. in Adv. Surface Science,
     Vol. 3 (Eds. J.F. Danielli, A.C. Riddiford, and M. Roseberg),
     Academic Press (1970), p. 121-168

(13) Steim, J.M., Tourtellotte, M.E., Reinert, J.C., McElhaney,
     R.N., and Rader, R.L. Proc. Natl. Acad. Sci. 63: 104 (1969).

(14) Benedetti, E.L., and Emmelot, P. in The Membranes, Vol. 4 of
     Ultrastructure in Biological Systems (Eds. A.J. Dalton and
     F. Haguenan), Academic Press (1968), p. 33-121.

(15) Fleischer, S., Fleischer, B., and Stoeckenius, W. J. Cell
     Biol. 32: 193 (1967).

(16) Napolitano, L., LeBaron, F. and Scaletti, J. J. Cell Biol.
     34: 817 (1967).

(17) Bar, R.S., Deamer, D.W., and Cornwell, D.G. Science 153: 1010
     (1966).

(18) Eylar, E.H., Madoff, M.A., Brody, O.W., and Oncley, J.L.
     J. Biol. Chem. 237: 1992 (1962).

(19) Winzler, R.J. in Red Cell Membrane (Eds. G.A. Jamieson and
     T.J. Greenwalt), J.B. Lippincott Co. (1969), p. 157-228.

(20)   Rosenberg, S. A., and Guidotti, G. in Red Cell Membrane
       (Eds. G. A. T. J. Greenwalt), J. P. Lippincott Co. (1969),
       p. 93-109.

(21)   Niehaus, W. G., Jr., and Wold, F. Biochim. Biophys. Acta
       196:170 (1970).

(22)   Lenard, J., and Singer, S. J. Science 159: 738 (1968).

(23)   Turner, J. D., and Rouser, G., unpublished results.

(24)   Cooper, R. A., and Jandl, J. H. J. Clin. Invest. 48: 906
       (1969).

(25)   Jacob, H. S. J. Clin. Invest. 46: 2083 (1967).

(26)   Murphy, J. R. J. Lab. Clin. Med. 65: 756 (1965).

(27)   Holden, J. T., Hild, O., Wong-Leung, Y. L., and Rouser, G.
       Biochem. Biophys. Res. Commun. 40: 123 (1970).

(28)   Rouser, G., unpublished results.

(29)   Siakotos, A. N., Rouser, G., and Fleischer, S. Lipids 4: 234
       (1969).

(30)   Therriault, D., and Rouser, G., unpublished results.

(31)   Rouser, G., and Yamamoto, A. in Handbook of Neurochemistry
       (Ed. A. Lajtha), Plenum Press (1969), p. 121-169.

(32)   Rouser, G., Kritchevsky, G., Yamamoto, A., Knudson, Jr., A. G.
       and Simon, G. Lipids 3: 287 (1968).

(33)   Eisenberg, S., Stein, Y. and Stein, O. J. Clin. Invest. 48:
       2320 (1969).

RECENT DEVELOPMENTS IN THE INVESTIGATION OF PURIFIED MYELIN

William T. Norton

Albert Einstein College of Medicine

Bronx, New York 10461

The first studies of isolated myelin began only about a decade ago. Therefore in the usual time scale all developments in this area are relatively recent. However in the field of neurobiology, events of ten years ago are ancient history, and myelin research has become an area where review chapters (1-8) are appearing with nearly the frequency of research papers. Needless to say, as in any field, as many new problems are raised as are solved. This review cannot touch on all of them and will be restricted to developments I feel are of special interest and related to the topic of this Institute. I am going to put special emphasis on the most basic problem, the preparation of myelin.

Isolation

As an editor and frequent referee I have occasion to see many manuscripts before publication. The most frequent and damaging criticism of myelin studies is that the preparative method and criteria of purity of the preparation may be inadequate. This is particularly important when differences in myelin are being sought, as in development, disease, species differences, mutant studies, etc. It is obvious that to study the properties of myelin it must be isolated in a degree of purity consistent with the aims of the experiment, and the type and amount of impurities should be known. Any of a number of published procedures (9-12) is capable of furnishing myelin of acceptable purity from normal, adult CNS tissue, however special cases may require special care.

All of the myelin isolation methods take advantage of one or both of two properties of myelin; large vesicle size and low

density. When brain is homogenized in sucrose solutions the myelin
layers swell, peel off the axons and reform in spherical vesicles
which are in the size range of nuclei and mitochondria, and therefore sediment with these two fractions during differential centrifugation. This swelling phenomenon is independent of the tonicity
of the sucrose (13) but does not occur in the presence of ions.
The myelin vesicles have the lowest intrinsic density of any subcellular fraction of brain because of their high lipid to protein
ratio. Myelin is less dense than 0.8 M sucrose, and will layer
on it in density gradient centrifugation procedures, whereas nuclei,
mitochondria and synaptosomes sediment through this solution.

The isolation methods fall into two groups, depending on
whether they were designed solely for myelin isolation (9,10,14)
or whether myelin is obtained as one of the products in a fractionation scheme designed to isolate all brain subcellular fractions
(15-17). These latter schemes might be expected to give preparations of lower purity, since extensive purifications of each fraction are not consistent with the aims of the technique. On the
other hand, the former schemes probably result in a somewhat lower
yield, and run the risk of giving a product not completely representative of myelin in situ. Most investigations of purified
myelin have used either the Laatsch, et al. (9), Autilio, et al.
(10) or Cuzner, et al. (11) methods, or modifications of them.
Allowing for analytical and species differences, the composition of
myelin prepared by these methods does not differ much. However, as
noted above, for certain special investigations, the amount of non-
myelin contamination might easily be enough to invalidate the
results.

The method worked out by Dr. Autilio and myself (10) has been
used extensively, sometimes in situations for which it was not
suitable. This method was designed for use with white matter of
large animals and was not evaluated with whole brains of small
animals. Even when used as recommended, with white matter as a
starting material, there is a deficiency in the method which could
lead to misinterpretations. The crude myelin fraction accumulates
at the interface of 0.32 M sucrose and 0.656 M sucrose during centrifugation. The myelin is then purified by a repetition of this
step, an osmotic shock and centrifugation on a continuous density
gradient. These conditions were worked out empirically before the
composition and physical properties of myelin were known. We now
know that the 0.656 M sucrose used in this method has a density
nearly equal to the mean myelin density in sucrose. Adams and Fox
(18) have shown that myelin distributes in a relatively wide band
in a sucrose density gradient. Samples taken throughout this band
show a continuously changing lipid/protein ratio. Thus, in the
method of Autilio, et al. (10) considerable myelin is discarded,
and that which is preferentially selected must have a higher than
average lipid/protein ratio.

With any of the common isolation techniques a high purity myelin fraction may be obtained from white matter (which is 50-60% myelin) or from whole brain of adult rats where myelin may be 15% of the dry weight. However it is more difficult to obtain clean preparations from immature whole brain that has perhaps only 2% myelin. Between 15 and 180 days the amount of myelin per rat brain increases 1500%, whereas the brain weight increases only 50-60%. In the extreme situation therefore, there is the possibility that the absolute amount of contamination of a myelin preparation with extraneous membrane particles could be nearly the same at all ages. Contamination of 60 mg of adult myelin with 4 mg of non-myelin membranes might go undetected, but the same amount of contamination with 3.5 mg of myelin would affect the data drastically.

For a study of development in the rat we (19) devised a new isolation technique which we hoped would answer most of these criticisms. This method proved to give myelin of constant purity, independent of the age of the animal, using our criteria of nucleic acid, ATPase and ganglioside sialic acid content. This method is also suitable for use with any brain tissue and although it has not been published in detail before, it has been made available to several laboratories and been used successfully in other studies (20-24).

The method as used for preparing myelin from brains of rats >30 days of age is given below. Modifications for younger animals are also given. The steps are written for a swing-out rotor having three 60 ml tubes (Spinco SW 25.2), and all steps are done at 0-4°.

Norton and Poduslo Method for Myelin Isolation.

1. A 5% homogenate of whole brain in 0.32 M sucrose (3 brains/100 ml) is layered over 0.85 M sucrose (25 ml per tube) and the tubes are centrifuged at 75,000 g for 30 min.
2. The myelin layer at the interface is removed to a clean tube, dispersed in distilled water and centrifuged at 75,000 g for 15 min. (myelin from 1 brain per 60 ml). The supernatant is discarded.
3. The myelin pellet is dispersed again in water and centrifuged at 12,000 g for 10 min. The supernatant is discarded.
4. Step 3 is repeated.
5. The three myelin pellets are homogenized in 100 ml 0.32 M sucrose, layered over 0.85 M sucrose and centrifuged as in step 1.
6. The myelin layers at the interface are collected, washed with water to remove sucrose, and can be freeze-dried for storage.

If 15 day old rats are used, take four brains in step 1, repeat step 1 with four more brains (8 brains in all), and combine two myelin layers per tube in step 2. In steps 3 and 4 use two tubes/8 brains, and use 1 tube/8 brains in step 5. For 20 day old rats, steps 1 and 2 are the same as for 15 day old rats. In steps 3 and 4 use 3 tubes/8 brains and use 2 tubes/8 brains in step 5. All dispersions must be thorough and should be done in an homogenizer.

The amount of brain per tube is kept at a constant low concentration in the first step, which effectively furnishes a quantitative yield of crude myelin, free from heavier particulates. Steps 2, 3 and 4 are designed to osmotically shock the myelin, freeing any trapped axoplasm and, during centrifugation, to sediment the larger myelin particles leaving smaller membranous material suspended in the supernatant. The final discontinuous gradient removes any heavier particulate material still remaining. The modifications allow the small amount of myelin in the younger animals to be concentrated without jeopardizing purity or yield. Higher purity preparations should be achieved by repeating step 1 with the crude myelin obtained in that step, and by using a continuous gradient in step 5.

A theoretically superior method to the above was devised by myself and Dr. Alan Davison (25) during a study of the nature of early myelin. This new method is shorter, and furnishes higher yields of myelin of presumably equivalent purity, although preparations made by this technique have not been as extensively characterized. Steps 1 and 2 are the same as in the Norton and Poduslo technique detailed above. Step 2 is repeated to complete the osmotic shock treatment and the resultant myelin pellet is dispersed in 0.3 M CsCl and purified by centrifugation on a continuous gradient of CsCl varying from 1.3 M to 0.3 M. The myelin forms a granular band centering at about 0.8 M CsCl. In this procedure any heavier material (including microsomes) forms a band or a precipitate below myelin. Although discontinuous gradients could probably be used, there is a danger of trapping material in the compact, cohesive myelin layers that are formed in such gradients. The losses incurred in steps 3 and 4 of the Norton and Poduslo technique are avoided in the CsCl method.

Criteria of purity of myelin preparations were difficult to set in the beginning because there was no a priori way of knowing what the intrinsic myelin constituents were. The types of criteria for myelin should be the same as for any other fraction: typical ultrastructure, the absence or minimization of markers characteristic of other particles and the maximization of markers characteristic of myelin. Most people have used the obvious criterion of electron microscopic appearance. Isolated myelin retains the typical 5-layered structure and repeat period of 120 A seen in situ. However the methods of isolation make it probable that the major contamination would be microsomes. The difficulty of identifying small membrane vesicles in a field of myelin membranes, and the well known sampling problem inherent in EM work, make this characterization unreliable after a certain purity level has been reached. Useful markers for contamination are succinic dehydrogenase, ATPase, glucose-6-phosphatase, 5'-nucleotidase and nucleic acid; all of which are very low in purified myelin (9,12,14,19,26).

Markers characteristic of myelin are fewer and some used previously are now to be regarded with suspicion. It is presumed that the myelin will be purified by a final density gradient centrifugation step (preferably using a continuous gradient). This establishes the density and therefore the lipid/protein ratio within limits. Many people have used the solubility, or near solubility, of the final product in 2:1 $CHCl_3:CH_3OH$ as an indicator of purity. We have found that most preparations are $\sim$ 95% soluble in this solvent but, depending on the history of the preparation, some, which satisfy all other criteria for purity, including identical lipid/protein ratios, were found to have as much as 10-15% insoluble material. Therefore complete or nearly complete solubility (this measure should be quantitative) is a good sign, but large amounts of insoluble residue may not indicate large amounts of impurity. In recent years the enzyme 2',3'-cyclic nucleotide-3'-phosphohydrolase has been shown to be myelin-specific (27-29). This enzyme should prove very useful in devising new myelin isolation methods and in assaying myelin contamination in other fractions. With increasing knowledge of myelin proteins and with the development of new methods for their separation and measurement, determination of the myelin protein pattern by, for example, acrylamide gel electrophoresis should be an excellent indicator of myelin purity.

If ions are included in the initial homogenization medium, it is probable that the crude myelin obtained will include considerable axoplasm. The axoplasm can be removed in subsequent steps provided one is aware of how extensive this impurity may be, and designs the procedures accordingly. We have found that if white matter is homogenized in a Dounce homogenizer using a buffered medium containing sucrose and 100 mM KCl (or less $CaCl_2$), then the crude myelin layer obtained in step 1 of the Norton and Poduslo method is actually largely composed of segments of intact myelinated axons. This discovery has been used to devise a method for preparing a myelin-free axon-enriched fraction (30). One surprising finding is that the axon fraction has a high content of cerebroside, a lipid generally considered to be a myelin marker.

Development

In 1965 Horrocks, et al. (31) reported that mouse brain myelin changes composition during development. This report was greeted

with skepticism by some, mainly because it was felt that the
preparative method was not adequate to give microsome-free material,
and that the myelin preparations were not controlled for this
possible impurity. In an effort to give a definitive answer to
this question, the method discussed earlier was developed. We found
that myelin prepared from rat brains of 15 days to 425 days of age
had a low and constant level of impurities (as judged by our criteria). (see table 1).

Table 1 - Criteria of Purity of Myelin Preparations

Age, days	15	20	30	60	144	190	425
% nucleic acid	0.26	0.58	0.24	0.20	0.17	0.18	0.13
Total ATPase, μMP/mg/hr.	0.81	2.17	0.84	0.32	0.29	--	--
Ganglioside sialic acid μg/100 mg	48.7	44.7	39.6	39.5	35.6	51.7	68.0

Moreover the yield of myelin was highly reproducible at any one
age; the isolated myelin had the same total lipid content at all
ages; and the same percentage of total brain galactolipid (40%) was
recovered in the myelin fraction at all ages. We were confident
that our myelin preparations were of high and constant purity and
that any compositional changes found would be unequivocal. Our
results (table 2) confirmed the general trends found by Horrocks,
et al. (31) and those reported later by Davison, et al. (32). As
 myelination proceeds the amount of galactolipid increases and the
amount of lecithin and desmosterol decrease. The other lipids do
not change significantly.

Table 2 - Rat Myelin Composition During Development

Age, days	15	20	30	60	144	190	425
Lipid, % dry wt.	74.2	70.4	72.1	72.5	70.4	70.1	68.0
Sterol	25.1	25.7	25.7	25.9	27.8	27.9	27.4
Desmosterol	3.0	1.8	1.5	1.0	0.3	0.2	0.2
Galactolipid	21.4	23.6	26.2	30.5	33.3	30.7	31.7
Phospholipid	50.4	48.3	48.0	44.0	43.7	43.5	44.9
PE	18.4	18.1	17.5	16.8	15.1	17.1	17.7
PC	16.4	15.6	13.6	12.4	10.7	10.7	11.5
Sph	3.4	3.5	3.3	3.4	3.2	3.1	3.1
PI	1.4	1.4	1.0	1.3	1.2	1.1	1.2
PS	7.0	6.9	7.4	6.5	6.7	7.1	7.6
Plasmalogen	14.1	13.1	13.7	14.4	13.9	13.7	14.3

(the figures for lipid are % of dry wt and for desmosterol are %
of total sterol. All other figures are wt % of total lipid)

Horrocks (33), Cuzner, et al. (12) and Eng and Noble (34)
have now published amplified studies. Eng, et al. (35) have also
shown that the ratio of basic protein to proteolipid protein increases in human myelin during development. Recently Einstein
and coworkers have published work on maturing rabbit brain showing
dramatic changes in composition of both lipids (36), and proteins
(37). They have used a Norton and Poduslo method, but unfortunately modified so as to possibly nullify its virtues. It may be that
their preparations from very young animals are mixtures of myelin
and "myelin-like" fractions.

The presence of the "myelin-like" fraction in crude myelin
from young animals is a complex issue which has confused studies
of myelin development. In Davison's earlier work it was apparent
that the composition of myelin from young (15 day old) rats was
quite different from our results reported above. Our young myelin,
while having less cerebroside than adult myelin, had considerably
more than that reported by Davison's group. In a collaborative
effort, Davison and I found that the difference lay in the isolation methods. Steps 3 and 4 of the Norton and Poduslo method removed a fraction retained by Davison's group in the final myelin
fraction. This fraction, which layers out with myelin in the
early steps of the procedure, could also be removed from myelin by
the CsCl gradient described above. This second fraction, which may
comprise 50% of the total crude myelin from a 15 day rat, is very
low in crude adult myelin. It has the ultrastructural appearance
of large microsomes and has quite a different lipid composition
from myelin. Banik and Davison (26) have extended this work and
shown that the myelin-like fraction from young rats resembles
myelin in its low enzyme activity and is different from brain
microsomes. The current thinking is that it may represent an
intermediate or transition form between oligodendroglial plasma
membrane and myelin, one which is present in the brain during
myelinogenesis. It is interesting that the adult quaking mouse, a
mutant characterized by a halt in myelination at an early stage,
has an amount of the myelin-like fraction comparable to that in a
15 day old normal mouse (38).

Proteins

The lipid composition of myelin is very well documented. In
recent years more attention has been directed toward the protein
part of myelin. There are now a number of methods for studying
myelin proteins by electrophoresis (39-41), solvent fractionation
(42,43) and gel filtration (44). It is apparent that CNS myelin
has a very simple protein pattern compared with other membrane
fractions. There are two major proteins, the classical proteolipid
protein and the basic protein. In addition there is a fraction,
which is probably heterogeneous called "Wolfgram protein". There

are some species differences, e.g., both rat and mouse myelin have
two basic proteins.

The basic protein has been extensively studied because of its
encephalitogenic activity. It is also water soluble following acid
extraction, and thus amenable to study by conventional protein
techniques. Eylar and coworkers have recently published a series
of detailed and thorough chemical and physical studies of this unusual protein. The protein from bovine brain has 169 residues and
a molecular weight of 18,000 (45). It is highly basic (isoelectric
point >12) and highly unfolded. Its lack of tertiary structure
makes it resistant to denaturation at 100°, in 8 M urea or at pH
10 (46). A series of peptides have been produced, some with EAE
activity and some without, however all the active peptides contain
the single tryptophan residue. This work has culminated in the
isolation of a nine residue peptide and the synthesis of an eleven
residue peptide including the same sequence, both of which have
high encephalitogenic activity (47). Eylar has also determined the
complete sequence of the parent protein.

It would be of considerable value to have a rapid method for
separating and studying all myelin proteins. We (38) have adapted
the Maizel technique (48) of acrylamide gel electrophoresis in
buffers containing sodium dodecylsulfate (SDS), which appears to be
ideal for this purpose. Whole myelin or delipidated myelin can be
readily dissolved in SDS solutions (without urea) and electro-
phoresed to give, in the case of bovine myelin, four major bands, two
of Wolfgram protein, one of proteolipid protein and one of basic
protein. Mouse and rat myelin yield two basic protein bands in
this system.

New Directions and Unsolved Problems

I can only make a brief list of some of the obvious areas for
exploration, and questions that should be clarified. Why are there
still discrepancies in the myelin composition of the same species
reported by different groups? Are these the result of different
analytical techniques or different preparative methods? What
constituents are lost from myelin during isolation? Myelin fractions from PNS, cord and brain are all different, are there differences in subregions of each that might be related to ontogeny? Do
different microscopic regions of the myelin sheath have different
metabolic rates?

One research area that has been badly neglected is the use of
myelin to study the fundamental nature of lipid-protein interactions. For example, it might be assumed that the membrane protein
determines in large part the lipid composition. We know that
myelin lipid composition in certain diseases (22-24, 49-51) and

during development is different from that of the normal adult, but
we know little about the differences in protein composition. Are
certain lipids always associated with specific proteins? Lastly,
our discovery of cerebroside in axons (30) has again raised the
question of how much the axon contributes to myelin and the nature
of the relationship between myelin and the oligodendroglial cell
membrane. The recent development, by Fewster and Mead (52) and
ourselves (53,54), of methods for the bulk isolation of oligodendroglia may furnish a new approach to the study of myelinogenesis.

References

(1)   Davison, A. N. in <u>Fortschritte der Padologie</u> (Ed. F. Linneweh)
      Berlin, Springer-Verlag, Band <u>II</u>, p. 65, 1968 (in English)
(2)   O'Brien, J. S. in <u>Developing Brain</u> (Eds. H. E. Himwich and
      W. A. Himwich) Springfield, C. C. Thomas, in press.
(3)   Smith, M. E. Adv. Lipid Res., <u>5</u>:241 (1967).
(4)   Mokrasch, L. C. in <u>Handbook of Neurochemistry</u> Vol. 1 (Ed.
      A. Lajtha) New York, Plenum Press, p. 171.
(5)   O'Brien, J. S. in <u>Handbook of Neurology</u> Vol. 18 (Eds. B. Vinken
      and G. Bruyn) Amsterdam, North Holland Publ. Co., in press.
(6)   Davison, A. N., and Peters, A. (Monograph on myelin, to be
      published by C. C. Thomas) Springfield, 1970.
(7)   Norton, W. T. in <u>The Cellular and Molecular Basis of Neurologic
      Disease</u> (Eds. E. S. Goldensohn and S. M. Appel) Phila.,Lea and
      Febiger, in press.
(8)   Norton, W. T. in <u>Basic Neurochemistry</u> (Eds. R. W. Albers, B.
      Agranoff, R. Katzman and G. J. Siegel) Boston, Little, Brown,
      in preparation.
(9)   Laatsch, R. H., Kies, M. W., Gordon, S., and Alvord, E. C., Jr.
      J. Exp. Med., <u>115</u>:777 (1962).
(10) Autilio, L. A., Norton, W. T., and Terry, R. D. J. Neurochem.,
      <u>11</u>:17 (1964).
(11) Cuzner, M. L., Davison, A. N., and Gregson, N. A. J. Neuro-
      chem., <u>12</u>:469 (1965).
(12) Cuzner, M. L., and Davison, A. N. Biochem. J., <u>106</u>:29 (1968).
(13) Worthington, C. R. and Blaurock, A. E. Biochim. Biophys. Acta,
      <u>173</u>:427 (1969).
(14) Shapira, R., Binkley, F., Kibler, R. F., and Wundram, I. J.
      Proc. Soc. Exp. Biol. Med., <u>133</u>:238 (1970).
(15) DeRobertis, E., Pellegrino de Iraldi, A., Rodriguez de Lores
      Arnaiz, G., and Salganicoff, L. J. Neurochem., <u>9</u>:23 (1962).
(16) Whittaker, V. P. Biochem. Soc. Symp., <u>23</u>:109 (1963).
(17) Eichberg, J., Jr., Whittaker, V. P., and Dawson, R. M. C.
      Biochem. J., <u>92</u>:91 (1964).
(18) Adams, D. H., and Fox, M. E. Brain Res., <u>14</u>:647 (1969).
(19) Norton, W. T., Poduslo, S. E., and Suzuki, K. <u>Abs., First
      Meeting Int. Soc. Neurochem.</u>, Strasbourg, p. 161 (1967).
(20) Suzuki, K., Poduslo, S. E., and Norton, W. T. Biochim. Biophys.

Acta, <u>144</u>:375 (1967).

(21) Suzuki, K., Poduslo, J. F., and Poduslo, S. E. Biochim.
      Biophys. Acta, <u>152</u>:576 (1968).

(22) Kamoshita, S., Rapin, I., Suzuki, K., and Suzuki, K. Neurology,
      <u>18</u>:975 (1968).

(23) Suzuki, K., Suzuki, K., and Kamoshita, S. J. Neuropath. Exp.
      Neurol., <u>28</u>:25 (1969).

(24) Kamoshita, S., Aron, A. M., Suzuki, K. and Suzuki, K. Am. J.
      Dis. Child., <u>117</u>:379 (1969).

(25) Norton, W. T., and Davison, A. N., in preparation.

(26) Banik, N. L., and Davison, A. N. Biochem. J., <u>115</u>:1051 (1969).

(27) Kurihara, T., and Tsukada, Y. J. Neurochem., <u>14</u>:1167 (1967).

(28) Kurihara, T., and Tsukada, Y. J. Neurochem., <u>15</u>:827 (1968).

(29) Olafson, R. W., Drummond, G. I., and Lee, J. F. Can. J. Bio-
      chem., <u>47</u>:961 (1969).

(30) Norton, W. T., and Turnbull, J. M. Fed. Proc., <u>29</u>:472 Abs.
      (1970).

(31) Horrocks, L. A., Meckler, R. J., and Collins, R. L. in <u>Varia-
      tion in Chemical Composition of the Nervous System</u> (Ed. G. B.
      Ansell) Oxford, Pergamon, p. 46, 1966.

(32) Davison, A. N., Cuzner, M. L., Banik, N. L., and Oxberry, J. M.
      Nature, <u>212</u>:1373 (1966).

(33) Horrocks, L. A. J. Neurochem., <u>15</u>:483 (1968).

(34) Eng, L. F. and Noble, E. P. Lipids, <u>3</u>:157 (1968).

(35) Eng, L. F., Gerstl, B., Pratt, D. V., and Tavaststjerna, M. G.
      <u>Abs. First Meeting Int. Soc. Neurochem.</u>, Strasbourg, p. 62
      (1967).

(36) Dalal, K. B., and Einstein, E. R. Brain Res., <u>16</u>:441 (1969).

(37) Einstein, E. R., Dalal, K. B., and Csejtey, J. Brain Res., <u>18</u>:
      35 (1970).

(38) Norton, W. T., and Morell, P., unpublished observations.

(39) Mehl, E. in <u>Macromolecules and the Function of the Neurone</u>
      (Eds. Z. Lodin and S. P. R. Rose) Amsterdam, Excerpta Medica,
      1968.

(40) Cotman, C. W., and Mahler, H. R. Arch. Biochem. Biophys., <u>120</u>:
      384 (1967).

(41) Mehl, E., and Wolfgram, F. J. Neurochem., <u>16</u>:1091 (1969).

(42) Eng, L. F., Chao, F. C., Gerstl, B., Pratt, D., and
      Tavaststjerna, M. G. Biochemistry, <u>7</u>:4455 (1968).

(43) Gonzalez-Sastre, F. J. Neurochem., <u>7</u>:1049 (1970).

(44) Autilio, L. A. Fed. Proc., <u>25</u>:764 (1966).

(45) Hashim, G. A., and Eylar, E. H. Arch. Biochem. Biophys., <u>135</u>:
      324 (1969).

(46) Eylar, E. H., and Thompson, M. Arch. Biochem. Biophys., <u>129</u>:
      468 (1969).

(47) Eylar, E. H., Caccam, J., Jackson, J. J., Westall, F. C., and
      Robinson, A. B. Science, <u>168</u>:1220 (1970).

(48) Maizel, J. V. Science, <u>151</u>:988 (1966).

(49) Norton, W. T., Poduslo, S. E., and Suzuki, K. J. Neuropath.
      Exp. Neurol., <u>25</u>:582 (1966).

(50) Norton, W. T., and Poduslo, S. in <u>Variation in the Chemical Composition of the Nervous System</u> (Ed. G. B. Ansell) Oxford, Pergamon, p. 82, 1966.

(51) O'Brien, J. S., and Sampson, E. L. Science, <u>150</u>:1613 (1965).

(52) Fewster, M. E., Scheibel, A. B., and Mead, J. F. Brain Res., <u>6</u>:401 (1967).

(53) Norton, W. T., Poduslo, S. E., and Turnbull, J. M. <u>Proc. 1st Meeting Am. Soc. Neurochem.</u>, Albuquerque, p. 22 (1970).

(54) Raine, C. S., Poduslo, S. E., and Norton, W. T., submitted to Brain Res.

[22] ...

[23] ...

ON Ca^{++} TRANSPORT ACROSS THE MITOCHONDRIAL MEMBRANE:THE ROLE OF THE CHEMICAL NATURE AND PATTERN OF MITOCHONDRIAL PHOSPHOLIPIDS

Carlo Stefano Rossi, Lodovico Sartorelli,
Ernesto Carafoli (+)
Department of Medical Biochemistry, Faculty of
Medicine, University of Nairobi,P.O.Box 30197,
Nairobi, Kenya

As membrane transport is defined the event by which a molecule or ion passes across a barrier (= membrane), not by passive diffusion, but by a process in which the movement,its directionality, its rate and extent are governed by specific devices, rather than by the gradient established across the barrier. In a number of cases this sort of transport involves expenditure of energy, which in the cell can be recovered from the oxidoreduction reactions of the respiratory chain, or from the hydrolysis of ATP.

Let us for example consider the mechanism by which in a nerve cell an excitation is transmitted from the periphery to the receiving center. This event is accompanied by a process in which an electrochemical potential is built up across the barrier, i.e. between the internal compartment of the cell and the external environment. This potential is merely due to an ion moving device (pump) that guarantees an higher activity of K^{+} ions inside the cell, as compared with the activity of the same ion in the surrounding medium.

In particular, it is well known that the chemical transmission of a nerve impulse depends upon a permeability change of the membrane due to the liberation of the chemical transmitter. These phenomena occur at the level

(+) Visiting Professor of Medical Biochemistry,University of Nairobi. Pr. Address: Institute of General Pathology, University of Modena, Italy.

of the so called chemical synapses. The ions involved in such a process are Ca^{++}, Mg^{++}, Na^+ and K^+, and the change in permeability seems to be somewhat coupled to a conformational modification of the post-synaptic membrane, which results in an opening up a cation selective channel. The event is paralleled by the destruction of the chemical transmitter due to a specific membrane bound enzyme (see the scheme here below, Figure 1).

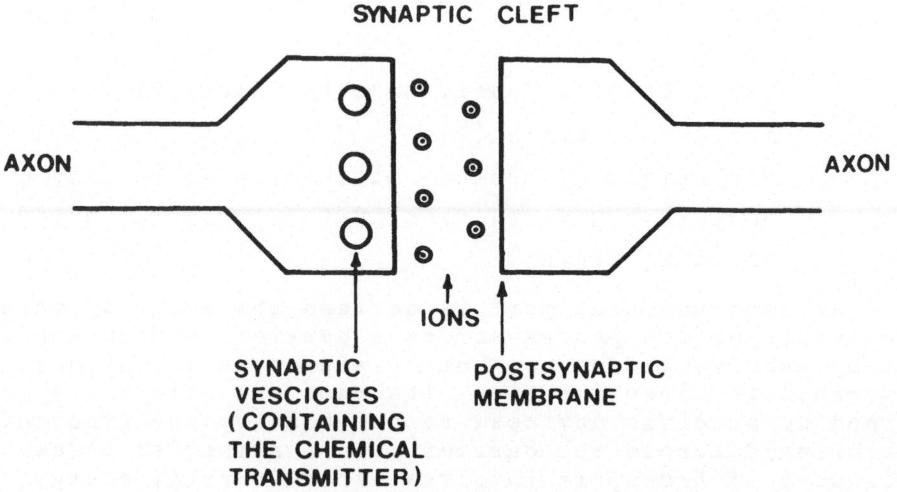

SYNAPTIC CLEFT

AXON

AXON

IONS

SYNAPTIC
VESCICLES
(CONTAINING
THE CHEMICAL
TRANSMITTER)

POSTSYNAPTIC
MEMBRANE

The most extensively investigated chemical transmitter is acetylcholine; acetylcholinesterase is the enzyme located in the post-synaptic membrane, which promotes the change of membrane permeability during the hydrolysis of the substrate. It is interesting that this enzyme seems to be widely distributed in the mitochondrial membrane.

Thus in the nerve cells the chemical neurotransmitters evoke changes in permeability, which in turn discharge the electrochemical potential established by the ion pumping machinery. Something similar occurs in the mitochondrion where specific compounds (valinomycin, gramicidin) are capable of increasing the permeability of the membrane and thus to abolish the gradient.

By considering these similarities, it is perhaps permissible to consider the ion translocating device of the mitochondrion as a model for the permeability change cycle occurring in the synapse during the chemical tran-

---

We wish to thank Dr. L. Galzigna of the Department of Medical Biochemistry, University of Nairobi, for the advice and consultation.

smission. In this presentation we will thus deal in detail with the mechanisms of ion movements across the mitochondrial membrane; although both monovalent and divalent cations, as well as anions, are transported across
the mitochondrial membrane, we shall concentrate on Ca^{++},
the transport of which has been studied in great detail.
Emphasis will be put on the role of phospholipids in the
ion transport activity of the mitochondrial membrane.

Figure 2 summarizes the most important observations
on the so called energy-linked Ca^{++} transport: it is
shown that this event requires the expenditure of the
energy pressure conserved in mitochondria by either the
operation of the respiratory chain, or by the hydrolysis
of $ATP^{(+)}$. The respiration dependent event is blocked by

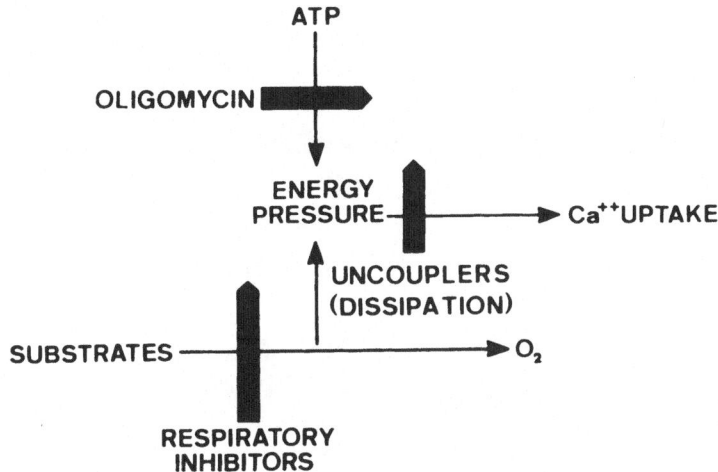

Figure 2. Energetics of calcium transport in mitochondrial systems.

respiratory inhibitors and not by oligomycin, whereas
the ATP driven process is insensitive to respiratory poi-
sons and is inhibited by oligomycin. Both processes are
completely prevented by uncouplers of oxidative phosphorylation. Although the nature of the energy pressure is
still a matter of controversy, it is important to say
that it is generally accepted that there are two levels
of energy conservation, and that only at the second level inorganic phosphate comes into the picture. It is
also generally believed that uncoupling agents discharge
the energy pressure by dissipation at the first step of
the sequence, whereas oligomycin inhibits at the step

(+) ATP = adenosinetriphosphate

where phosphate enters the sequence. The above mentioned mode of action of inhibitors clearly suggests that
the energy-dependent movements of Ca^{++} do not involve
the phosphorylated level of energy conservation.

        When the electron flux through the respiratory
chain is the driving force for ion movements, the uptake of Ca^{++} is accompanied by a transient stimulation of
respiration (1,2,3), as shown in Figure 3. The amount
of extra-oxygen consumed, and the amount of Ca^{++} taken
up are stoichiometrically related: about 2 Ca^{++} ions
are translocated as a pair of electrons is passed across
each of the three energy conserving sites of the respiratory chain. Concomitant to the respiratory jump, there is also a reversible shift of the redox state of respiratory carriers towards a more oxidized state (4). At
the end of the respiratory jump the extramitochondrial
Ca^{++} concentration has been reduced to less than 1 μM.

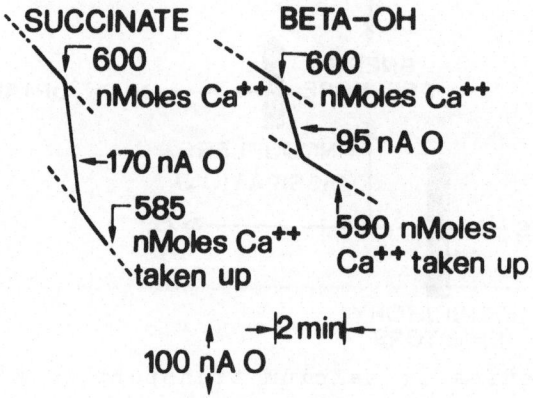

Figure 3. Stoichiometry of oxygen consumption and Ca^{++}
uptake in mitochondria. Oxygen is measured by a Clark
electrode, and Ca^{++} isotopically. The medium contains
80 mM NaCl, 10 mM tris-Cl, pH 7.4, 10 mM Na succinate
or Na-βhydroxybutyrate, as respiratory substrates, and
2 mg mitochondrial protein per ml. Temp.21°C (5,6).

        Under these conditions several successive additions
of Ca^{++} may be made, each producing a transient stimulation of oxygen uptake, followed by an abrupt return to
the initial rate. The mitochondria remain undamaged and
retain complete respiratory control. Another aspect of
this kind of transport is illustrated in Figure 4. The
movements of Ca^{++} are coupled to the opposite translocation of H^+ ions; in the absence of permeant anions(such

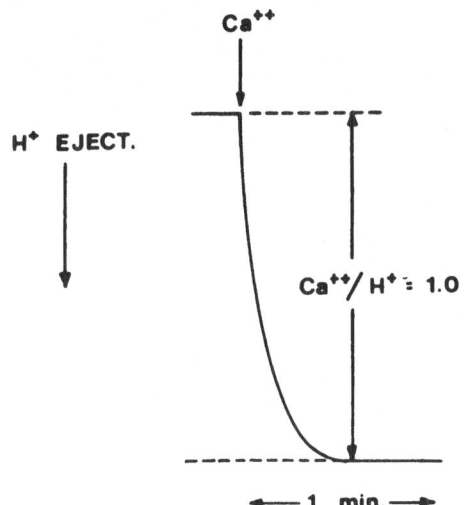

Figure 4. Acid-base changes in the suspending medium during Ca++ uptake (followed with a glass electrode). Experimental conditions as in Figure 2, except for the concentration of tris-Cl (4 mM).

as inorganic phosphate or acetate) the Ca++/H+ ratio is about 1 (7,8).

As shown by Rossi and Lehninger (5), following the completion of a Ca++ induced respiratory stimulation, the mitochondria are still capable of oxidative phosphorylation and of respiratory control. Conversely, following an ADP + induced stimulation of respiration, the addition of Ca++ produces a normal respiratory jump with the usual Ca++: oxygen stoichiometry. Furthermore, when Ca++ and ADP are added simultaneously, Ca++ is taken up first, and then ADP is phosphorylated to form ATP. The two processes are alternative. These observations indicate that mitochondria will accumulate Ca++ in preference to forming ATP. All these facts may have important physiological implications. The mitochondria remain undamaged and functionally intact, and retain respiratory control during Ca++ uptake, and can thus be considered organelles with a dual function: the formation of ATP, and the regulation of the concentration of Ca++ in the cytosol (Figure 5).

+ ADP = adenosinediphosphate

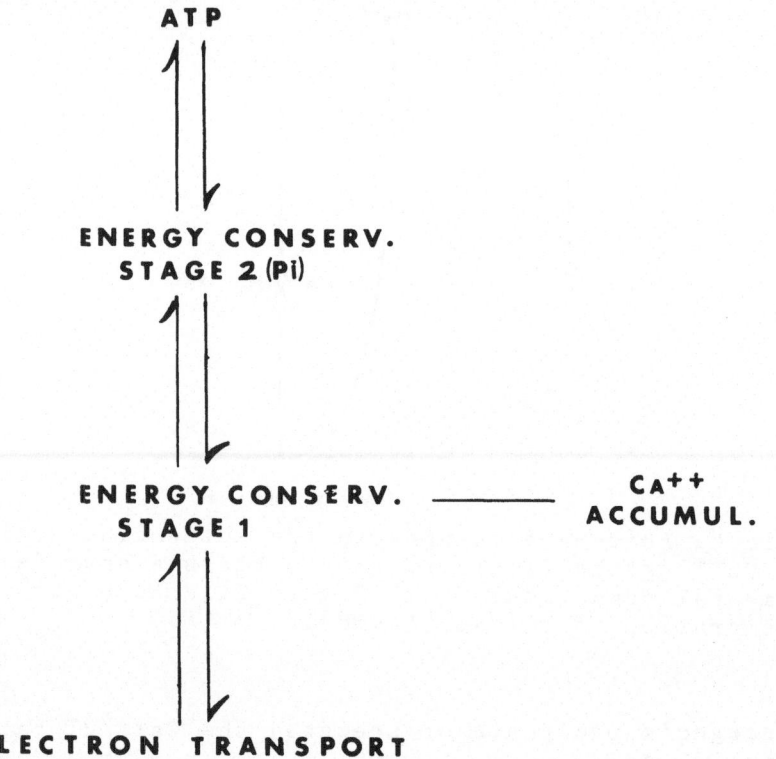

Figure 5. Energy-linked translocation of Ca^{++}, and oxidative phosphorylation of ADP (= ATP formation), as alternative processes.

        In addition to the transport of Ca^{++} strictly dependent upon the metabolism of the mitochondrion, other Ca^{++} binding processes have been recently reported. Azzone and his group (9) have observed the binding of large amounts of Ca^{++} in the complete absence of oxidative metabolism or ATP hydrolysis: up to a maximum of 50 nmoles Ca^{++} can be bound in a process which shows a rather low affinity constant (10^{-4}M). From the work of Azzi and Scarpa (10) it is suggested that the process requires the mitochondrial phospholipids. In fact, the local anesthetic butacaine, a reagent which is known to displace Ca^{++} from phospholipids micelles, severely inhibits this metabolism-independent Ca^{++} binding. This effect has suggested to us a detailed analysis of the role of phospholipids in both the energy-dependent and the metabolism independent Ca^{++} movements (see below).

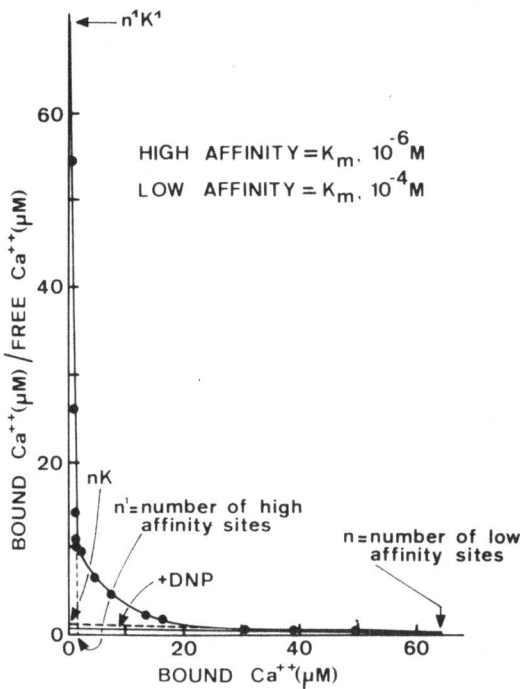

Figure 6. Scatchard plot of Ca^{++} binding by mitochondria in the absence of metabolism (redrawn from Lehninger et al. 12, 13).

    In 1969 a third type of Ca^{++} binding was discovered by Reynafarje and Lehninger (12). When extremely small amounts of Ca^{++} are presented to the mitochondria, the ion is bound to the mitochondria in a process which shows a very high affinity constant. A plot of the data obtained under these circumstances is shown in the Figure 6. The Scatchard plot is biphasic, and therefore suggestive of two independent classes of binding sites: one class has some 40-50 binding sites per mg protein, with an apparent half-saturation at about 100 µM Ca^{++}. This reaction has been called *Low Affinity Binding*. The other class has only 1-2 binding sites per mg protein, and its affinity for Ca^{++} is very high, and the half-saturation is about 0.1 µM. This reaction has been called *High Affinity Binding*. The high affinity reaction is completely abolished by uncouplers, a fact which indicates that this process is somewhat involved in the energy-linked uptake.

    The above mentioned observations prompted the pos-

tulation of a specific carrier molecule capable of moving Ca^{++} across the mitochondrial membrane. A carrier
has been postulated also by Mela and Chance (14, 15),
who found that extremely low concentrations of La^{+++} inhibited the initial phases of energy-linked Ca^{++} uptake.
We have recently found that La^{+++} inhibits the high affinity binding; as shown in Figure 7, the Scatchard plot
in the presence of La^{+++} is clearly monophasic, and the
class of binding sites completely abolished is that having high affinity for Ca^{++}.

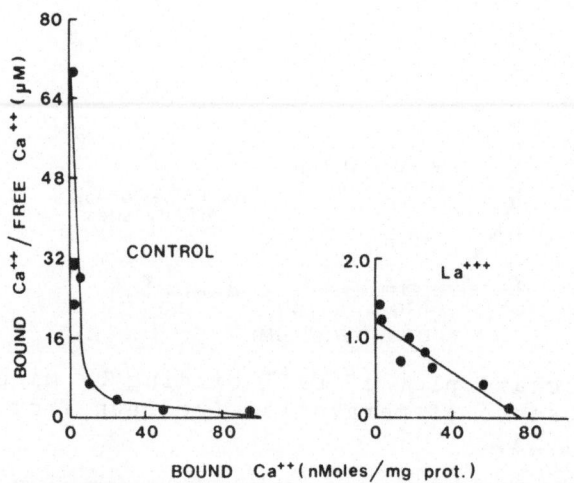

Figure 7. Effect of La^{+++} (1 µM) on the high affinity
binding. Experimental conditions as in Figure 5.

    In agreement and in addition to the findings of o-
ther Laboratories, we have reported that the breakdown
of mitochondrial phospholipids is paralleled by the de-
cay of respiratory control in mitochondria aged or treated with phospholipase A (15, 16). The loss of respiratory control under these conditions is not due to the
formation of lysocompounds derived from the hydrolysis
of phospholipids. In fact, when mitochondria are exposed to increasing concentrations of lysophosphatidylcho-
line, they retain complete respiratory control and thus
show normal ADP/O ratio.

    In Figure 8 it is shown a typical chromatogram of
intact freshly isolated mitochondria as compared with
aged and phospholipase treated organelles. In this experiment mitochondrial phospholipids were separated by

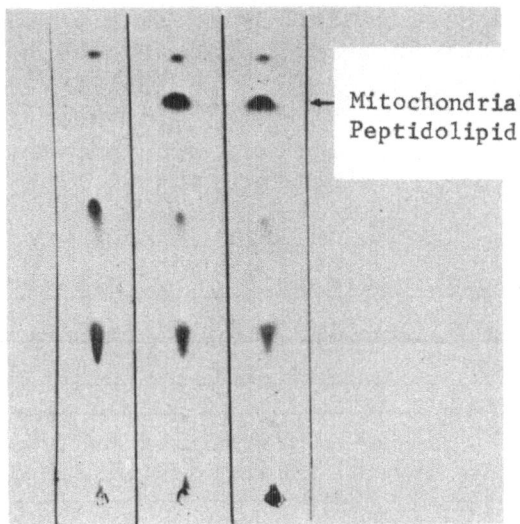

A. Fresh isolated mitochondria
B. Aged mitochondria (24 hr at 0°C)
C. Phospholipase treated mitochondria

Figure 8. Direct chromatography of rat liver mitochondria on silica gel, according to Curri et al. (18).
a) intact, fresh isolated mitochondria; b) mitochondria
incubated in 250 mM sucrose for 30 min. at 20°C; mitochondria incubated with phospholipase A (0.1 µg per mg
mitochondrial protein) for 5 min. at 20°C, with gentle
stirring.
1: phosphatidylcholine; 2: phosphatidylethanolamine;
3: mitochondrial peptidolipid (MPL).

a procedure proposed in our Laboratory (17). According
to this method, the direct thin layer chromatography of
mitochondria without any previous extraction step, re-
solves the phospholipid fractions into their components
as easily as does thin layer chromatography of chloro-
form-methanol extracts of mitochondria. In the same Figure it is seen that a spot running ahead of phosphatidylethanolamine (plus phosphatidylserine) has been detected only in aged and phospholipase A treated mitochondria. This mitochondrial peptidolipid (18) (MPL) is
similar in composition to the material extracted by Remmert and Lehninger (19) from the "Releasing Factor";
the two compounds differ only in their respective fatty
acid composition. From time to time speculations have
been made as to the contribution of these materials in

releasing the respiration from the control by ADP, and
the relevance of these compounds to the mechanism of
translocation of Ca^{++} across the mitochondrial membrane is under detailed investigation in our Laboratory.

In an attempt to specify and correlate the binding
of Ca^{++} with the phospholipid composition of the mito-

Table 1

Phospholipid Composition of Mitochondria

Isolated from Different   Mammalian Tissues

and from Blowfly Flight Muscle

| | Mitochondria from | | | |
	Liver	Heart	Kidney	Blowfly
phosphatidyl-choline	43	41	40	12
phosphatidyl-ethanolamine	30	38	38	65
phosphatidyl-inositol	3	3	3	10
cardiolipin	17	19	19	15
(the values are expressed as % of the total lipid P)				

chondrial membrane, we have fractionated the phospholipids from a wide variety of mammalian tissues, as well
as from blowfly flight muscle mitochondria. Blowfly flight mitochondria are of particular interest, in view of
the highly unusual properties of Ca^{++} transport recently described by Carafoli et al. (20). The separation of
phospholipids into different components has revealed no
radical differences between mitochondria prepared from
mammalian liver, kidney or heart. As expected by quot-
ing mean result of the determinations in our and other
laboratories (21), the phosphatidylcholine : phosphatidylethanolamine ratio is slightly above 1. On the other
hand, Carafoli et al. (20) have recently reported a completely different phospholipid composition of mitochondria isolated from blowfly flight muscle. In these organelles the above ratio is about 0.2. The summary of all
these data is reported in Table 1. The most striking difference concerns phosphatidylethanolamine, which is far
more abundant in blowfly than in mammalian mitochondria.
Carafoli et al. (20) have reported that metabolism-inde-

pendent Ca^{++} binding is almost completely absent in blowfly flight muscle mitochondria; this indicates that phosphatidylethanolamine is unlikely to play any role in this type of binding in these mitochondria. A different situation may however prevail in mitochondria from different sources (22).

A summary of the main properties of the different Ca^{++} binding mechanisms taken by a recent review by Carafoli and Rossi (23) and modified somewhat to include some data presented in this paper, is shown in Table 2. It is clear that energy-linked uptake and high affinity binding, share many properties. On the other hand, low--affinity binding, and energy-linked movement differ in many respects.

Table 2

Comparative Properties

of the Different Ca^{++} Binding Mechanisms

	requir. for resp. or ATP hydrol.	aff. (K_m)	Inhib. by	Biol. distr.	sensit. to phospholip. A
Met.-dep.	yes	$\sim 10^{-6}$M	DNP,La^{3+}	absent from:yeast blowfly muscle(+)	+
High-aff. binding	no	$\sim 10^{-6}$M	DNP,La^{3+}	absent from:yeast blowfly muscle	+
Low-aff. binding	no	$\sim 10^{-4}$M	Butac.	absent from: blowfly muscle	+++

(+) Under very special conditions active uptake of Ca^{++} can be demonstrated in yeast and blowfly muscle mitochondria.

A soluble protein component capable of binding Ca^{++} with a very high affinity constant and most likely responsible for the high affinity reaction, is presently under investigation in Lehninger's Laboratory (24). Its proper functioning apparently requires the integrity of the phospholipid structure of mitochondria. Figure 9

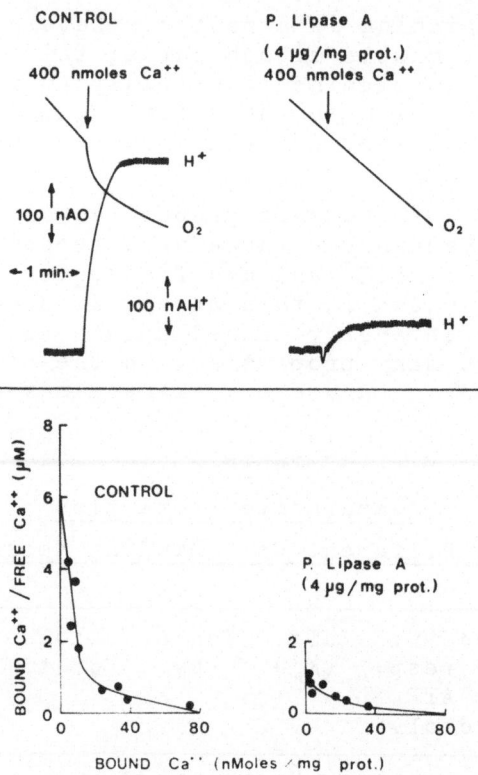

Figure 9. Effect of phospholipase A on high affinity
Ca^{++} binding and energy-linked Ca^{++} uptake by mitochondria. Conditions as in Figure 3. The incubation with
phospholipase A was carried out at 21°C for 5 min., in
250 mM sucrose + 0.5% BSA, with gentle stirring. At the
end of incubation, aliquots of the suspension containing the mitochondria were added to the reaction medium.
Specific activity of phospholipase A was 200 U/mg.

shows that a brief incubation of mitochondria with phospholipase A eliminates the high affinity binding of Ca^{++}.
It is interesting that the abolishment of the high affinity reaction is accompanied by the disappearance of the
energy-linked translocation.

As for the problem of the relationships among the
different Ca^{++} binding mechanisms, there are several lines of evidence that relate the high-affinity binding
reaction with the energy-dependent translocation of Ca^{++}.
The two systems reveal the same affinity for the substrate, and are both sensitive to the same inhibitors.

As shown in Figure 10, the disappearance of the e-

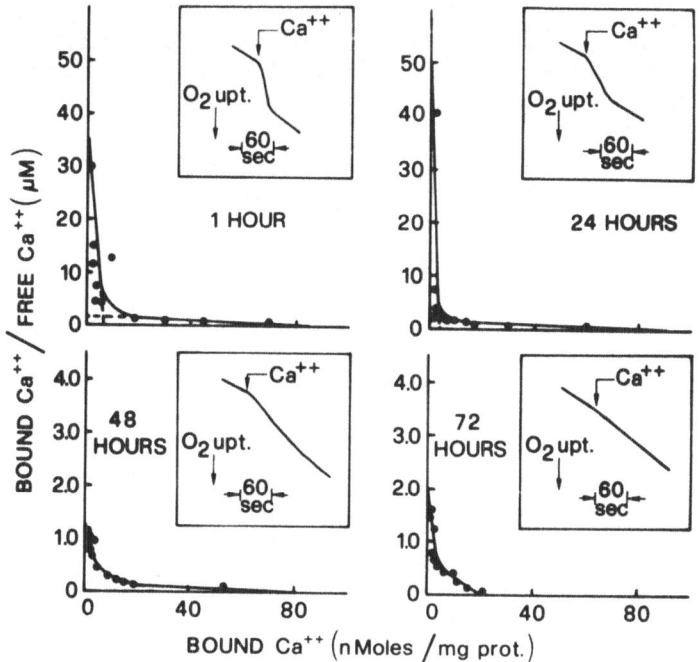

Figure 10. Energy-linked Ca++ uptake, and high-affinity binding in aged mitochondria. Conditions as in Figure 3. Mitochondria were aged at 2°C for 48 hours.

nergy-linked translocation in aged mitochondria is accompanied by the absence of the high-affinity binding sites. Finally, we have recently found that high affinity reaction and energized translocation share a common distribution within the mitochondrial structure (inner membrane).

It is thus reasonable to conclude that the high-affinity reaction, representing a carrier molecule, plays a role in the energy-dependent Ca++ uptake. Work is now in progress on the chemical nature of the Ca++ carrier, its mode of action, and its functional link with other ion transport systems in the mitochondrial membranes.

This research was supported by a grant to C.S.R. from J.S. Karmali, Ltd, Leitz East Africa, Nairobi, Kenya.

Acknowledgements. Some of the experiments described in this paper were carried out with the collaboration of Godfrey Maina of the Department of Medical Biochemistry, University of Nairobi, Nairobi, Kenya. The collaborations of drs. Roberto Tiozzo and Paolo Gazzotto of the Institute of General Pathology of the University of Modena is also gratefully acknowledged.

REFERENCES

1. Lehninger,A.L., Carafoli,E., and Rossi,C.S., Adv. Enzymol. 29: 259 (1967).

2. Rossi,C.S., and Lehninger,A.L. Biochem.Z. 338: 698 (1963).

3. Greenawalt,J.W., Rossi,C.S., and Lehninger,A.L., J. Cell Biol. 23: 21 (1964).

4. Chance,B., J.Biol.Chem. 240: 2729 (1965).

5. Rossi,C.S., and Lehninger,A.L., J.Biol.Chem. 239: 3971 (1964).

6. Chance,B. Ed. Energy Linked Functions of Mitochondria, Academic Press, New York (1963).

7. Sris,N.E. Soc.Sci.Fennica, Comment.Biol. 28:1 (1963)

8. Gear,A.R., Rossi,C.S., Reynafarje,B., and Lehninger, A.L., J.Biol.Chem. 242: 3403 (1967).

9. Rossi,C., Azzi, A., and Azzone,G.F., J.Biol.Chem. 242: 951 (1967).

10. Scarpa, A., and Azzi,A., Biochim.Biophys.Acta 150: 473 (1968).

11. Scarpa,A., and Azzone,G.F., Europ.J.Biochem. 12: 328 (1970).

12. Reynafarje,B., and Lehninger,A.L., J.Biol.Chem. 244: 584 (1969).

13. Lehninger,A.L., Rossi,C.S., Carafoli,E., and Reynafarje,B. in Mitochondria Structure and Function (eds.L.Ernster,Z.Drahota,Academic Press, 1969).

14. Mela,L., Arch.Biochem.Biophys. 123: 286 (1968).

15. Mela,L., and Chance,B., Biochem.Biophys.Res.Comm. 35: 556 (1969).

16. Rossi,C.R., RossiC.S., Sartorelli,L., Siliprandi, D., and Siliprandi,N., Arch.Biochem.Biophys. 99: 214 (1962).

17. Rossi,C.R., Sartorelli,L., Tatò,L., and Siliprandi, D., Arch.Biochem.Biophys. 107: 170 (1964).

18. Curri,S.B., Rossi,C.R., and Sartorelli,L. Thin Layer Chromatography (Ed.G.B.Marini-Bettolo), Elsevier Amsterdam (1964).

19. Remmert,L.F., and Lehninger,A.L., Proc.Natl.Acad. Sci. 45: 1 (1959).

20. Carafoli,E., Saktor,B., and Lehninger,A.L. in press

21. Fleischer,S., Fleischer,B., Stoekenius,W., J. Cell Biol. 32: 193 (1967).

22. Scarpa,A., and Azzone,G.F., Biochim.Biophys.Acta 173: 78 (1969).

23. Carafoli,E., and Rossi,C.S., Proceedings of the 1st Intern. Symp. on Cell Biol. and Cytopharmacol., Venice, 1969, Raven Press, N.Y., in press.

24. Lehninger,A.L., and Carafoli,E., Fed. Proc. 28: 669 (1969).

BRAIN NUCLEOTIDES AND EXCITATORY PROCESSES

V. Bonavita   and   F. Piccoli

Department of Neurology, University of Palermo

Palermo, Italy

The trivial analogy in the structure of barbiturates and pyrimidines has suggested a series of studies, which were undertaken in our Laboratory since 1961 (Bonavita et al., 1961) to investigate on the role of some free nucleotides in cerebral activity.

Studies on non-polymerized nucleotides appear as a useful method to obtain data on the energetic balance of the tissue and also, though not directly, on the biosynthesis of structural components to which free nucleotides participate (see Mandel and Weill, 1962).

The experiments reported in the present review are mainly concerned with the attempt of correlating changes in the turnover of some nucleotides with brain activity. They give also an unexpected insight into the possible compartmentation of pyrimidine nucleotides in the brain of adult animals.

It was on the basis of a first series of neurochemical experiments (see Piccoli et al., 1969, 1970a and Guarneri et al., 1970) that two years ago we could suggest the presence, in the nervous tissue, of two metabolic pools of UMP. Fig. 1 gives a graphic representation of the two pools. Pool "A", whose precursor is orotate (see Guarneri et al., 1970) is a large pool, with a relatively slow turnover, in which the conversion of UMP into CMP seems to take place (Piccoli et al., 1969, 1970b); it is also the pool that receives UMP from RNA (Piccoli et al., 1969). The existence of a second, small pool,

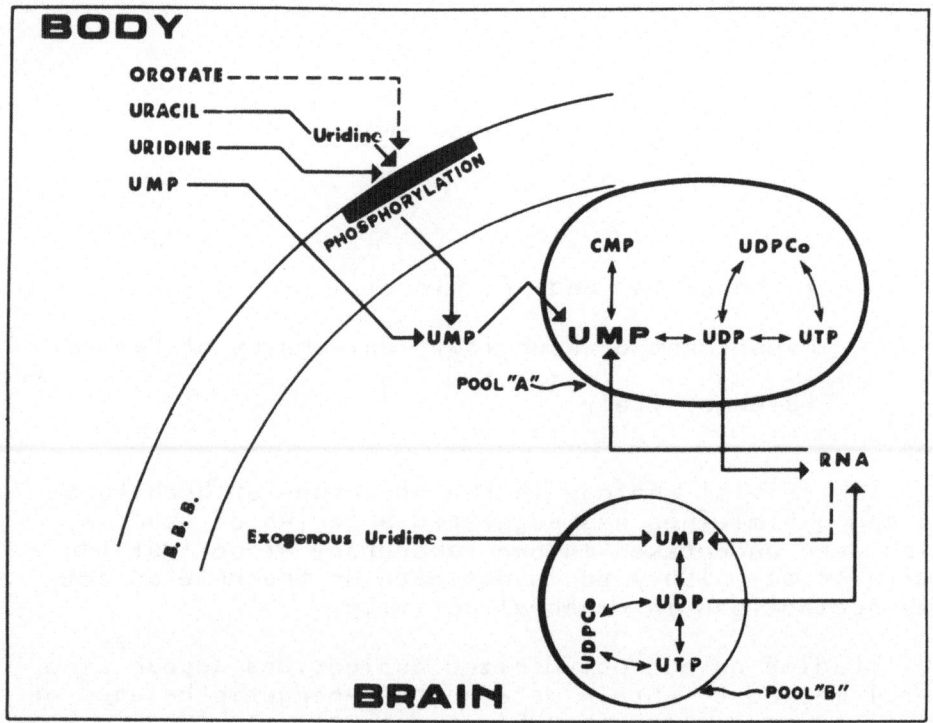

Fig. I - A grafic representation of the two pools of
          UMP. For description, see text.

with a high rate of turnover (pool "B" in Fig. I) has
been detected by injecting 3H-uridine intracisternally
(Piccoli et al., 1969).

     The finding of such a metabolic compartmentation of
UMP and uracil-containing nucleotides has acted as
the major stimulus for the investigation of the function
of UMP and the selective role of each of the two pools.
Three different experimental approaches have been chosen
for this purpose. Brain concentration and turnover of
pyrimidine nucleotides have been analyzed: i) after
electrical stimulation; ii) during pharmacological sleep;
iii) after bilateral eye enucleation in the rat.

                    o   o   o   o

     i) When labelled uridine is injected intracisternal-
ly to the rat, UTP is the compound exhibiting the highest specific activity at the earliest time; UDPCo (UDP-
acetylglucosamine plus UDP-glucose) and UMP both follow

UTP, UMP reaching a slightly higher specific activity than UDPCo by 60 min after injection. CMP has a definitely lower specific radioactivity which attains its maximum not earlier than 2 hours (Fig. 2) (Piccoli et al., 1969).

A maximal electroshock seizure is responsible for abrupt changes in the brain level and turnover of several nucleotides (see Bonavita, 1967). Table I summarizes the quantitative changes of only uracil derivatives at the earliest times after electroshock. As indicated also in Fig. 3, changes of UTP duplicated those of ATP with minor differences (see Bonavita, 1967). The extra UTP measured at 5 min after electroshock should derive from extra UMP, measured in the convulsant brain not accounted for by the decrease of UTP and UDPCo. Since post-paroxysmal metabolic events are related to recovery processes, it is interesting to recall that the UTP level in the brain of the immature rat is much higher than in the adult tissue

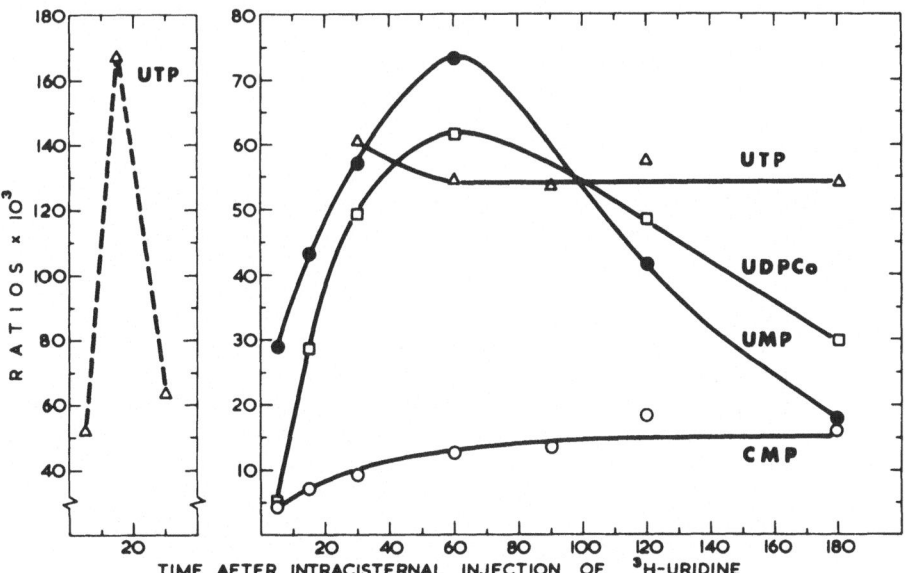

Fig. 2 - Radioactivity of UTP, UDPCo, UMP and CMP from the acid-soluble fraction of rat brain after intracisternal injection of 3H-uridine. Values are expressed as ratios between the specific radioactivity of each nucleotide and the total radioactivity of the acid-soluble fraction.
From Piccoli et al., 1969.

Table I - Concentration of uridine phosphates in rat
          brain before and after an electroshock seizure

	UMP	UDPCo	UTP
Control (A)	1.8	25.5	12.3
5 sec after ESK (B)	16.8	18.3	8.0
15 sec after ESK (C)	13.8	19.25	10.4
(B) - (A)	+ 15.0	- 7.2	- 4.3
(C) - (A)	+ 12.0	- 6.25	- 1.9

Values are expressed as micromoles in 100 grams of
wet tissue.

(Jacob and Mandel, 1966). One wonders which metabolic
role of UTP plays a major part in the need of the
transitory post-convulsive excess of brain UTP. UTP
and UDPCo did not balance the increase of UMP during
the development of the convulsant activity: it can be
assumed that a UMP-containing compound is split during
convulsant activity. The possibility that one or more
species of ribonucleic acids are the source of this UMP
is supported by preliminary evidence obtained in our laboratory (Bonavita and Guarneri, unpublished data).

     Analysing the effect of electroshock on the turnover
of pyrimidine nucleotides (Fig. 4 to 7), one should under-
line the similarity of changes of specific radioactivity
of CMP and UMP: a sudden fall of specific radioactivity
during the seizure, a quick recovery to control values
in the case of CMP, and a slower recovery for UMP. When
UTP was investigated in the brain during electroshock, a
different pattern was observed. The specific radioactivity decreased during convulsant activity when this occurred 15 min after the injection of the labelled precursor
(experiment A in Fig. 6), did not change significantly

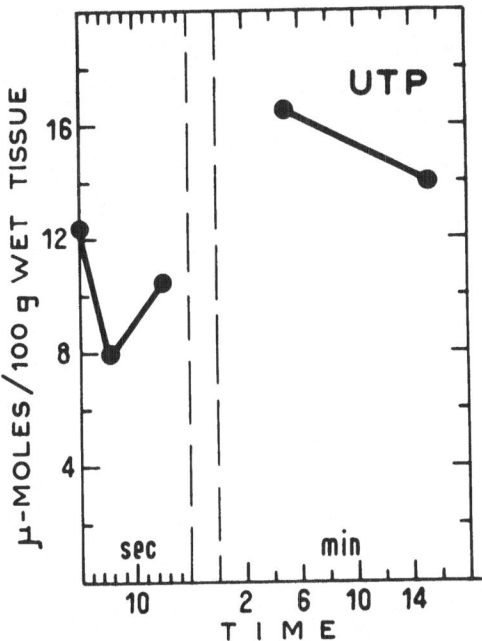

Fig. 3 - Quantitative changes of UTP in the rat brain
after electroshock. The statistical analysis
(Student's t) has shown that the concentration
of UTP is significantly different from that of
controls (at the 0.1 level of probability) up
to 5 min after electroshock.
From Piccoli et al., 1969.

if the electroshock was given at 30 min (experiment B),
increased at 60 min (experiment C). This last experiment
was the only one performed when UTP was in equilibrium
with its precursor.

As already noted, experiments with labelled uridine
contribute to knowledge of the physiological turnover
of pyrimidine nucleotides as well as of the metabolic
changes related to convulsant and post-paroxysmal events.

With reference to the physiological turnover of
brain nucleotides, the high and early specific radioactivity of UTP is the most intriguing observation we
have reported on. The finding is in apparent contrast
with the postulated sequential phosphorylation of uridine.
A possible explanation is that brain UMP exists in two
separate pools with different turnover rates, and that
the smaller pool with a high turnover rate is the pre-

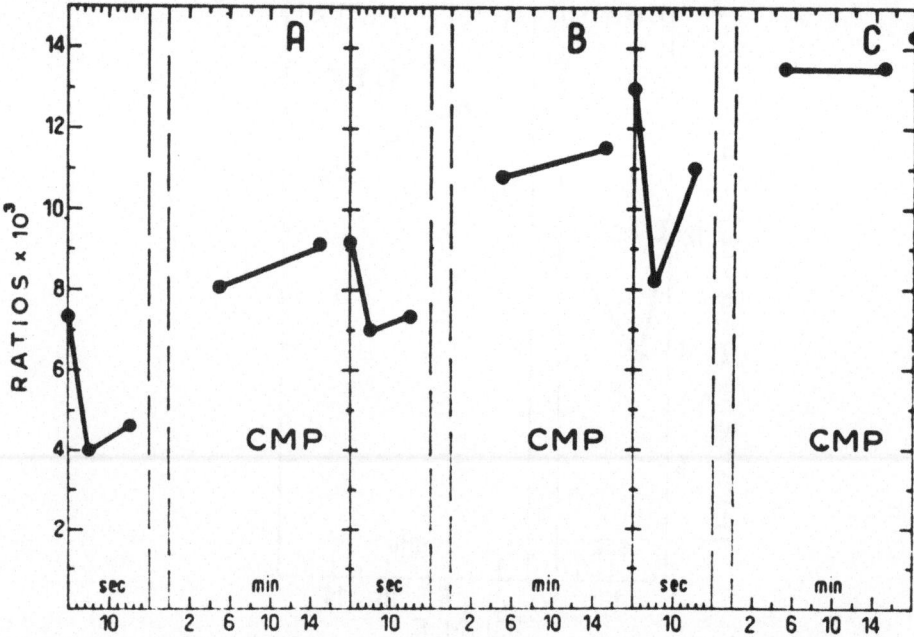

Fig. 4 - Changes of the radioactivity of free CMP in
the rat brain after electroshock. In experiments A, B and C, the electroshock was given
at 15,30 and 60 min respectively, after the
intracisternal injection of 3H-uridine. Values
are expressed as in Fig. 2.
From Piccoli et al., 1969.

cursor of UTP from uridine.

Another interesting finding is the relatively slow
turnover of CMP as compared with uracil derivatives. If
the existence of two pools of UMP is postulated, CMP
would presumably derive from the pool of lower turnover.

While quantitative measurements did not show significant changes of the CMP concentration in the convulsant
brain, experiments with labelled uridine revealed an
abrupt change of CMP turnover during the seizure (see
Fig. 4). Also, the specific radioactivity of CMP returned
to its control value much earlier than UMP. To explain
this observation, it may be assumed that the long-lasting
subnormality of UMP specific radioactivity in electro-
shocked animals is due to a dilution of UMP deriving from
the dephosphorylation of UTP with UMP of lower specific
radioactivity from a different source. It is evident from

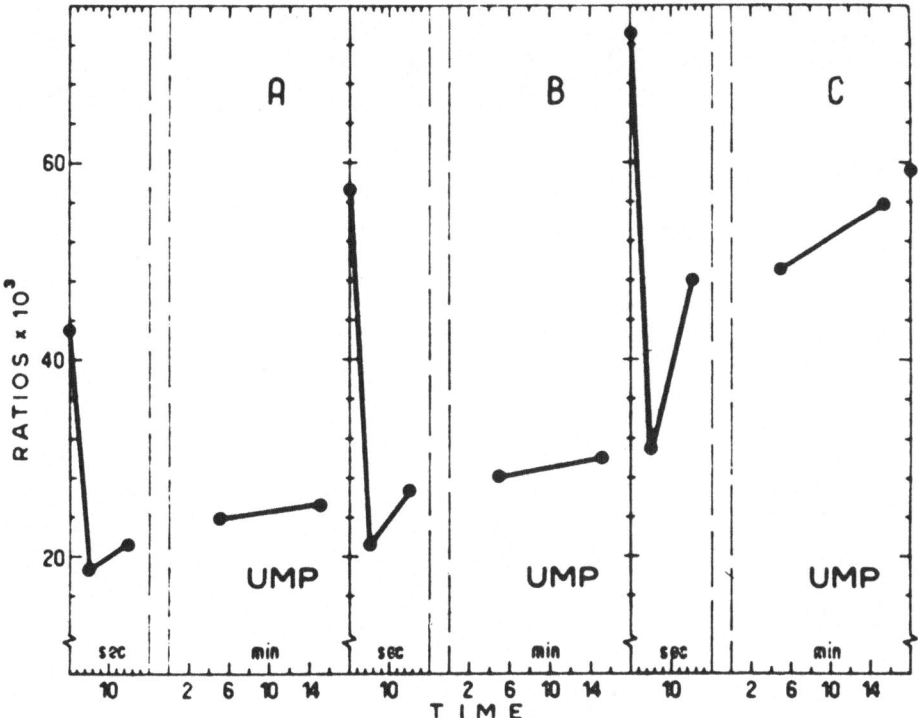

Fig. 5 -   Changes of the radioactivity of free UMP in
           the rat brain after electroshock. For details,
           see Fig. 4.
           From Piccoli et al., 1969.

Fig. 5 that any decrease of UMP labelling related to convulsant activity is greater at 15 or 30 min after uridine
injection, i.e. when the specific radioactivity of UMP is
normally increasing (Fig. 5).

     The prolonged maintenance of high levels of UTP after
a seizure is in keeping with the finding of a specific
radioactivity still above normal at 15 min after electroshock (Fig. 6). Since UTP measured after convulsant activity arises from a sequential phosphorylation of UMP, it
must be explained why the specific radioactivity of the
triphosphorylated nucleotide is higher than normal at the
time of the subnormal specific radioactivity of UMP. Such
a discrepancy suggests again that UMP exists in two separate pools only one of which is in equilibrium with UTP,
i.e. UMP from pool "B".

     Data on UDPCo are also quite interesting (Fig. 7).

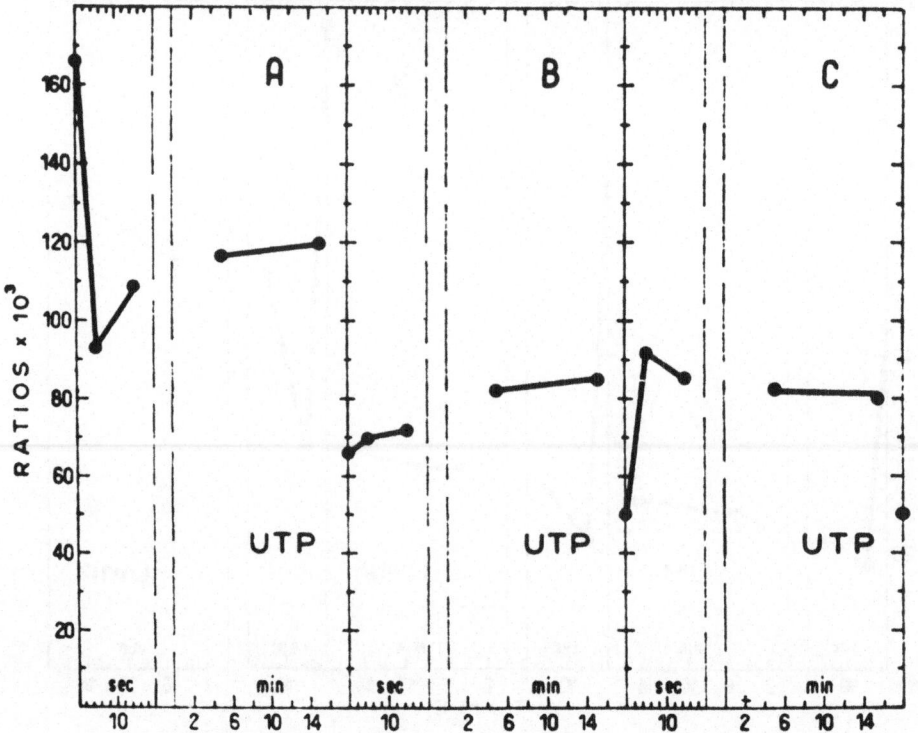

Fig. 6 -   Changes of the radioactivity of free UTP in
           the rat brain after electroshock. For details,
           see Fig. 4.
           From Piccoli et al., 1969.

Turnover studies show that their specific radioactivity
is normal at 15 min after the electroshock, while the
quantitative measurements reveal the persistence of a
significant decrease of the compounds at the same time.
The easiest explanation for such a divergence is that
UDPCo are not resynthetized at a high rate since UTP is
preferentially funnelled towards a different recovery
process.

                        o o o o

     ii) A different line of investigation on the role
of free nucleotides in brain activity has been the study
of their turnover in the rat brain during pharmacological
sleep (Piccoli et al., 1970b).

     Changes of levels and turnover of purine and pyrimi-
dine nucleotides were measured during sleep induced by a
single injection of a short-acting barbiturate, sodium

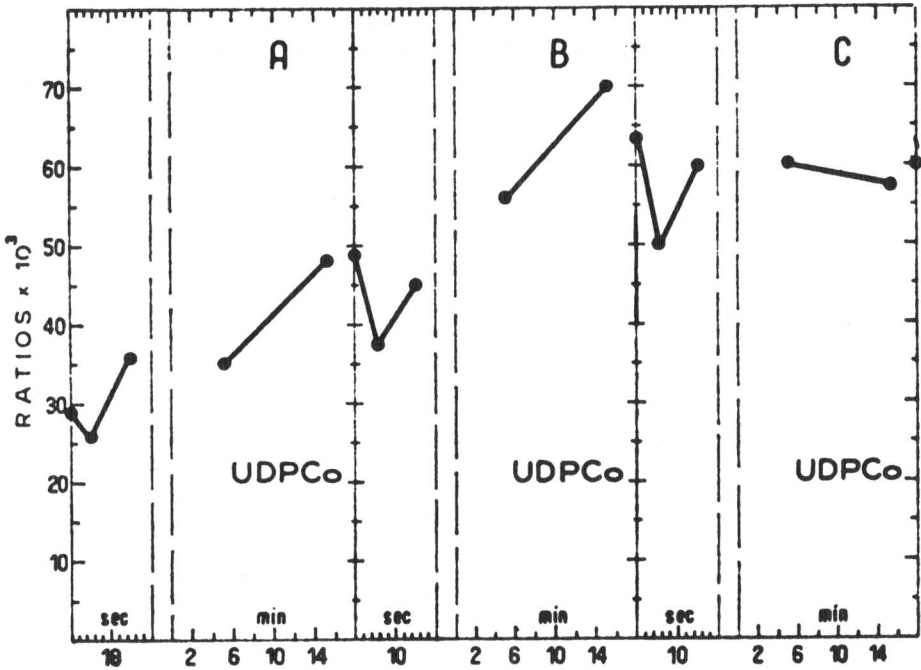

Fig. 7 -   Changes of the radioactivity of free UDPCo
in the rat brain after electroshock. For
details, see Fig. 4.
From Piccoli et al., 1969.

thiopental, or of sodium gamma-hydroxybutyrate, which
is responsible for a sleep behaviourally similar to spontaneous sleep.

Table II shows the quantitative changes of UMP, GTP
and ATP in the sleeping brain. The analysis of the specific radioactivity of UTP, UDPCo and UMP leads to the
conclusion that the turnover of these compounds is slowed
down during sleep induced by both the drugs used. Fig. 8
shows that CMP exhibited a definite increase of incorporation of radioactivity. Both sodium thiopental and
sodium gamma-hydroxybutyrate induced similar changes in
the compounds analyzed, thus suggesting that changes of
levels and turnovers measured after the injection of the
two drugs are related to sleep itself.

Sleep seems to be characterized by an increase of
the cerebral level of UMP and by a preferential conversion

Table II -   Quantitative changes of brain
nucleotides in the sleeping rat

Comp.	Control	10 min		60 min		120 min	
		A	B	A	B	A	B
UMP	1.8	4.2	3.2	4.9	4.1	4.6	4.4
ATP	147.0	112.4	150.0	115.0	148.2	111.5	139.6
GTP	29.5	23.3	22.6	25.8	26.5	25.0	28.4

A: after sodium thiopental   B: after sodium gamma-
                                         hydroxybutyrate

of UMP into CMP. Since the level of UTP and UDPCo did
not show any change during sleep to balance the increase
of UMP, the conclusion may be that an extrasynthesis of
UMP takes place in the sleeping brain. As an alternative
pathway, one could postulate that an UMP-containing compound is split during sleep. Since the specific radioactivity of UMP decreases greatly (Fig. 9), mainly during
the sleep induced by gamma-hydroxybutyrate, it seems to
be likely that UMP from the large pool (pool "A", see
above) is increased. As a consequence of such a peculiar
distribution of UMP between its two pools, the turnover
of UTP (Fig. 10) and UDPCo (Fig. 11) is slowed down and
the turnover of CMP is increased. The alternative hypothesis of a decrease of UMP phosphorylation, UMP being funnelled into CMP, contrasts with the lack of a decrease
of UTP concentration.

Even considering the peculiar relationship between
UMP and CMP in the sleeping brain, as compared to awake
controls, one could say, as a general statement, that
there is no qualitative difference between sleep and
wakefulness. Such a statement appears as a return, in
neurochemical terms, to Sherrington's hypothesis on the
depressed activity of the sleeping brain (Sherrington,
1947). At any rate, the decrease of turnover of uridine

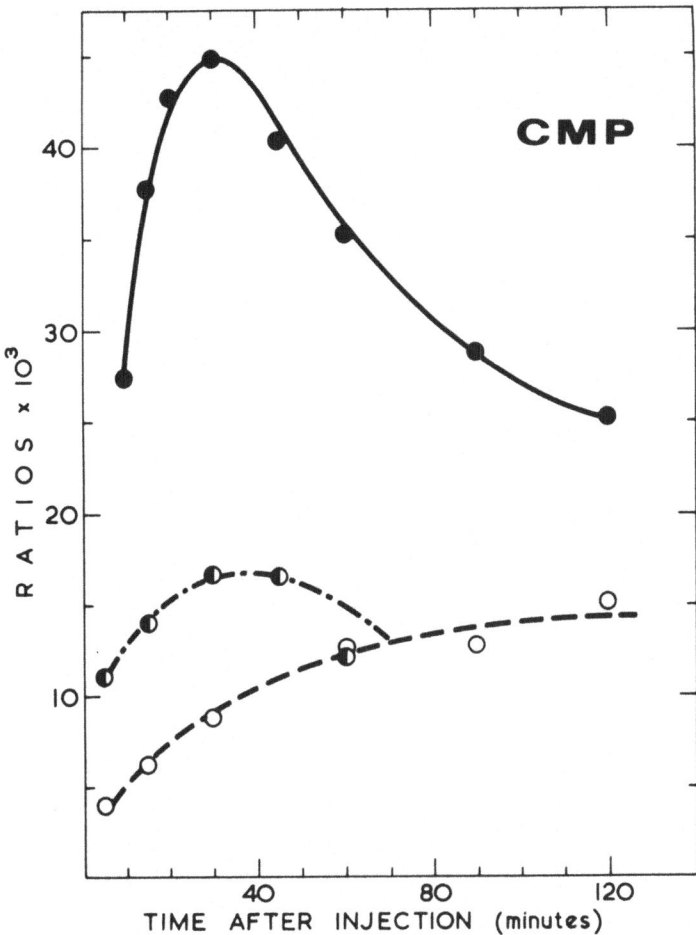

Fig. 8 - Changes of the radioactivity of free CMP in the rat brain during pharmacological sleep. Dotted line and open circles: control; dotted line and semi-closed circles: after sodium gamma-hydroxybutyrate; solid line: after sodium thiopental. Values are expressed as in Fig. 2. From Piccoli et al., 1970b.

derivatives points out to a possible mechanism of recovery during sleep, and seems to be related to the potentiation by uridine derivatives of thiopental induced sleep (Bonavita et al., 1964).

o o o o

iii) The results described above have several limitations. The convulsive outburst is a phenomenon not lo-

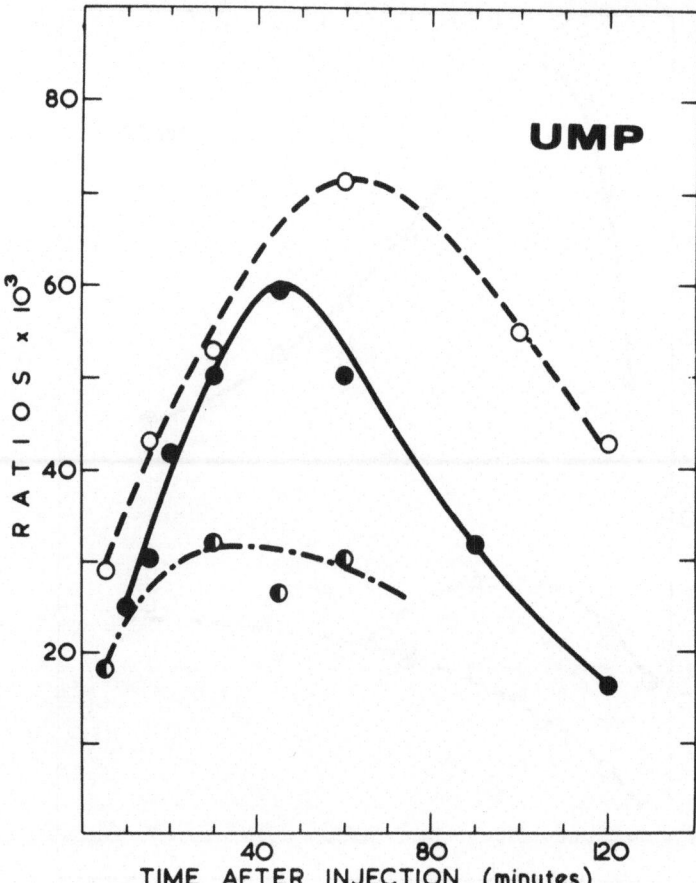

Fig. 9 - Changes of the radioactivity of free UMP in
         the rat brain during pharmacological sleep.
         For details, see Fig. 8.
         From Piccoli et al., 1970b.

calized to defined areas of the brain; drug induced
sleep may be criticised, because drugs can be responsible
for effects on brain tissue, not related to sleep itself.
For studies seeking to investigate metabolic changes in
relation to functional activity of the brain, investigations on sensorial areas, i.e. visual cortex after
light stimulus or sensorial deprivation, have appeared
as a better approach.

     The following series of results concerns the changes
in concentration and turnover of free nucleotides from
the occipital lobe of the rat, at different periods of
time after bilateral eye enucleation (Guarneri et al.,

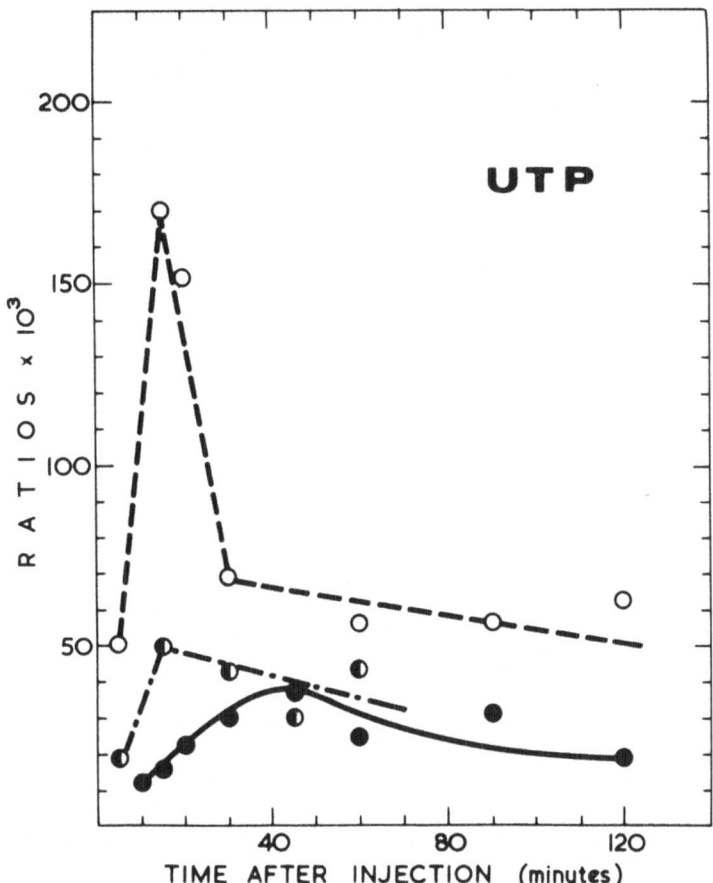

.Fig. 10 - Changes of the radioactivity of free UTP in
the rat brain during pharmacological sleep.
For details, see Fig. 8.
From Piccoli et al., 1970b.

1970). In this series of experiments, the turnover of pyrimidine nucleotides was studied after intracisternal
injection of 14C-orotate.

Before any comment on the experimental results one
should emphasize that, when utilizing this pyrimidine
precursor, the sequence of labelling was consistent with
the postulated pathway of sequential phosphorylation of
uracil-containing nucleotides. Thus, the compound exhibiting the highest specific activity at the earliest time
was UMP followed by UTP, UDPCo and CMP (Fig. 12), uridine
not being an intermediate of the metabolic pathway from
orotate to UMP.

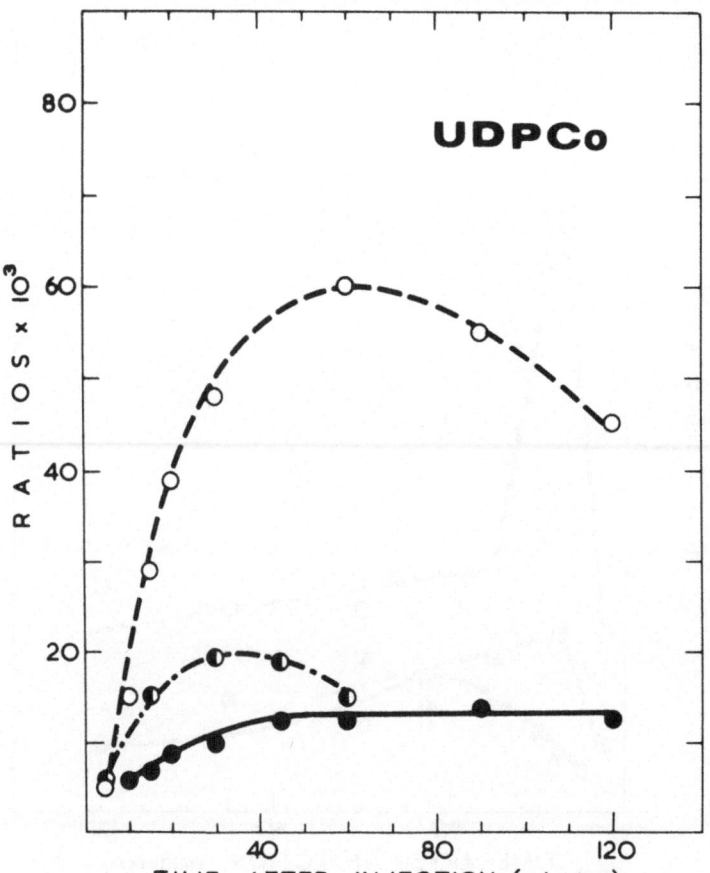

Fig. 11 - Changes of the radioactivity of free UDPCo in
the rat brain during pharmacological sleep.
For details, see Fig. 8.
From Piccoli et al., 1970b.

Table III summarizes the quantitative changes of
UTP and GTP from the occipital and frontal lobes of the
rat at various times after bilateral enucleation. Among
the nucleotides isolated, UTP and GTP were the only two
compounds that underwent significant modifications in
these experimental conditions, but the turnovers of UMP,
CMP and UTP were all slowed down in the brain areas under
investigation. The specific radioactivity of UDPCo (Fig.
13) shows a clear increase both in the occipital and
frontal areas at the earliest times after enucleation.

The general decrease of turnover of UMP (Fig. 14)
and UTP (Fig. 15) does not seem to require a particular

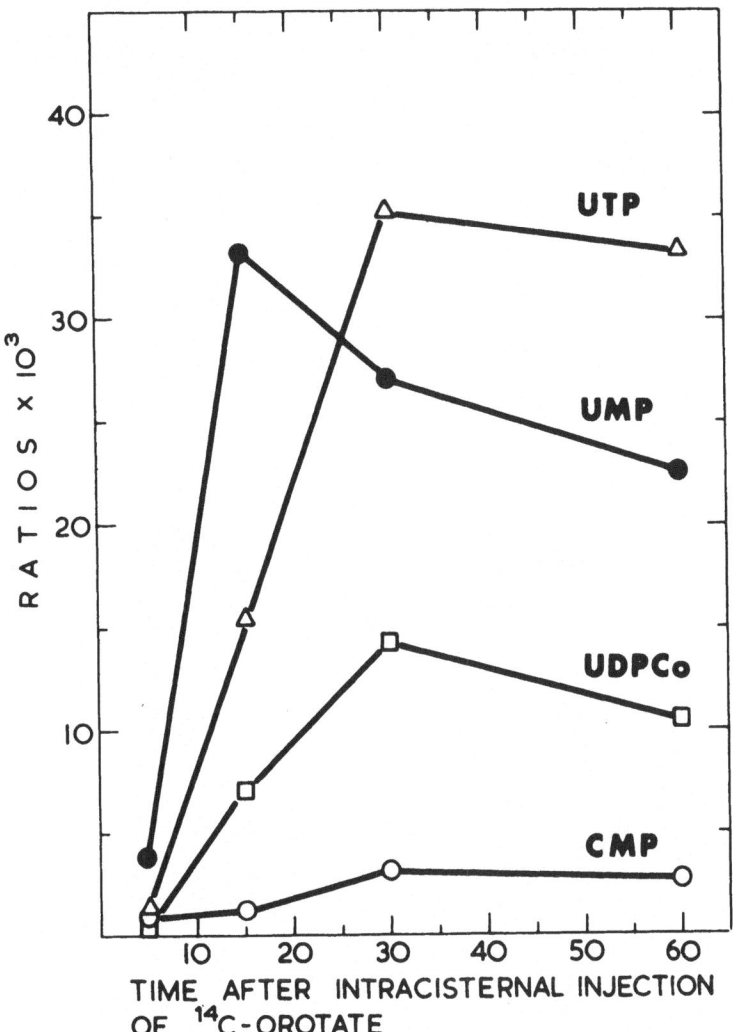

Fig. 12 - Radioactivity of UMP, UTP, UDPCo and CMP from
           the acid-soluble fraction of rat brain after
           intracisternal injection of 14C-orotate.
           Values are expressed as in Fig. 2.
           From Guarneri et al., 1970.

explanation. Conversely, the early and persistent increase of UTP concentration and the increased specific
radioactivity of UDPCo deserve some comment. As it is
known, one of the types of enzymatic catalysis to which
UDPCo participate is the lengthening of the glycogen
chain by addition of terminal glucose. It is also known
that in the depression of cerebral activity by prolonged
anesthesia an accumulation of glycogen in brain occurs

Table III -   Quantitative changes of brain nucleotides
              after bilateral enucleation of the rat.

Comp.	Control		7 hrs		25 hrs		100 hrs	
	F	O	F	O	F	O	F	O
UTP	13.8	11.3	20.1	18.1	22.0	18.2	12.0	12.0
GTP	28.5	24.5	21.2	12.0	24.5	16.4	25.4	25.0

F: frontal lobe   O: occipital lobe

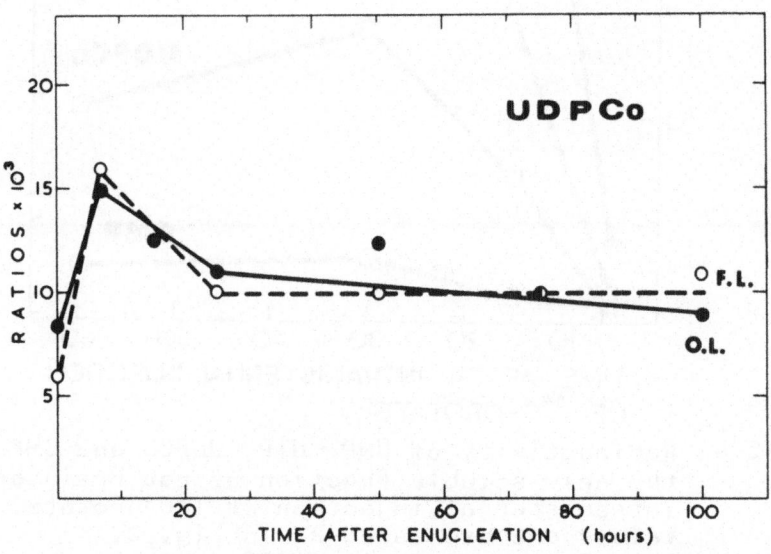

Fig. 13 - Changes of the radioactivity of free UDPCo
           from different areas of the rat brain after
           bilateral eye enucleation.
           F.L. = frontal lobe; O.L. = occipital lobe.
           Animals were killed 30 min after intra-
           cisternal injection of 14C-orotate. Values
           are expressed as in Fig. 2.
           From Guarneri et al., 1970.

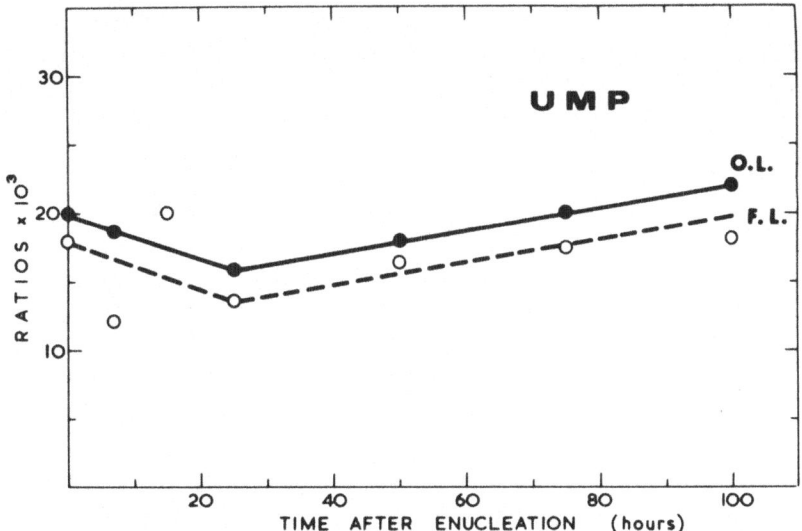

Fig. 14 - Changes of the radioactivity of free UMP
from different areas of the rat brain after
bilateral eye enucleation. For details, see
Fig. 13.
From Guarneri et al., 1970.

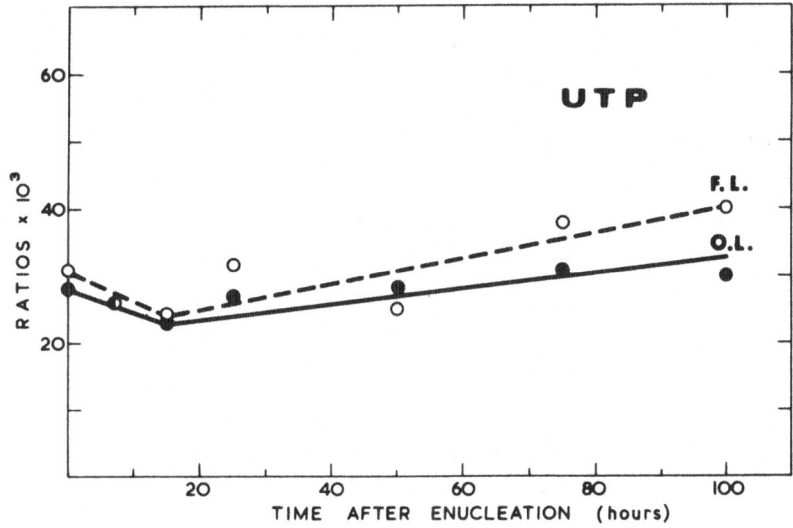

Fig. 15 - Changes of the radioactivity of free UTP
from different areas of the rat brain after
bilateral eye enucleation. For details, see
Fig. 13.
From Guarneri et al., 1970.

(Nelson et al., 1968). The quantitative increase of UTP
participating in the production of UDPCo (Munch-Peterson
et al., 1953) and the increase of the turnover of brain
UDPCo after bilateral enucleation support the hypothesis
that a glycogen increase occurs even after sensorial
deprivation.

The role of GTP in providing energy for moving
m-RNA relative to the ribosome and positioning the amino-
acyl-t-RNA seems quite probable (Mandel, 1970), and
it may thus be postulated that the decrease of GTP
after enucleation is somehow related to a decreased synthetic rate of proteins and RNA (Bondy and Margolis,
1970).

Wegener (1970), Bondy and Margolis (1970) and
Margolis and Bondy (1970) have found that light stimulation or deprivation produce, in different animals, metabolic changes diffused to the whole hemisphere affected.
These and our own results demonstrate the need for both
input and integration in order to develop optimal activity in single brain areas.

o o o o

By following three different experimental lines we
have discussed some aspects of brain activity involving
free nucleotides. The results we have presented suggest
quite strongly the existence of a specific role of uracil-containing nucleotides in developing and suppressing
the activity of nervous tissue.

UMP seems to play indeed an interesting role in
brain excitability. Several findings support this hypothesis: i) UMP as well as uridine is an effective anti-
convulsant agent when injected intraperitoneally to animals (Bonavita et al., 1964); ii) cerebral concentration
of UMP increases by many times (7 to 8) during the arrest
of the convulsive seizure (Piccoli et al., 1969); iii)
pharmacological sleep activates the large pool of UMP increasing 2 to 4 times the cerebral concentration of the
nucleotide and shifting its metabolic pathway toward CMP
(Piccoli et al., 1970b); iiii) the diffuse increase of
UTP concentration in brain areas of animals submitted to
bilateral eye enucleation and the simultaneous apparent
decrease of UTP turnover (Guarneri et al., 1970), suggest
that the turnover of UMP from the large and slow pool is
accelerated in sensorial deprivation, though no evidence
supporting such a conclusion has been obtained in experiments with labelled orotate. The increase of turnover of
UDPCo is the only finding in favour of the hypothesis.

REFERENCES

(1)   Bonavita, V., Scarano, E., and Zito, M. Acta Neurol. 16: 493 (1961).

(2)   Mandel, P., and Weill, J.D. J. Physiol. (Paris) 54: 190 (1962).

(3)   Piccoli, F., Camarda, R., and Bonavita, V. J. Neurochem. 16: 159 (1969).

(4)   Piccoli, F., Camarda, R., and Bonavita, V. In preparation (1970a).

(5)   Guarneri, R., Piccoli, F., Camarda, R., and Bonavita, V. In preparation (1970).

(6)   Piccoli, F., Guarneri, R., Camarda, R., and Bonavita, V. In preparation (1970b).

(7)   Bonavita, V. in Atti XVI Congresso di Neurologia, Vol. 4, p. 31, II Pensiero Scientifico, Roma (1967).

(8)   Jacob, M., and Mandel, P. in Protides of the biological Fluids (Ed. H. Peeters) p. 63, Elsevier, Amsterdam (1966).

(9)   Sherrington, C. The Integrative Action of the Nervous System. Yale University Press, New Haven (1947).

(10) Bonavita, V., Monaco, P., and Tripi, E., Acta Neurol. 19:215 (1964).

(11) Nelson, S. R., Schulz, D. W., Passonneau, J.V., and Lowry, O. H. J. Neurochem. 15: 1271 (1968).

(12) Munch-Peterson,A., Kalckar, H. M., Cutolo, E., and Smith, E. E. B. Nature 172: 1036 (1953).

(13) Mandel, P. in Handbook of Neurochemistry, Vol. 5 (in press) (Ed. A. Lajtha). Plenum Press (1970).

(14) Bondy, C. S., and Margolis, F. L. Exptl. Neurol. 27: 344 (1970).

(15) Wegener, G. Exptl. Brain Res. 10: 363 (1970).

(16) Margolis, F.L., and Bondy, C.S. Exptl. Neurol. 27: 353 (1970).

THE BIOCHEMISTRY OF MYELINOGENESIS IN THE CENTRAL NERVOUS SYSTEM

ALAN N. DAVISON

DEPARTMENT OF BIOCHEMISTRY

CHARING CROSS HOSPITAL MEDICAL SCHOOL, LONDON W.C.2

The myelin sheath is built up of many lipoprotein lamellae wrapped around the axon in a continuous sheet of membrane leading from the plasma membrane of the oligodendroglial cell. Immediately before myelination begins, movement of the formative oligodendroglial cells occurs and the glial cell promulgates wrapping, at first, loosely round the axon. At this early stage lipid droplets containing cholesterol esters (1) accumulate in the brain serving, perhaps, as a reservoir of sterol and fatty acids for the needs of the developing tissue. In the rat cholesterol and phospholipid biosynthesis reaches a maximum rate by about 10 days after birth but the most rapid synthesis of sulphatides and cerebroside is further delayed by more than a week. This sequence of change correlates well with the accumulation of lipids in the developing brain (2,3) but not with the histology. Thus in the kitten optic nerve although myelin rings can be easily detected 16 days after birth, the typical myelin constituent is virtually absent, even after 27 days much less galactolipid is detectable than would be anticipated from the morphology (4). During the same period the phosphoglyceride fatty acids have a relatively shorter chain length and are more saturated than those found in the adult nerve. Osmiophilia of the optic nerve myelin is absent in the initial stages of myelination (4).

Early Myelin

Myelin isolated from the developing rat brain has been found to differ substantially from that of the mature central nervous system. The crude myelin has a higher phospholipid :

cholesterol ratio and less galactolipid is present than in the
same fraction prepared from adult brain. It has been found
(5) that after osmotic shock, the crude myelin can be separated
into two membrane fractions. The first appears in the electron
microscope as compact myelin lamellae but the second so called
myelin-like fraction appears to be composed of single membrane
vesicles similar to those of microsomal subfractions or synaptic
membranes. The amount of the latter fraction is maximal in rats
of about 16 days old whereas the quantity of myelin increases
exponentially up to a year after birth. This suggests that the
myelin-like fraction is implicated in the process of myelination.
Enzyme marker studies show that both myelin fractions contain
the cyclic phosphohydrolase and aminopeptidase thought to be
characteristic of myelin, but other marker enzymes are absent.

In order to further characterise the myelin fractions,
proteins were separated from each of the subcellular fractions
of developing brain by disc electrophoresis on polyacrylamide
gel according to the technique of Takayama et al (6). The
separated protein bands showed similar profiles for purified
myelin, the second myelin-like, the microsomal and the nuclear
fractions. The protein bands from the synaptic endings and free
mitochondria fractions were quite different from that for myelin
(7) as has already been shown by Cotman & Mahler (8) in their
studies on the adult rat brain. Since it has been claimed that
basic encephalitogenic protein is a unique constituent of myelin
the distribution of this protein was examined after extraction
with acid. It appeared from electrophoresis under alkaline and
acid conditions (9) that the characteristic basic encephalitogenic
protein was only present in purified myelin.

The identity of the myelin-like fractions

The possibility that the myelin-like form is due to
contamination of a small myelin fraction with membrane derived
from another subcellular system seems to be unlikely as judged
by enzyme marker, protein patterns and the cytological evidence
(10). It seems unlikely that the myelin-like material is light
myelin described by Eng & Noble (11) for our sample was obtained
from fresh brain and the myelin-like fraction was more dense
than myelin. Moreover the myelin-like fraction does not contain
the basic encephalitogenic protein and is therefore unlikely
to be a form of degraded myelin.

It has previously been suggested that the myelin-like
fraction is principally derived from oligodendroglial plasma
membrane (12) but the proposal that myelin derived from the glial
plasma membrane pushes round the axon has been always difficult

one to accept. Analysis shows that the myelin-like fraction resembles plasma membranes in its lipid composition (5,10). It therefore seems possible that the myelin-like material could have been regarded as a precursor membrane system which would act as a framework for the synthesis of mature myelin, but our isotope experiments do not support this contention. Thus, when ^{14}C acetate or ^{14}C leucine was injected into developing rats no evidence was seen of any precursor - product relationship between the myelin-like and myelin lipids or proteins (5).

There is evidence that the composition of both myelin and the myelin-like fractions changes in the maturing brain. We therefore propose that the myelin-like material is a form of myelin in continuity with the glial plasma membrane and with compact myelin. 'Early' myelin therefore includes a mixture of both the myelin-like and myelin fractions. It would appear that compact myelin contains cerebroside and the basic encephalitogenic protein possibly localized in the inter and intraperiod lines of the myelin sheath.

Current studies on the metabolism of sterols, and proteins of the 'early' myelin fractions suggest that these fractions undergo relatively rapid turnover in the developing compared to the adult brain. How this process occurs is not known neither is there any information about the site of basic protein synthesis. There is evidence to suggest that cerebroside and sulphatide biosynthesis occurs in the endoplasmic reticulum and that the galactolipid is transferred to newly forming myelin as a lipoprotein complex. Synthesis of the myelin complex would thus be localised to the membrane itself. It is even possible that new membrane units may be inserted into the already compact myelin lamellae. Such a concept would explain the observation that during maturation the proportion of newly synthesised myelin-like membranes would decrease and the composition of the compact myelin become closer to that in the adult.

Myelin Metabolism

Previous experiments in which $4-^{14}C$ cholesterol and various precursors were injected into the myelin lipids was followed by a persistence of radioactivity for remarkably long periods (13,14,15). This work was interpreted as indicating the relative metabolic stability of myelin in comparison to other parts of the body. Similar studies on the metabolism of myelin proteins (16-20) also indicated a slow turnover of these constituents with a half-life of 35 days or more. However while these latter results remain unassailed there is some doubt about the interpretation of the radioactive lipid experiments. Firstly following labelling mitochondrial sterol was unexpectedly

found to be as apparently metabolically stable as that of myelin
(21,22). Secondly Rawlins Hedley-White, Villegas and Uzman (23)
have found radioautographic evidence for reutilization of [14]C
cholesterol in the sciatic nerve long after administration of the
labelled sterol. Thirdly, it has also been difficult to explain
the degree of incorporation of isotopes into myelin lipid of
the adult brain if as was postulated the lipids were metabolically
stable (19,24,25).

7-Dehydrocholesterol in Developing Myelin

If anti-cholesterolaemic drugs are injected into developing
rats large amounts of sterol precursors accumulate in the brain
and myelin (26,27). Within a few weeks after injection of AY9944
the 7-dehydrocholesterol which formed up to 50% of the total
myelin sterol largely appears as cholesterol. This work has now
been extended. Dr. N.L.Banik has injected radioactive acetate
into young rats after drug treatment and shortly before
myelination commences, so that 7-dehydrocholesterol in the myelin
is extensively labelled. It has been found that as the
dehydrocholesterol disappears so the equivalent amount of
radioactive cholesterol is detected in the myelin, suggesting
a localized conversion of dehydrocholesterol to the more
saturated sterol. However, we have only been able to detect
7-dehydrocholesterol reductase activity in the microsomes of
young brain and not in the purified myelin. These observations
therefore suggest that in the developing rat brain the
7-dehydrocholesterol is reduced by the smooth endoplasmic
reticulum of the cell and that cholesterol molecules are
reincorporated into the myelin.

In further experiments Dr. N.L.Banik has incubated myelin
containing labelled lipid with whole brain homogenate at 37°C
for 30 min. On isolation of the microsomal fraction from the
combined in vitro preparation about 10% exchange of ^{14}C lipid
has been detected into the hitherto unlabelled microsomes.
Similar results have been obtained with 7-dehydrocholesterol
containing myelin and in incubation experiments with labelled
microsomes with subsequent isolation of myelin. Graham & Green
(28) also reported a 40% exchange of ^{14}C cholesterol from
lipoproteins to myelin in in vitro experiments but their
incubations were for long period and presumably on adult
animals. The possibility of lipid exchange between different
membranes is of course well established. Thus there is dynamic
interchange between serum cholesterol and erythrocyte membrane
sterol (29). Wirtz & Zilversmit (30) have demonstrated similar
exchange in liver mitochondrial membrane and microsomes in vitro
However this observation should be regarded with caution since
the experiments with nervous tissue are difficult to control

particularly as myelin seems to be easily degraded on storage.

Conclusion

The recent experiments on lipid metabolism make it necessary to re-examine the whole question of myelin stability as judged by lipid metabolism, particularly in the mature animal. It may be that exchange of sterol, as reported here, is a feature of 'early' myelin and that cholesterol in the multilamellae of the adult brain is less available for interchange.

Drs. H.C.Agrawal, N.L.Banik, M.L.Cuzner, R.F.Mitchell and Martha Spohn were concerned in much of the original work reported in this paper.

References

1. Adams, C.W.M., and Davison, A.N. J. Neurochem. 4:282 (1959)

2. Cuzner, M.L., and Davison, A.N. Biochem. J. 106:29 (1968)

3. Wells, M.A., and Dittmer, J.C. Biochemistry 6:3169 (1967)

4. Banik, N.L., Blunt, M.J., and Davison, A.N., J. Neurochem. 15:471 (1968).

5. Agrawal, H.C., Banik, N.L., Bone, A.H., Davison, A.N. Mitchell, R.F., and Spohn, M. Biochem. J. in press (1970)

6. Takayama, K., MacLennan, D.H., Tzagoloft, A., and Stoner, C.D. Arch. Biochem. 114:223 (1964)

7. Mehl, E., and Wolfgram, M., J. Neurochem. 16:1091 (1969)

8. Cotman, C.W., and Mahler, H.R., Arch. Biochem. 120:384 (1967)

9. Martenson, R.E., Deibler, G.E., and Kies, M.W., J. biol. Chem. 244:4268 (1969)

10. Banik, N.L., and Davison, A.N. Biochem. J. 115:1051 (1969)

11. Eng L.F., and Noble, E.P., Lipids 3: 157 (1968)

12. Davison, A.N., Cuzner, M.L., Banik, N.L., and Oxberry J.M., Nature, Lond. 212:1373 (1966)

13. Davison, A.N., p.178 in Applied Neurochemistry (Ed. Davison, A.N., Dobbing J.) 1968

14. Davison, A.N., p.547 in Handbook of Neurochem. Vol.3
    (Ed. Lajtha, A) 1970

15. Smith, M.E., in Adv. in Lipid Res. vol. 5 (Eds. D.Kritchevsky
    and R. Paoletti) (1967)

16. Furst, S., Lajtha, A., and Waelsch, H. J. Neurochem $\underline{2}$:216 (1958)

17. Davison, A.N., Biochem. J. $\underline{78}$:272 (1961)

18. Metzger, H.P., Cuenod, M., Grynbaum, A. & Waelsch, H.
    J. Neurochem. $\underline{14}$:183 (1967)

19. Smith, M.E., Biochim. biophys, Acta $\underline{164}$:285 (1968)

20. Gaitonde, M.K. and Martenson, R.W. J. Neurochem. $\underline{17}$:551 (1970)

21. Cuzner, M.L., Davison, A.N., and Gregson, N.A. Biochem.J.
    $\underline{101}$:618 (1966)

22. Khan, A.A., and Folch-Pi, J.P., J. Neurochem. $\underline{14}$:1099 (1967)

23. Rawlins, F.A., Hedley-Whyte, E.T., Villegas, G., Uzman, B.G.,
    in 2nd int.meet. of the int.soc. for Neurochemistry (Eds.
    R. Paoletti, R. Fumagalli G. Galli) (1969).

24. August, C. Davison, A.N., and Williams, F.M., Biochem. J.
    $\underline{81}$:8   (1961)

25. Davison, A.N., and Gregson, N.A., Biochem. J. $\underline{98}$:915 (1966)

26. Banik, N.L., and Davison, A.N., J. Neurochem. $\underline{14}$:594 (1967)

27. Fumagalli, R., Smith, M.E., Urna, G and Paoletti, R.
    J. Neurochem. $\underline{16}$:1329 (1969)

28. Graham, J.M. and Green, C. Biochem. J. $\underline{103}$:16C (1967)

29. Rouser, G., Nelson, G.J., Fleischer, S., and Simon G.,
    p.5 in Biological Membrane (Ed.by Chapman, D) London :
    Academic Press 1968

30. Wirtz, K.W.A., and Zilversmit, D.B., J. biol. Chem. $\underline{243}$:3596
    (1968)

THE BLOOD BRAIN BARRIER

Hugh Davson

University College, London, W.C.1., England

The concept of a barrier between blood, on the one
hand, and the brain on the other derives from the
classical studies of Ehrlich and of Goldmann on vital
staining; it was noted that a large variety of acid
dyes, of which trypan blue was one, would pass out of
the blood into the tissues of the body making them
voloured; the brain and spinal cord stood out in contrast to the rest of the tissues in that, with the exception of highly localized regions, such as the area
postrema, they remained unstained. In the thirties and
forties of this century some doubt was cast on this
concept of a barrier between blood and brain; thus King
in 1938 (1) argued that the essential phenomenon was one
of failure to stain, and this could have been due to the
absence of suitable connective tissue to take up the
stain. In this respect, however, he was ignoring one
half of the experimental evidence offered by Goldmann
(2, 3) to support the concept; thus Goldmann in his
"second experiment" injected trypan blue into the cerebrospinal fluid and showed that, this time, the brain
did become stained. Thus it is not for lack of tissue
to take up the stain that the brain and spinal cord are
unable to do so, hence Goldmann's evidence does, in fact,
constitute a sound basis for the concept of a barrier
between blood and the central nervous parenchyma. An-

other attack on the concept was prompted by the applica-
tion of the electron microscope to the study of the
minute anatomy of the central nervous system; earlier
light-microscopic studies had relied, of course, on the
staining of the cellular elements, and the techniques
available had never been adequate to stain effectively
the two classes of cells at the same time, namely, the
neurons and the glia. The osmium fixation-staining
technique of electron microscopy revealed both types of
cell at the same time, and the picture presented was one
of densely packed cells and processes, so that it appeared
that there was little or no extracellular space between
the cellular elements. Since the substances employed
in demonstrating the barrier were essentially substances
that remained extracellular, failing to penetrate tissue
cells, it was argued that the apparent barrier between
blood and the tissue was due to the fact that there was
so small an extracellular space that the amounts of
material taken up by brain to bring this extracellular
space into equilibrium with the blood plasma were so
small as to escape notice. This argument found acceptance
by many workers in the field, but like the argument of
King mentioned above it was based on ignoring Goldmann's
second experiment; thus Goldmann had shown that trypan
blue was taken up by the tissue if presented by way of
the cerebrospinal fluid so that there was, indeed, a
space into which it could diffuse, and the more quantitative studies of Stern and Gautier (4, 5, 6) had shown
with a variety of compounds, such as iodide, thiocyanate,
ferrocyanide, and so on, that these could establish
measurable concentrations in the brain when presented
through the cerebrospinal fluid, but not when presented
by way of the blood. It was considered, however, by
Davson and Spaziani (7) that a strictly quantitative
approach was necessary to reestablish the concept of the
blood-brain barrier, and these authors measured the uptake of brain <u>in vivo</u>, after maintaining a steady level
of the substances examined for several hours. The results
are shown in Fig. 1, and indicate that the uptakes are
measurable, but small, corresponding, in the case of
sucrose, PAH and ^{131}I to spaces of 2-3%. Pieces of brain
from freshly killed animals were exposed to Ringer-Locke
solutions containing these substances, and analysed at

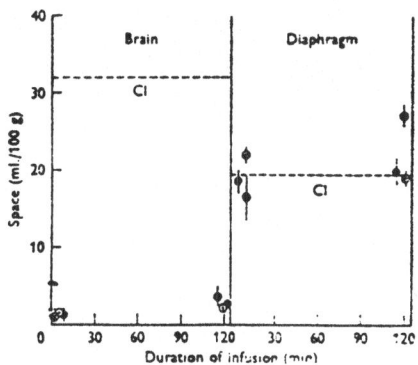

Fig. I. Penetration of PAH, ⊗, sucrose, ●, and ^{131}I, ⊙,
  into brain and diaphragm mucle of rabbits infused with
the test substance for 6 or 120 min. The ordinate refers
to the volume of tissue water (per 100 g tissue) that
contains the same concentration of the test substance as
in a plasma dialysate. Each point represents the average
of determinations in 4-10 animals; S.E. shown as vertical
lines. (Davson & Spaziani, J. Physiol.)

different time-intervals. Now, the uptake was rapid
and progressed far beyond the 2-3% space found in vivo,
showing that if these substances remained extracellular
the extracellular space was in the region of 12 per cent
or more. Thus the very small volumes of distribution of
such solutes as sucrose and inulin cannot be ascribed to
a low extracellular space in central nervous tissue; if
this were so, there should be no difference between the
volumes of distribution according as the solute was
presented by one or other route. The use of excised
tissue can obviously give rise to practical objections,
so that an in vivo approach along the same lines would
be preferable; this was done by Davson, Kleeman and
Levin (8) who presented the solutes through the cerebrospinal fluid by barbotage, and measured a volume of
distribution of sucrose of 12% for the spinal cord of the
rabbit, and later, in a much more refined study, Rall,
Oppelt and Patlak (9) perfused the ventricles of dogs
with artificial cerebrospinal fluid, containing [14]C-
sucrose, for long periods and measured the uptake in
small blocks of tissue taken at successive distances from

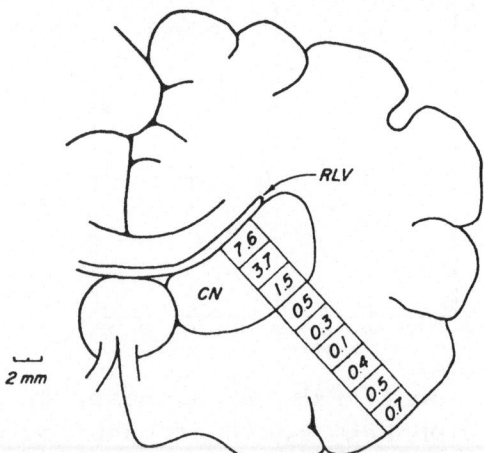

Fig. 2. Showing inulin concentration as per cent of
perfusate in tissue blocks from caudate nucleus outward.
CN = caudate nucleus, RLV = right lateral ventricle.
(Rall, Oppelt & Patlak, Life Sciences)

the ventricular wall, as illustrated by Fig. 2; the
mathematical analysis of the results indicated a probable
extracellular space of some 12 per cent; they also
indicated that there was little or no barrier to diffusion
across the ependymal wall. Finally, Van Harreveld,
Crowell and Malhotra (10) showed that the electron microscopical appearance of the brain parenchyma as a densely
packed mass of cells with little or no extracellular
space was largely artefactual, because of the rapid
swelling of the tissue that takes place post mortem;
when they froze the surface of the brain in situ and
subsequently fixed this by freeze-substitution, quite
definite extracellular spaces were observed.

    Thus the concept of a blood-brain barrier seems
now to be well reinstated, this time on a sound
quantitative basis in so far as the volume of the extracellular space, so important for kinetic considerations,
can be taken to be in the region of 10-15 per cent of
the weight of the tissue.

Having established what is only the bare fact of a barrier between blood and nervous tissue we must now proceed to discuss some of its interesting features and, finally, its significance for the economy of the brain.

LIPID SOLUBILITY

When the penetration of a variety of solutes into brain is studied, it immediately emerges that the feature that determines the ease with which a substance may cross the barrier is its lipid-solubility; thus Davson (11) showed that ethyl alcohol, with a high lipid-solubility, passed so rapidly from blood into the brain that it had equilibrated within less time than that required for isolation of the tissues; with ethyl thiourea the rate was rapid but measurable; with the less lipid-soluble thiourea the rate was less. Thus the blood-brain barrier is similar to the barrier separating an individual cell from its environment, since it has been known from the classical studies of Overton that lipid-solubility is a dominant factor in determining ease of penetration of cells. This suggested that the capillaries of the nervous tissue were fundamentally different from those of other tissues, such as muscle, since the kinetics of escape of materials from these other capillaries showed that lipid-solubility did not exert a dominant influence, rather the main route of penetration was through quite large water-filled pores, so that molecular size was the important parameter. The capillaries of the central nervous tissue are covered by the protoplasmic end-feet of the astrocytes associated with them; and it seemed reasonable to attribute the special selectivity of the nervous tissue capillaries to this layer, especially as the endothelial linings of the nervous capillaries seemed identical with those of muscle even in the electron microscope. Evidence in favour of this was the observation of Clemente and Holst (12) that X-ray treatment of the brain caused a breakdown of the barrier at the same time as it caused degenerative changes in the astrocytes associated with the capillaries. It is only recently that the true morphological basis for the blood-brain barrier has been elucidated, thanks to the development

of the horseradish peroxidase technique by Karnovsky (13).
Horseradish peroxidase is a protein of relatively low
molecular weight (40,000) that consequently can escape
from the capillaries of most tissues of the body but
cannot, of course, cross the blood-brain barrier. The
molecules may be made visible in the electron microscope
by their reaction with H_2O_2 and 3-3'-diaminobenzidine,
the product reacting with osmium to give dense electron-
staining. Karnovsky (13) found, in cardiac capillaries,
that the marker was concentrated in the intercellular
clefts, indicating that this was the main, if not the
exclusive, pathway out of the capillary. When Reese and
Karnovsky (14) examined brain capillaries, they found no
escape of horseradish peroxidase into the tissue; however,
the marker did not accumulate between the basement membrane and the glial end-feet, as would be expected were
the latter the functional barrier. The marker was retained in the lumen of the capillary, its passage through
the cell clefts being blocked by what have been called
zonulae occludentes, or tight junctions, i.e. regions
where the membranes of adjacent cells have fused and
thus made it necessary for solutes to pass through the
cells rather than along the intercellular clefts. This
requirement obviously accounts for the importance of
lipid-solubility in the kinetics of the barrier, indicated above. Certain regions of the brain, notably the
area postrema, the median eminence, tuber cinereum, etc.,
were early found to be deficient in a blood-brain
barrier, in the sense that they became stained after
intravenous injections of trypan blue; examination of
the capillaries in these regions showed that they lacked
tight-junctions of the type seen in the rest of the brain.

We have seen that the blood-brain barrier may be
circumvented by injecting subjstances, such as trypan
blue or inulin, into the cerebrospinal fluid, in which
case they are able to diffuse into the central nervous
parenchuma apparently without appreciable hindrance at
the ependyma or the pia-glial surface of the brain.
Brightman and Reese (15) have applied the horseradish
peroxidase technique to the study of the ependyma and
pia-glia; as might have been expected from the physiological studies, horseradish peroxidase passed from

ventricles and subarachnoid space into the adjacent
nervous tissue, passing through intercellular clefts;
earlier studies had indicated that these clefts were
sealed by tight junctions, but the more exacting study
of these junctions demanded by this finding of a ready
passage of horseradish peroxidase revealed that the
sealing of the clefts, described earlier, was only
apparent; the adjacent membranes came close together
but there were, indeed, channels between them; this
type of junction was called a gap-junction.

BLOOD-CEREBROSPINAL FLUID BARRIER

The classical workers in the field of the blood-
brain barrier observed that the cerebrospinal fluid
shared in the protection from the blood afforded by the
blood-brain barrier; thus trypan blue does not enter
the cerebrospinal fluid after intravenous injection.
Stern and Gautier (4, 5, 6) likewise found that substances that failed to enter the brain failed to enter
the cerebrospinal fluid, and Stern and Gautier actually
argued that, when a substance did enter the brain, it
entered exclusively by way of the cerebrospinal fluid;
the fluid was in a series kinetically with the brain-
parenchyma, and passage across the blood vessels of this
parenchyma was excluded absolutely. The choroid plexuses,
which were regarded as the source of the cerebrospinal
fluid, were regarded as analogous with the placenta,
which controlled the passage of material from maternal
to foetal tissue. The theory of Stern and Gautier (4, 5,
6) is only of historical interest now, but their work and
their thinking have been of value, in so far as they drew
attention to the relation between the blood and brain,
the blood and cerebrospinal fluid, and finally the cerebrospinal fluid and brain. Modern work has emphasized
this triangular relationship and has shown, in fact, that
the blood-brain and blood-cerebrospinal fluid barriers
cannot profitably be studied in isolation. This point
may be illustrated by the schematic compartmentation of
the blood-brain system in Fig. 3 which forms the basis
of modern thinking on the subject.

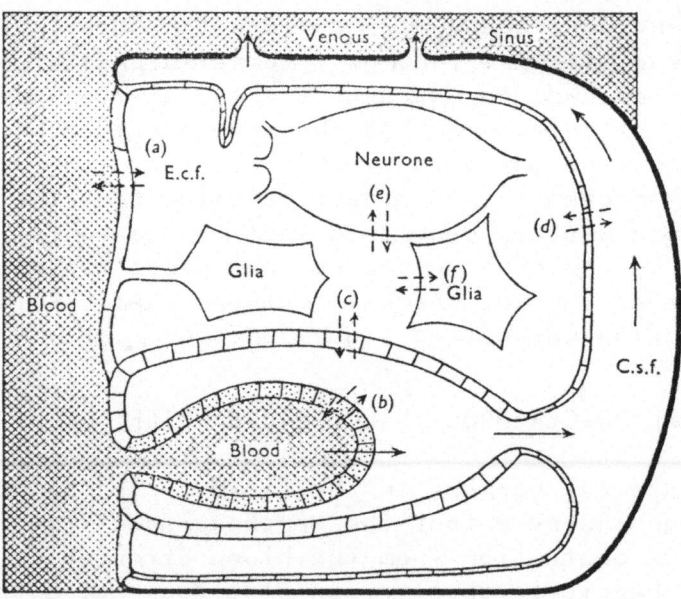

Fig. 3. Diagram of fluid comparments of the blood-brain-
cerebrospinal fluid system. Continuous arrows represent
proven directions of fluid flow. Interrupted arrows
indicate where diffusion of water and solutes may occur
between the different compartments, namesly: (a) across
the blood-brain barrier, between brain capillaries and
extracellular fluid; (b) across the epithelia of the
choroid plexuses; (c) across the ependyma; (d) across
the pia-glial membranes; (e) and (f) across the cell
membranes of neurones and glial cells. Thick line
represents the arachnoid-dural enclosure of the system.
(Davson and Bradbury, S.E.B. Symp.)(18).

     An important step forward in the quantitative
description of the relation between the three compartments was made by Wallace and Brodie (16, 17); these
authors injected Br^-, I^-, and CNS^- into the blood and
estimated their concentrations in both cerebral tissue
and cerebrospinal fluid after different periods, comparing these values with the concentrations of Cl^- in
the same samples. If the Br^-, for example, came into
rapid equilibrium with the tissue fluid of the brain,
we should expect to find the ratio of Cl^- to Br^- the

same in both plasma and tissue after a short period; if,
on the other hand, diffusion equilibrium were slow of
achievement, the ratio Cl^-/Br^- would be higher in the
tissue than in the plasma. Studies of these ratios
showed that they were invariably greater than unity even
after many hours; i.e., they showed that Br^- penetrated
slowly into the tissue spaces. The striking finding,
however, was that the ratio in the tissue was always the
same as that in the cerebrospinal fluid, and this seemed
to indicate, as Stern and Gautier (4,5, 6) had suggested,
that the cerebrospinal fluid was acting as an intermediary
for the supply of the anions to the brain. Their later
study, on the other hand, led them to reverse this
relationship, and they suggested that the brain might
well be the source of the material reaching the cerebrospinal fluid, i.e., the brain might be forming the
cerebrospinal fluid, at least in part.

KINETIC STUDIES

The work of Wallace and Brodie (16, 17), who studied
both blood-brain and blood-cerebrospinal barriers, was
extended by Olsen and Rudolph (19) to Na^+ and Br^-, and
to a number of other solutes by Davson (11). It was
shown that with ^{24}Na the kinetics of uptake by cerebrospinal fluid and extracellular fluid of brain were remarkably similar, as illustrated by Fig. 4 where the
curve represents the uptake by c.s.f. and the plotted
points refer to brain. The congruence was interpreted
as the result of separate transport across the blood-
c.s.f. barrier in the choroid plexuses and across the
blood-brain barrier, presumably the capillaries of the
central nervous parenchyma. Studies with more lipid-
soluble solutes, such as thioureas and ethyl alcohol,
revealed that equilibration of the brain could occur
much more rapidly than equilibration between blood and
cerebrospinal fluid, as illustrated by Fig. 5, and the
approximate equality of the rate of equilibration in
the case of ^{24}Na revealed by Fig. 4, was therefore
regarded as a special case. It was emphasized that the
rate of equilibration of a lipid-soluble substance with
the cerebrospinal fluid would be helped by exchanges

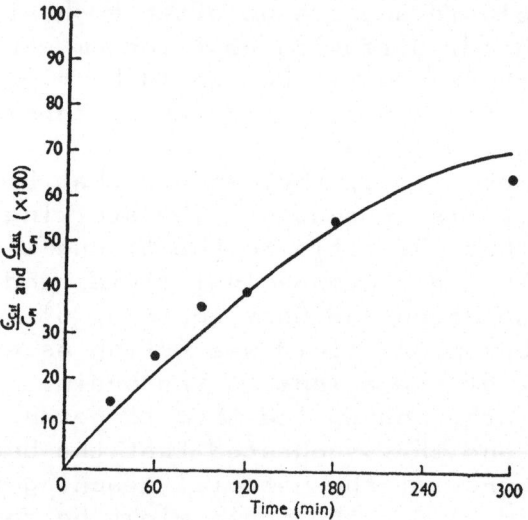

Fig. 4. Penetration of ^{24}Na into c.s.f. (smooth curve) and brain extracellular H$_2$0 (points).

Ordinates: $\dfrac{(C_{csf}) \text{ or } (C_{ext})}{(C_{pl})} \times 100$

C_{csf} = concentration in c.s.f.; C_{ext} = brain extra-cellular H$_2$0; C_{pl} = concentration in plasma.

Abscissae: Time in min.   (Davson, J. Physiol.)

between brain and cerebrospinal fluid since, as Fig. 5 shows for ethyl thiourea, exchanges between brain and cerebrospinal fluid will favour net flux into the cerebrospinal fluid (Davson, 20). The virtual absence of net exchanges of ^{24}Na between the brain and cerebrospinal fluid, deducible from the similar rates of equilibration with plasma, suggested that the rate of formation of cerebrospinal fluid might be measurable by the rate of turnover of ^{24}Na in the cerebrospinal fluid; thus it was argued that ^{24}Na would enter the cerebrospinal fluid in the freshly secreted fluid at approximately the concentration in that of a dialysate of plasma; in the absence of net gains from the brain, the rate of equilibration would then be given by rate of formation of the fluid. The turnover constant for ^{24}Na, deduced from Fig. 4, was 0.0041 min^{-1} and would correspond to a

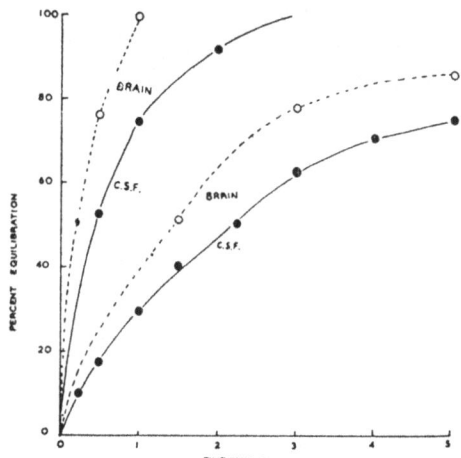

Fig. 5. Penetration of ethyl thiourea (upper curves) and thiourea (lower curves) into brain-H$_2$O and cerebrospinal fluid. (Davson, Kleeman, Levin, Drugs and Membranes).

replacement of 0.41 per cent of the volume per minute. The rabbit's cerebrospinal fluid has an average volume of 2.5 ml, so that this would correspond with a rate of secretion of 10.2 μl/min. This is sufficiently close to the measured rate, namely 10.1 μl/min (21) to suggest that the main influx of sodium into the cerebrospinal fluid is by way of the primary secretion of the choroid plexuses.

     The triangular relationship between the three compartments, blood, cerebrospinal fluid and brain, provides the clue to what had been a very puzzling situation. Thus there is a measurable permeability of the blood-brain barrier to such extracellular labels as inulin and sucrose and sulphate, yet the volume of distribution of these substances in the brain, when a steady concentration is maintained for many hours in the plasma is very low, of the order of 3-4% of the total weight. It might be thought that the volume would eventually rise to the true extracellular space which, we have seen, may be as high as 15 per cent of the weight of the tissue. The explanation lies in the circulating character of the

cerebrospinal fluid;   the concentration of, say, sucrose,
in the fluid at the steady state is only a small fraction
of that in plasma, because of the low permeability of
the choroid plexuses to this molecule and the fact that
the fluid drains away through unselective channels in
the arachnoid villi.   If the fluid were completely stagnant, then, of course, the steady-state concentration
would be the same as that in the plasma, but because of
the flow outwards and the restricted rate of penetration
inwards, a steady state is achieved with a low concentration in the cerebrospinal fluid.   It is this low concentration that provides a "sink" for diffusion of
sucrose from brain extracellular fluid so that, at the
steady state, this fluid likewise has a low concentration
compared with that in plasma.   Thus the volume of distribution will be smaller than the extracellular space.
Recent kinetic analyses of the situation have justified
this concept of a sink-action of the cerebrospinal fluid
(22).

Under steady-state conditions, then, the extracellular fluid may well be comparable in chemical composition with the cerebrospinal fluid, a hypothesis that
had been mooted by Wallace and Brodie (16, 17) in the
light of their studies on the exchange of bromide and
other anions between blood and the two compartments. The
composition of the cerebrospinal fluid is different from
that of the blood plasma or of a filtrate of blood plasma,
so that if this suggestion is correct, we must conclude
that the central neurones are surrounded by a medium
that is different from the medium surrounding other cells,
e.g. those of skeletal muscle, since it is reasonably
well established that the extracellular fluid surrounding
these is similar to a dialysate of blood plasma.   If this
suggestion is correct, then we must conclude that the
composition of the extracellular fluid of the brain is
controlled by active transport processes, or, if not,
that it is the cerebrospinal fluid that controls its
composition.   Thus the cerebrospinal fluid could be
secreted with its characteristic composition, with low
K^+, high Mg^{++}, high Cl^-, low HCO^-_3, low glucose, and so
on, and by virtue of the rapidity with which exchanges
occurred between the cerebrospinal fluid and the central

nervous parenchyma, the cerebrospinal fluid could impose
its own composition on that of the extracellular fluid
of the parenchyma. It was suggested that, on this basis,
the composition of the cerebrospinal fluid should change
as it passed from the ventricles, where it was first
secreted, to the subarachnoid spaces, where it was
finally drained away (23). An examination of the composition of the fluid drawn from different regions failed
to indicate any serious changes in composition; in fact,
the constancy of the composition was remarkable (23) so
that it was concluded that the extracellular fluid of
brain was, indeed, controlled by active processes,
separate from those controlling the composition of the
cerebrospinal fluid.

Additional support for this concept was provided by
examination of the concentrations of a variety of constituents of the cerebrospinal fluid drawn as near
simultaneously as possible from the various regions of
the dog's cerebrospinal system, in particular the cortical
subarachnoid fluid, which was chosen as one that was more
likely to indicate the composition of the extracellular
fluid of the brain parenchyma than the lumbar fluid,
since the latter, especially in the immobilized animal,
is probably very stagnant and may well reflect changes
due to the partial breakdown of the blood-cord barrier.
In general, the results indicated a remarkable uniformity
in composition, so far as Na^+, Ca^{++}, Cl^-, HCO_3^- and
glucose were concerned, but the changes in concentration
of K^+ that took place after the fluid had left the
ventricle were quantitatively significant and of considerable interest. The results are shown in Table I.
The cortical subarachnoid fluid has a lower concentration
of K^+ than the ventricular fluid, so that on its way
through the cranial subarachnoid space the fluid has
become less like a dialysate of plasma, in that its concentration, already well below that of a plasma dialysate,
has become even lower. Clearly, in this case, the extracellular fluid of brain not only has a considerably
lower concentration of K^+ than that in plasma or its
dialysate, but also less than that of ventriculated
cerebrospinal fluid. It cannot possibly be argued that
the cerebrospinal fluid is imposing its composition on

Table 1. *Mean concentration of K^+ (mequiv/kg H_2O) in plasma and cerebrospinal fluids of dogs and cats. Numbers in parentheses indicate numbers of animals from which a given fluid was taken (Bito & Davson, 1966)*

Plasma₁	Plasma₂	Plasma₃	Cist Mag	Cort Subarach	Ventr	Lumbar
Plasma$_1$	Plasma$_2$	Plasma$_3$	Cist Mag	Cort Subarach	Ventr	Lumbar
4.56 ± 0.12	4.57 ± 0.14	$3.97 + 0.12$	2.98 ± 0.06	2.65 ± 0.10	2.93 ± 0.08	3.22 ± 0.08
(16)	(16)	(12)	(15)	(8)	(15)	(8)

Cort Subarach/Cist Mag	0.86 ± 0.02	$P < 0.001$
Ventr/Cist Mag	0.98 ± 0.01	$P > 0.1$
Lumbar/Cist Mag	1.06 ± 0.01	$P < 0.001$

Plasma$_1$ is from blood withdrawn before anaesthesia; Plasma$_2$ from blood withdrawn immediately after anaesthesia, and Plasma$_3$ from blood withdrawn after all cerebrospinal fluid had been taken. P is the probability that the observed ratio would occur by chance.

that of the extracellular fluid of brain; if one fluid imposes its composition on that of the other, then it must be the brain extracellular fluid imposing its composition on that of the cerebrospinal fluid. In fact, studies on the composition of the freshly secreted fluid, collected under oil from the exposed choroid plexus, indicate that the cerebrospinal fluid, as freshly secreted, has its characteristic composition (24, 25, 26) which may be modified by exchanges with the extracellular fluid of brain. Thus both fluids must be regarded as specific secretions, the one by the choroidal epithelial cells and the other either by the cerebral blood capillary endothelial cells or by their associated astroglia. A variety of studies on the relation between cerebrospinal fluid and brain have amply confirmed the suggestion of Wallace and Brodie (16, 17) that the extracellular fluid of brain and the cerebrospinal fluid are similar in composition; for example we may mention the conclusion of Fencl, Miller and Pappenheimer (27) that the factor determining the control of respiration-rate was the pH of the cerebrospinal fluid rather than that of the blood plasma, and the measurements of the responses of the resting potentials of glial cells of Necturus to altered K-concentration of the blood (Cohen, Gerschenfeld and Kuffler (28)) which showed that the response was determined by the concentration in the cerebrospinal fluid, rather than that in the blood.

LYMPH FUNCTION

Finally, we may point to an important function of the cerebrospinal fluid as a substitute for a lymphatic system; however "tight" the blood-brain barrier may be normally, we must envisage the escape of small quantities of blood proteins into the brain tissue, and these can only escape back to the blood through a system that flows under hydrostatic pressure from the subarachnoid spaces into the vascular system. The free connexion between subarachnoid spaces and dural sinuses, revealed by a variety of studies both functional and histological on the arachnoid villi, permits this flow; in consequence, high concentrations of proteins are prevented from building up in the tissue.

REFERENCES

1. King, L.S. Res. Publ. Ass. nerv. ment. Dis. 18:150 (1938)

2. Goldmann, E.C. Beitr. klin. Chir. 64:192 (1909)

3. Goldmann, E.C. Abh. preuss. Akad. Wiss., Phys.-Math. Kl. No. 1, 1 (1913)

4. Stern, L., and Gautier, R. Arch. int. Physiol. 17:138 (1921)

5. Stern, L., and Gautier, R. Arch. int. Physiol. 17:391 (1922)

6. Stern, L., and Gautier, R. Arch. int. Physiol. 20:403 (1923)

7. Davson, H., and Spaziani, E. J. Physiol. 149:135 (1959)

8. Davson, H., Kleeman, C.R., and Levin, E. J. Physiol. 159:67 (1961)

9. Rall, D.P., Oppelt, W.W., and Patlak, C.S. Life Sci. 2:43 (1962)

10. Van Harreveld, A., Crowell, J., and Malhotra, S.K. J. Cell Biol. 25:117 (1965)

11. Davson, H.   J. Physiol.   129:111 (1955)

12. Clemente, C.D., and Holst, E.A.   Arch. Neurol.
     Psychiat.   71:66 (1954)

13. Karnovsky, M.J.   J. Cell Biol.   35:213 (1967)

14. Reese, T.S., and Karnovsky, M.J.   J. Cell Biol.   34:
     207 (1967)

15. Brightman, M.W., and Reese, T.S.   J. Cell Biol.   40:
     648 (1969)

16. Wallace, G.B., and Brodie, B.B.   J. Pharmacol.   65:
     220 (1939)

17. Wallace, G.B., and Brodie, B.B.   J. Pharmacol.   70:
     418 (1940)

18. Davson, H., and Bradbury, M.   Symp. Soc. Exp. Biol.
     19:349 (1964)

19. Olsen, N.S., and Rudolph, G.G.   Amer. J. Physiol.
     183:427 (1955)

20. Davson, H.   Physiology of the Ocular and Cerebro-
     spinal Fluids.   J. & A. Churchill, London (1956)

21. Bradbury, M., and Davson, H.   J. Physiol.   170:195
     (1964)

22. Welch, K.   Brain Research   16:453 (1969)

23. Davson, H. in The Cerebrospinal Fluid.   Ciba Foundn.
     Symp. London: Churchill. (1958)

24. Rougemont, J. et al.   J. Neurophysiol.   23:485 (1960)

25. Ames, A. et al.   J. Neurophysiol.   27:672 (1964)

26. Sadler, K., and Welch, K.   Nature   215:884 (1967)

27. Fencl, V., Miller, T.B., and Pappenheimer, J.R.
     Amer. J. Physiol.   210:459 (1966)

28. Cohen, M.W., Gerschenfeld, H.M., and Kuffler, S.W.
     J. Physiol.   197:363 (1968)

SECTION IV

NUTRITION AND BRAIN DEVELOPMENT

UNDERNUTRITION AND THE DEVELOPING BRAIN: THE USE OF ANIMAL MODELS TO ELUCIDATE THE HUMAN PROBLEM

John Dobbing

Department of Child Health, University of

Manchester, Clinical Sciences Bldg., York

Place, Manchester 13, England

INTRODUCTION

There is no longer any doubt that growth restriction due to malnutrition at certain ages is associated in many children with an irreversible deficit in higher mental function. The phenomenon is by no means confined to the underprivileged communities of developing countries, but can be found, although on a smaller scale, in similar communities in the most advanced country in the world. It may also be allied to the diminished ultimate potential of some low birth weight babies from even the most privileged homes. A constant feature appears to be the need to maintain a good growth rate at least until the eighteenth month of human postnatal life.[7] By contrast, the effects of growth retardation at later ages can apparently be reversed on restoring good dietary and other conditions.

It is important to appreciate that we cannot at present distinguish between malnutrition as an aetiological factor in the mental impairment, and all the other environmental hazards of deprived socio-economic and cultural conditions. The increased exposure to infective disease, and even the poor genetic endowment by which such children may also be afflicted, may also conspire in a cumulative fashion to reduce the child's eventual mental potential.[1] Thus it is very probable that malnutrition is only one facet, although a central one, of a highly complex aetiology.[2] Still less can we be certain that any such effect of early malnutrition is mediated by a modification of the normal physical development of the brain. To demonstrate this last point would require a knowledge which none of us possess, of the molecular or cellular basis of higher mental function.

With these reservations, however, it must be conceded that the effects on mental function described above bear some striking resemblances to the known effects of infantile malnutrition on the physical growth and development of the brain; and there is no good reason why the discipline of developmental neurochemistry should continue to neglect this topic which is of such sociological and humanitarian importance. The present paper represents a plea that neurochemists will begin to apply their already extensive knowledge and skills to the problem.

THE VULNERABLE PERIOD

Experimental paediatric pathologists[3,4,5] using simple experimental designs and comparatively crude techniques, have already demonstrated that there is a specific period of brain growth during which the organ is susceptible to growth restriction resulting in permanent physical deficit. One version of this idea relates vulnerability to that period of brain growth when cells are undergoing mitosis.[5] Brain cells comprising any particular region undergo a once-and-for-all growth-hyperplasia (discounting pathological reactions) at an early stage of development, corresponding well with the timing of the observed period of clinical vulnerability.[6] Restriction of general growth at this time reduces the rate of mitosis in a manner which is not recoverable, even when optimum growth rates are subsequently restored.[5]

However when this hypothesis is applied to the brain, it tends to ignore that the two main cell types, neurons and glia, multiply at quite different times. Are both periods of mitosis vulnerable? Furthermore it may be wrong to assume that glial (or even neuronal) cell number is a functionally significant parameter.[8] It seems much more likely that brain function is more dependent on a correct and orderly sequence of cellular migration and the formation of the appropriate

histological micro-architecture. Of central importance
to proper function is probably the growth of neuronal
processes (axons and dendrites) and the proper
establishment of their myriads of synaptic
interconnections. Both of these aspects of brain
growth post-date the achievement of neuronal cell
number, although they do overlap with the period of
glial multiplication known to pathologists as
myelination gliosis.[9] Indeed it may be significant
that most of the known experimental work up to the
present time which has produced permanent deficits of
cell numbers by undernutrition during the vulnerable
period, has imposed the experimental stress during the
later period of glial mitosis, after adult neuronal
numbers have been achieved, and the reported deficits
in cell numbers must almost certainly refer to glial
cells. Functional consequences have sometimes been
observed, and it seems more likely that these may be
the result of permanent reduction of post-mitotic
neuronal development, rather than of restriction of
glial cell number. Neurochemists could help to
resolve this question with a quantitative assessment of
the effects of vulnerable-period undernutrition on the
ultimate extent of later neuronal growth.

Such speculation led some years ago[4] to a rather
wider vulnerable period hypothesis, of which the cell
multiplication theory mentioned above forms only an
early part. This states in general terms that
vulnerability of the brain is greatest during the
transient period known as the "brain growth spurt".
It is thus suggested that vulnerability may be directly
related to the growth velocity of all the brain growth
processes. The neuronal mitotic period of the "growth
spurt" occupies only the early stages, and is followed
by the neuronal growth mentioned above as well as by
the glial mitosis which precedes and prepares for
myelination. Myelination occupies the later half of
the "growth spurt". This whole vulnerable period is
broadly encompassed by the period of fastest accumulation
of brain fresh weight. It is also a time when many
enzymes are being rapidly elaborated resulting in
dramatic increases in their activity,[10,11] and so it
raises the possibility that the developing enzymology
of the brain may also be directly and permanently
restricted at this time.

Much experimental evidence is being accumulated
which relates the degree of vulnerability to the velocity
of brain growth. In neurochemical terms all the

evidence is crude, but at least it is sufficiently
substantial to make a more sophisticated neurochemical
approach worth while.

THE BRAIN GROWTH SPURT

    If the whole growth spurt is a period of greatest
vulnerability, it is first necessary to define its
extent in quantitative and qualitative terms. The
growth curve of whole brain wet weight has a sigmoid
shape, and the brain growth spurt can be easily
represented as a first order velocity curve.[12]

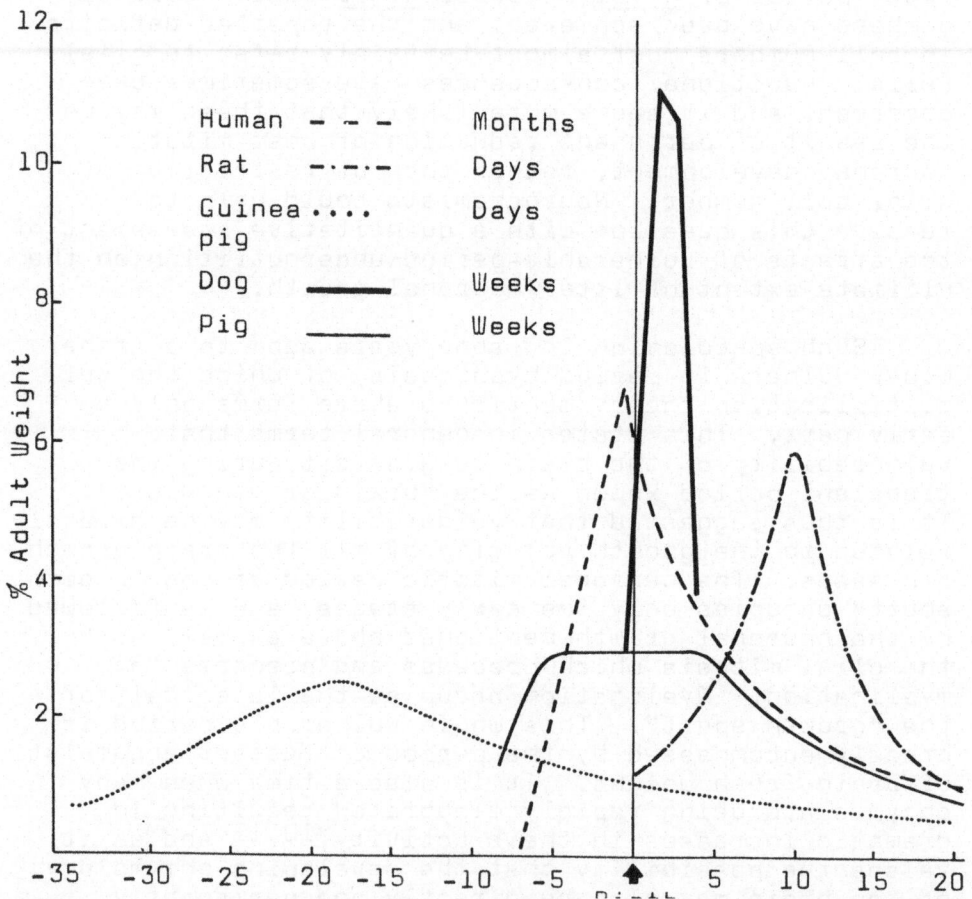

Figure I Rate curves of brain growth in relation to
birth in different species. Values are calculated at
different time intervals for each species.

Figure I shows the velocity curves of brain growth in various species, from which it is clear that there is an important species difference in its timing in relation to birth. Maximum vulnerability in the guinea pig should be during foetal life, and so it can be shown to be. In the rat it is postnatal, whereas in humans it is perinatal. The characteristics of human brain growth velocity will be illustrated in more detail in a later figure (Fig.II).

The principal events of whole brain growth are related to these velocity curves as follows. Neuronal cell division is complete to adult numbers before the growth spurt begins.[13] Oligodendroglial multiplication, quantitatively much more important than neuronal multiplication, occupies the first half of the growth spurt. Lipid accumulation, in which myelination is prominantly included, occupies the second half.[12] Lipid accumulation occurs partly at the expense of water content, leading to a progressive dehydration of the tissue.[14] The greatest contributor to growth in fresh weight, however, is probably growth in cell size, including the all-important special growth of neuronal processes and their interconnections, and this occurs throughout the main growth spurt.

Most of these events can be quantitatively defined even in whole brain using the velocity curves of DNA (nuclear) and lipid accumulation. For these purposes the comparatively few tetraploid cells can be ignored, and the accumulation rates of many lipids inferred from the estimation of cholesterol, to which they bear a known temporal relation. The need to employ such crude descriptions is imposed by the extremely laborious nature and questionable validity of much quantitative histology, although it must be conceded that much topographical detail is lost. It is difficult to distinguish different cell types (but see Figure V), and the migration of cells as well as the death of some of them which forms such a striking feature of brain development is unrepresented. It is therefore perhaps surprising that a recognisable developmental pattern emerges from analysing whole brain such as is seen in Figure II.

A little more meaning can be derived by dividing the brain into anatomical regions, but in a developing series this must be done according to anatomical land-marks and not by linear measurement. Thus the very different growth characteristics of the cerebellum can

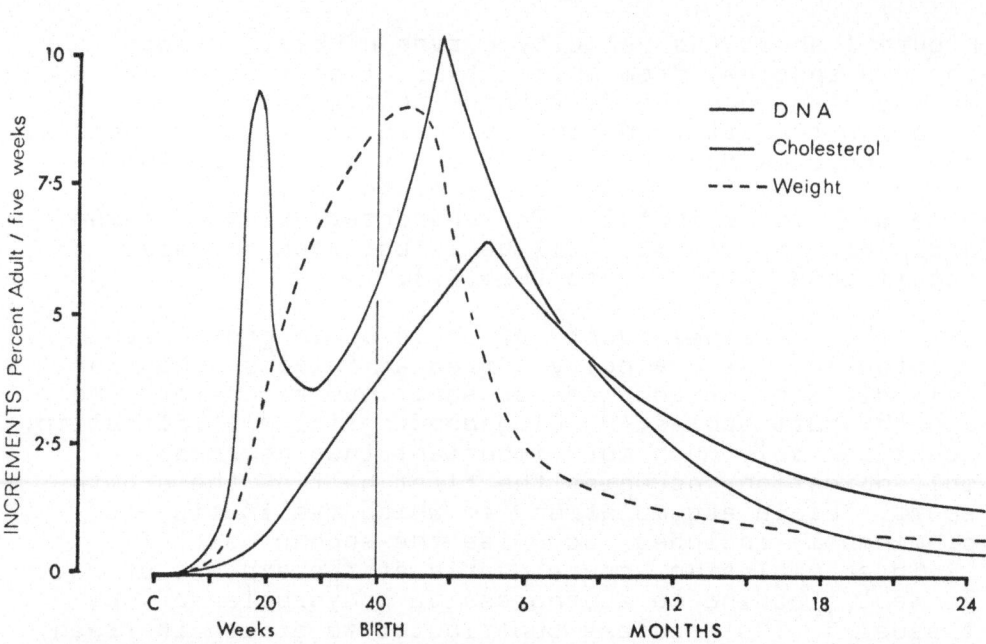

Figure II Velocity curves showing incremental rates of
DNA (two peaks), cholesterol (single peak) and fresh
weight in whole human brain. Data from analysis of 200
normal human brains[6]. Note the bimodel curve for DNA,
representing neuroblast followed by glial multiplication.

be distinguished from those of forebrain and stem.[15,16]
Unfortunately subdivisions into smaller regions makes
it almost impossible to derive benefit from DNA
estimations, since these are only of value when expressed
as totals per whole region. Cellularity, or DNA
concentration per unit volume or weight is more
influenced in a developing series by changes in the
denominator of the expression than by the DNA itself
and is consequently almost impossible to interpret.
Thus DNA concentration will fall in some regions as it
rises in others with increasing age, even though cell
division is occurring in both (Figs. III & IV). The
meaning of the different ways of expressing analytical
data has been discussed elsewhere[17].

TESTING THE VULNERABLE PERIOD HYPOTHESIS

It is not difficult to show that the brain is much
more easily influenced during the time of the brain
growth spurt than either before or afterwards. Much of,

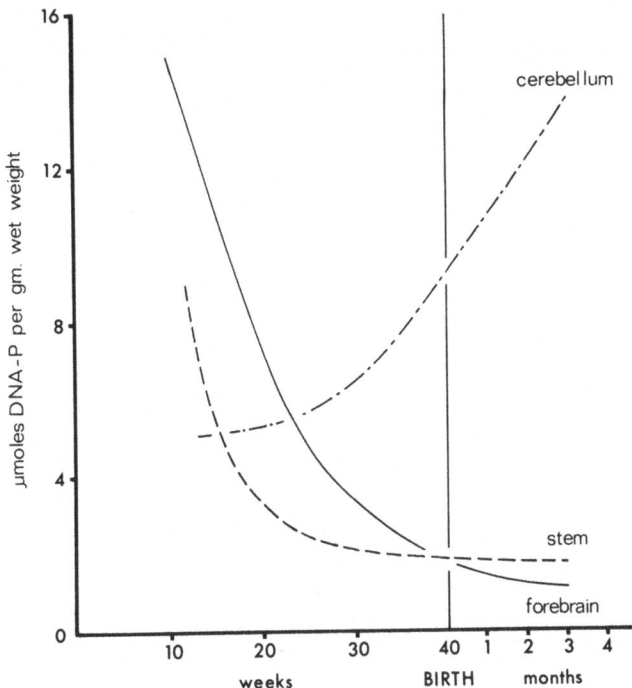

<u>Figure III</u> Changing concentration of DNA (per gram wet weight) in human forebrain, cerebellum and stem. Data from analysis of 200 human brains.[6]

the early results is summarised elsewhere[4] and it may now be added that the progress of behavioural ontogeny is depressed just as are the other gravimetric and analytical parameters. The age at which important milestones of reflex and other behavioural growth occur in undernourished animals is significantly later than in normal controls[18]. Growth retardation in the physical parameters manifests itself in ways which can appear contradictory unless the normal direction of progress is taken into account.

Thus although the undernourished brain weight is <u>less</u> than the controls during development, the <u>relative</u> brain weight, or the brain:body ratio is <u>higher</u>.[4] This is <u>not</u> a brain sparing phenomenon as is so often assumed. It simply reflects that the brain weight increases with age before the body weight; and the brain:body ratio thus declines with increasing age. A simple growth retardation must therefore inevitably

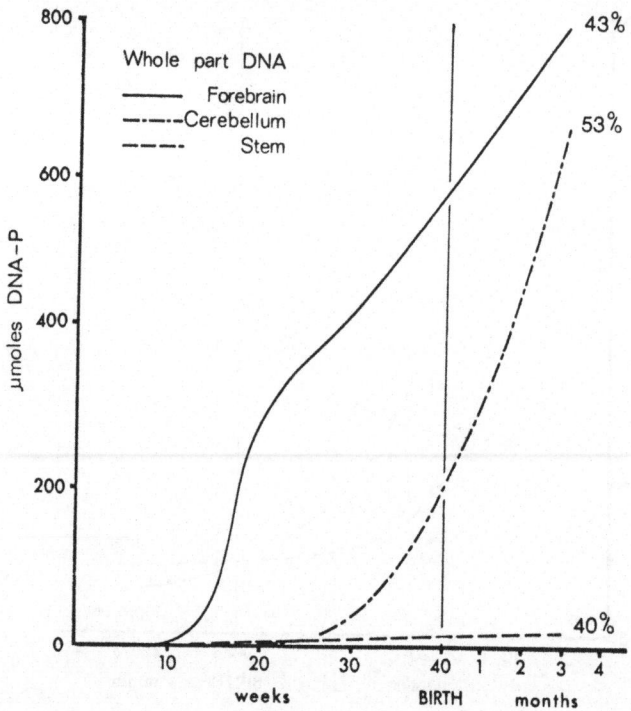

<u>Figure IV</u> Rising levels of whole part DNA in human
forebrain, cerebellum and stem. Data from analysis of
200 normal human brains[6]. Percentage figures denote
percent of adult values achieved at 3 postnatal months.

lead to a <u>higher</u> value for the brain:body ratio.
Eventually, in the adult, this same brain (again
contrary to earlier claims) can be shown to be
relatively, as well as absolutely smaller.[19] This
new finding in rats compares well with the similar
finding of small ultimate head circumference in
previously underfed children.[20]

     Similarly, during the developmental period, total
DNA per whole region increases as cells divide, but in
all regions, except the cerebellum the DNA <u>concentration</u>
or "cellularity" normally declines with increasing age
(see Fig.III). Thus, except for the cerebellum, the
brain retarded in its development by undernutrition
will have a smaller total quantity of DNA but a higher
concentration of the same material. Undoubtedly if it
were possible to examine much smaller and more specific
regions other similar false paradoxes would be

discovered. Analytical findings thus require great
care in their interpretation, and a quantitative as
well as qualitative knowledge of the normal movements
of each parameter with time must always be considered.

This caveat may apply with even greater force to
the effects on the developing enzymology of the brain.
It can easily be shown, however, that the activity of
several enzymes is significantly reduced by undernutrition only at the time of most rapid development,
although others are apparently unaffected.[10]

THE LONG TERM EFFECTS

Interesting as it may be to study the
contemporaneous effects of undernutrition on the
developing brain, it is presumably much more important
to examine those long-term results which persist in
spite of nutritional rehabilitation after the vulnerable
period. It may be argued that temporary retardation
of developmental processes is of little permanent
significance if it can be reversed. However this
raises completely uninvestigated aspects of the
problem. Educationalists are well aware of the
permanent effects on a child's prowess of even
temporary restrictions of progress which allow him to
get cumulatively behind his fellows. In the same sense
the orderly sequence of morphological and neurochemical
development of the brain may be sensitive to such
retardation, leading to an accumulating lag or deficit.

EXPERIMENTAL EVIDENCE FOR THE
VULNERABLE PERIOD HYPOTHESIS.

At present probably the most significant research
is that which attempts to detect deficits or differences
in adults who were undernourished at the time of the
vulnerable period. Accounts of these will be found
elsewhere and will only be summarised here.[17]

Perhaps the most important positive finding refers
as much to the whole body as to the brain. This is
the finding that if whole body growth retardation is
imposed at the time of the brain growth-spurt, neither
the whole body nor the brain are able to undergo
complete recovery on restoration of a normal diet.[4]
The relevant period is in the last half of pregnancy
in the guinea pig or the first three postnatal weeks

in the rat. (It is interesting that guinea-pig birth
weight can be reduced by as much as forty percent by
maternal undernutrition, contrary to the old, outworn
hypothesis of perfect fetal parasitism. Growth
restriction is not related to the foetal or postnatal
condition, but to the stage of development of a species
at those particular times.)

     Having imposed permanent stunting in this way,
the adult brains will show the following permanent
deficits:-

1) Small size. This will be somewhat smaller than is
    appropriate for the body weight.[19]

2) Deficits in total numbers of cells which may
    spuriously appear to parallel the reduction in
    weight.[4]

3) A selective cell deficit in the cerebellum compared
    with the remaining brain. This is probably related
    to the faster rate of growth and hence greater
    vulnerability of the cerebellum.[15,21]

4) A reduced concentration (per gram fresh weight) of
    myelin lipids.[4,22]

     As has been already stressed, there is no evidence
that any of these deficits are of functional significance.
The idea that they may be mentally deleterious is mere
hypothesis.

SIGNIFICANCE FOR HUMAN CHILDREN

     It is quite clear that similar findings can never
be experimentally demonstrated in human children,
although a few analyses of the brains of malnourished
children have already shown all the deficiences
mentioned above.[23]

     Assuming that the animal findings which can be
extrapolated from one species to another can also be
extrapolated to the human, it now becomes important
to define the timing and duration of the human brain
growth spurt.

     Preliminary results from about two hundred human
brains, ranging from ten gestational weeks to six
postnatal years[6] suggest the following conclusions:-

1. The human brain growth spurt begins at the end of

the second trimester, and ends at between eighteen
months and two years of post-natal life. This is
significantly later than the previous estimate of
five postnatal months.[24]

2. Human brain composition is about half-way towards
the adult state by about three postnatal months
(see Fig. IV).

3. The differences between cerebellar growth and that
of the remaining brain are the same as those
previously shown in animals, and could perhaps lead
to a similar differential vulnerability of the
cerebellum to growth retardation (see Figs. III & IV).

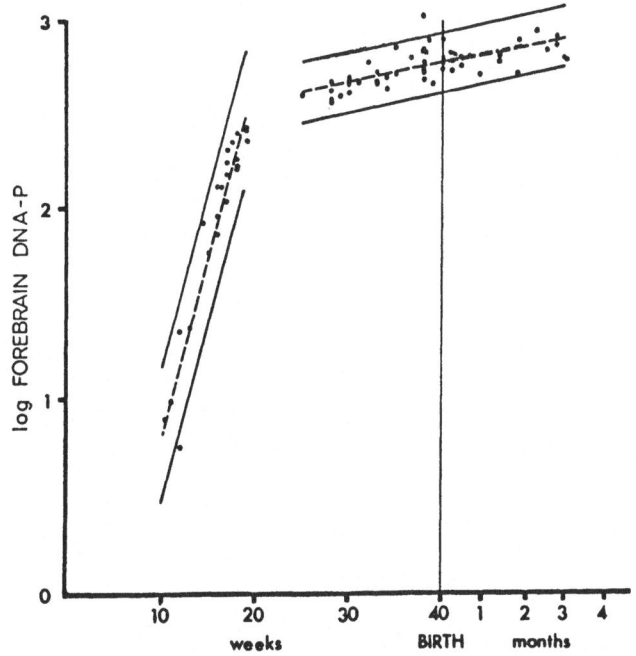

Figure V   Two phases of human forebrain DNA
multiplication. The data has been analysed in two
groups, from 0 - 22 gestational weeks and from 25
gestational weeks to 3 postnatal months. Two
exponentials are shown, representing neuroblast
followed by glial multiplication[6]. (Broken lines are
calculated regression lines, solid lines are 95%
confidence limits.)

4.   There is a period between fifteen and twenty-five
weeks of gestation when human neuroblasts complete

their mitoses, before the commencement of glial multiplication. In this regard also the human brain resembles the animal in its growth sequence (see Fig.V).

CONCLUSIONS.

Experimental studies of the long-term effects of undernutrition on the developing brain have shown results which emphasise the paramount importance of the timing of the undernutrition in relation to the brain "growth spurt". In humans this period is an extensive one, occupying the last half to one third of gestation and the first eighteen months to two years of postnatal life. This implies that permanent physical deficits in the brain could result from fetal growth retardation in the later months of pregnancy, and it underlines the importance of good postnatal growth in the prematurely born. Furthermore the first two years of postnatal life probably present a further hazard to the developing brain if malnutrition or any other growth-retarding influence is allowed to persist. Alternatively the postnatal period may also present an opportunity in our species for compensatory brain growth, provided good nutrition or the correction of other deleterious influences is diligently pursued at this early stage. These remarks are based on the double assumption (a) that the changes which can be permanently produced in developing animal brain are functionally significant; and (b) that they can be extrapolated to man by carefully matching comparable stages of brain growth, especially the complex of interrelated events known collectively as the brain "growth spurt".

ACKNOWLEDGEMENTS

I wish to acknowledge a grant from the Medical Research Council, with additional help from the National Fund for Research into Crippling Diseases, and the Spastics Society.

REFERENCES

1. Cravioto, J., Birch, H.G., de Licardie, E.R., &
   Rosales, L. Acta Paediat. Scand., 56: 71 (1967)

2. Scrimshaw, N.S. & Gordon, J.E. (Eds). Malnutrition
   Learning and Behaviour. Boston, M.I.T. (1968)

3. Dobbing, J. Biol. Neonat. 9: 132 (1966)

4. Dobbing, J. In Applied Neurochemistry (Eds.
   Davison, A.N. & Dobbing, J.) Oxford, Blackwell
   (1968)

5. Winick, M. & Noble, A. J. Nutr. 89: 300 (1966)

6. Dobbing, J. & Sands, J. (in preparation)

7. Cravioto, J. Amer. J. Dis. Child (In press) (1970)

8. Dobbing, J. Amer. J. Dis. Child. (In press) (1970)

9. Rorke, L.B. & Riggs, H.E. Myelination of the brain
   in the newborn. Toronto, Lippincott (1969)

10. Adlard, B.P.F. & Dobbing, J. Brain Res. (in press)
    (1971).

11. Chase, H.P., Dorsey, J. & McKhann, G.M. Pediatrics
    40: 551 (1967)

12. Davison, A.N. & Dobbing, J. In Applied Neurochemistry
    (Eds. Davison, A.N. & Dobbing, J.) Oxford, Blackwell
    (1968)

13. Flexner, L.B. In Biochemistry of the Developing
    Nervous System (Ed. Waelsch, H.) New York,
    Academic Press (1955)

14. Dobbing, J. & Sands, J. Brain Res. 17: 115 (1970)

15. Dickerson, J.W.T. & Dobbing, J. Proc.Roy.Soc.Med.
    59: 1088 (1966)

16. Dickerson, J.W.T. & Dobbing, J. Proc. Roy. Soc. B.
    166: 384 (1967)

17. Dobbing, J. In Developmental Neurobiology (Ed.
    Himwich, W.A.) Springfield, Thomas (1970)

18. Smart, J.L. & Dobbing, J. Brain Res. (in press) (1971)

19. Dobbing, J. & Sands, J. (In preparation)

20. Stoch, M.B. & Smythe, P.M. S.Afr. Med. J. 41: 1027 (1967)

21. Chase, H.P., Lindsley, W.F.B. & O'Brien, D. Nature, 221: 554 (1969)

22. Culley, W.J. & Lineberger, R.D. J. Nutr. 96: 375 (1968)

23. Winick, M. & Rosso, P. Pediat. Res. 3: 181 (1969)

24. Dobbing, J. & Sands, J. Nature 226: 639 (1970)

INFLUENCE OF NEONATAL UNDERNUTRITION ON THE DEVELOPMENT OF RAT

CEREBRAL CORTEX: A MICROCHEMICAL STUDY

Norman H. Bass

Departments of Neurology and Pharmacology, University of

Virginia School of Medicine, Charlottesville, Virginia

Neurochemists have assembled a large body of data on developing whole brain in many species, including man. Although of great value, these observations do not permit close morphologic correlations because developing brain matures at different times in various regions. Information derived from analyses of whole brain, therefore, is a summation of different events, and the specific changes underlying structural transformation of any one region may be obscured. Therefore, we have confined our observations to the somatosensory area of developing rat cerebrum, combining microchemical techniques with histologic examination to study the differentiation of neuronal cytoplasm, proliferation of axons and dendrites, and the formation of synapses and myelin (1,2).

Flexner (3) defined a "critical period" for development of cerebral cortex about the 10th day of postnatal life in the albino rat. Maturation of the isocortex, however, continues well beyond this period so that many morphological, biochemical and physiological changes occur between 10 and 50 postnatal days of age. Four major changes reported in this study of rat cerebrum are: (1) the extensive proliferation of large neuronal processes accompanied by the appearance of ribosomal organelles as cortical thickness doubles to normal adult width between birth and 10 postnatal days, (2) further neuronal differentiation characterized by proliferation of fine axo-dendritic fibers and formation of synapses, between 15 and 25 postnatal days, (3) the continuous migration of neuroglial cells from the subependymal zone through white matter into cerebral cortex between 10 and 50 days of age, and (4) myelination occurring at a maximum rate between 30 and 50 postnatal days (Figure 1). The maturation of rat isocortex therefore is a complex process,

MATURATION OF THE SOMATOSENSORY AREA OF RAT CEREBRAL CORTEX DURING POSTNATAL DEVELOPMENT

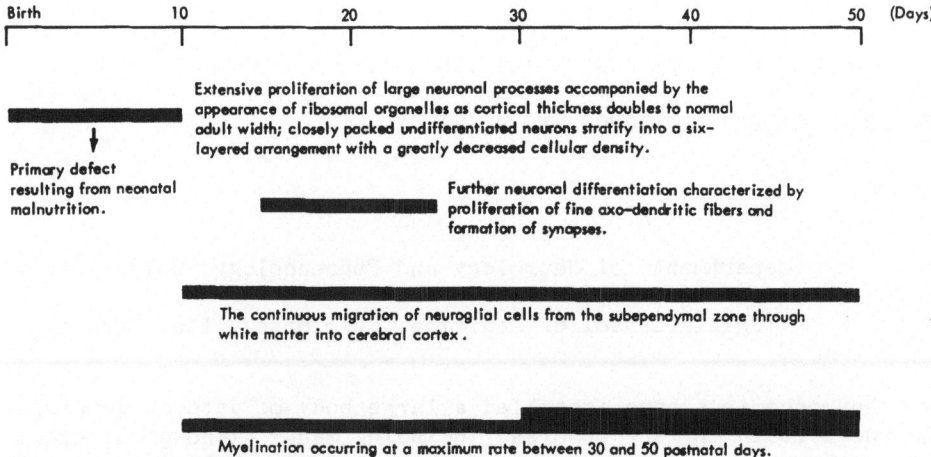

Figure 1. Schematic representation of 4 processes contributing to the normal maturation of the somatosensory area of rat cerebral cortex during the first 50 days of postnatal development. The primary defect of cortical development in animals subjected to neonatal malnutrition occurs during the first 10 days of postnatal life.

involving not a single "critical period" but many stages, both critical and interdependent.

During the past few years, laboratories have presented evidence from animal and human research suggesting that under-nutrition during early postnatal life may lead not only to physical impoverishment but to mental retardation. Early studies of undernutrition on growth of the albino rat indicated the importance of the so-called "sparing effect"; that is, of all organs of the body, the brain is least affected by inanition. This widely accepted, but only partially correct conclusion is based on obser-vations in rats undernourished after weaning at 21 postnatal days of age. Recent morphological and biochemical investigations (6,7,8) in addition to the detailed report of the present study (9,10) have suggested that rats subjected to undernutrition during the first 21 days of life show permanent cerebral defects. Hence, during the first 3 weeks of postnatal development, prior to the body growth spurt, the brain appears the most susceptible to the effects of nutritional deprivation.

<u>Methods</u>: Two methods, introduced in 1918 by Sugita (11) have been commonly used to produce neonatal malnutrition in albino rats. The first method, further elaborated by Widdowson and McCance (12), was to increase litter size to 20 in comparision with a control litter of 6 to 9 pups. This method results in a large variation in growth of individuals in the experimental litter, dependent on competition of the young for the 12 maternal feeding stations. The second method, employed by Culley and Mertz (13), is to separate the young of a litter from the nursing mother for 18 to 20 hours each day. This method results in a mortality of greater than 50% because of severe starvation. It should be noted that both experimental methods create many important uncontrolled variables: over-crowding of the litter, changes in the quality and quantity of milk production by the lactation dam, inadequate temperature regulation in the maternal nest, and unassessible emotional factors.

In the present study, neonatal undernutrition is produced by a modification combining features of both previously reported methods. Litters are increased to 12 and the young separated from the nursing mother for 8 hours a day (9 a.m. to 5 p.m.) during the first 21 postnatal days. Although many variables are still poorly controlled, this new technique results in a more uniform depression of growth rate and allows full survival in experimental litters. Rehabilitation is begun by then allowing those mothers to nurse their pups continuously until 35 days of age. After weaning, the animals were segregated by sex, placed in separate cages and fed ad libitum on a standard rodent diet. Rats treated in this way show a 40% reduction in body growth at 20 postnatal days, followed by recovery to 92% of normal body weight by 40 days of age (Figure 2). Control litters, reduced to 6 to insure maximal nutrition, were continuously nursed by their mothers and then were weaned at 21 postnatal days. The variability or "runting" produced by submaximal growth of a few pups in a normal "control" litter of 12 was avoided in favor of small litters of 6 in which optimal nutritional conditions were provided for all pups. In this way, we have been able to contrast the effects of neonatal malnutrition and subsequent rehabilitation with an ideal standard of maturation of body and brain.

The procedures described by A. Pope and H. Hess for sampling a frozen block for quantitative microchemical assays and for relating the results to the microscopic appearance of the tissue were described previously (1,2). Both nutritionally deprived and control animals were decapitated at birth and at 10, 20, 30, 40, and 50 days of age, and the brains removed immediately. Blocks of right somatosensory cortex were fixed in 10% formalin for histological examination. The frozen hemisphere was transferred to a Harris microtome cryostat maintained at -12 degrees C., where a cylinder 2 mm. in diameter was punched perpendicular to the cortical surface of the block. The frozen cylinder was cut with a rotary micro-

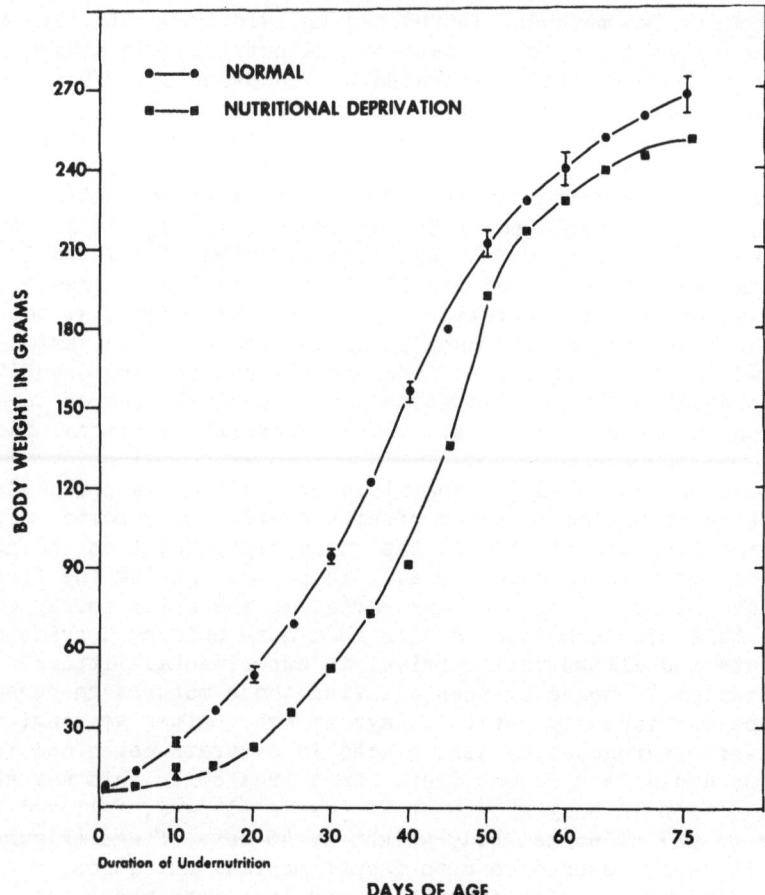

Figure 2.   Growth curve of normal male albino rats compared to those
undergoing nutritional deprivation during early postnatal life.
Duration of undernutrition was from birth to 21 postnatal days.
Each point is mean for 30 animals. Bracketed vertical lines on
normal curve represent standard error of mean.

tome into serial horizontal frozen sections 40 microns in thickness.
Sections were used in consecutive groups (usually 9) of 6 sections
each. Sections of each group were desiccated over calcium chloride,
weighed on a Mettler ultra-microbalance and placed in tubes grouped
into two anatomic categories: cortex and subcortical white matter.
Each group of slices was extracted with chloroform-methanol-water
and partitioned with aqueous KCl. Quantitative histochemical
indices of histologic fine structure were determined: DNA as an
index of total cells, RNA as an index of ribosomal organelles and
cytoplasmic volume, ganglioside sialic acid as a neuronal membrane
component, and cerebrosides, cholesterol, and proteolipid protein

as indices of myelin. Fresh weight, dry weight, water content and
cortical thickness were also determined.

Results and Discussion: The somatosensory cortex of the
normal newborn rat is composed mostly of undifferentiated neurons
tightly packed in vertical columns. The axodendritic fibers are
sparse, neuroglial cells are few in number, and extracellular spaces
are extensive. During the first 10 days of normal postnatal development, neurons enlarge, the pattern of nuclear chromatin matures,
and cytoplasmic organelles and cellular processes increase in size
and number. Cellular packing density decreases greatly and the
cortical thickness doubles. Neurons stratify to create a six-
layered neocortex. This rapid increase in cortical width and the
isocortical stratification in the first 10 postnatal days are
attributed to the extensive proliferation of large diameter
axodendritic processes (14).

At 10 days of age, cortical thickness reaches adult size and
a six layered cytoarchitectonic pattern is apparent: Silver
staining fibers are absent and only small numbers of neuroglia are
present. A large source of undifferentiated cells is seen around
the ependymal wall of the lateral ventricle and these cells have
been shown to migrate postnatally through white matter into
cerebral cortex and become mature glia (15). By 50 days of age, an
adult histologic appearance is seen: there is an increase in cell
density and silver staining fibers in both cortex and white matter
and a decrease of migrating glial cells around the ependymal surface.

DNA per cubic millimeter of fresh frozen tissue has been used
as an index of cellular packing density in developing rat cerebrum
(Figure 3). In normal cerebral cortex, the highest cellular density
is found at birth when DNA values are 20% greater than those of the
50 day old normal adult. Between birth and 10 days of normal
development, as cortical thickness increases and differentiating
neurons stratify into 6 layers, DNA levels decrease by 90%.
Between 10 and 50 postnatal days, DNA progressively increases as
glial cells migrate into cortex. In rats subjected to neonatal
malnutrition, cortical thickness at 10 postnatal days is decreased
25%, layering is poorly defined, and DNA concentration is increased
by 45% compared to controls. This increase in cellular density is
maintained between 10 and 30 days of age despite increase of cortical width to normal values by 20 postnatal days. Histologic
examination reveals many undifferentiated neurons, a severely
decreased neuropil, and a paucity of mature glial cells. In subcortical white matter of normal animals, cellular density increases
in steps during postnatal development. DNA concentrations in
undernourished rats are similar to controls except for significant
increases in cellular density at 10 and 30 days. These findings
suggest that the rate of normal postnatal glial cell migration into

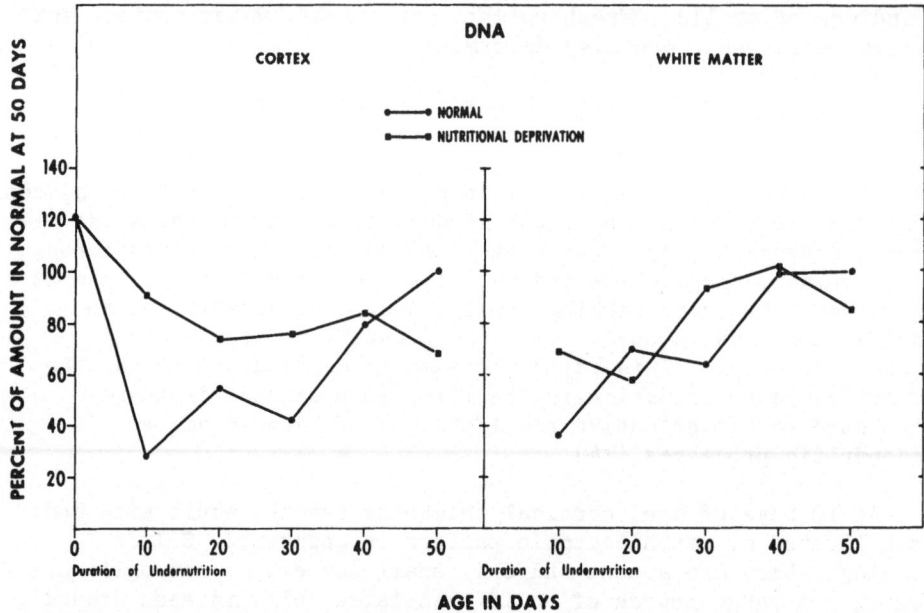

Figure 3.   DNA in developing cortex and white matter of normal and
nutritionally deprived rats.   All values are expressed as percentage of DNA in normal specimens at 50 postnatal days.

cortex has been retarded in the first 40 days.   Undifferentiated
glial cells destined to migrate into cortex then accumulate in
white matter.   Some of this accumulation is near the ependymal
surface deeper than our sample of white matter, hence not fully
shown by our DNA values in white matter.   The final decreased
cellular density in the cerebrum of the 50 day old rat subjected
to malnutrition, therefore, is the product of retarded cortical
growth with incomplete isocortical stratification, decreased
proliferation of large diameter dendritic processes of undifferentiated neurons, and diminished rate of glial cell migration.

     Total ganglioside sialic acid, a neuronal membrane component,
normally has peak values in cortex and white matter at 20 postnatal
days of age (Figure 4).   This peak attributed to the major
disialoganglioside (GD1A) concentration, is correlated with the
formation of synaptic (16) junctions and axodendritic proliferation
(17), as well as with maturation of cortical electrical activity
(18) and behavior (19).   By the 21st postnatal day, food-seeking
behavior allows weaning to take place, and also can be related to
the accumulation of ganglioside sialic acid.

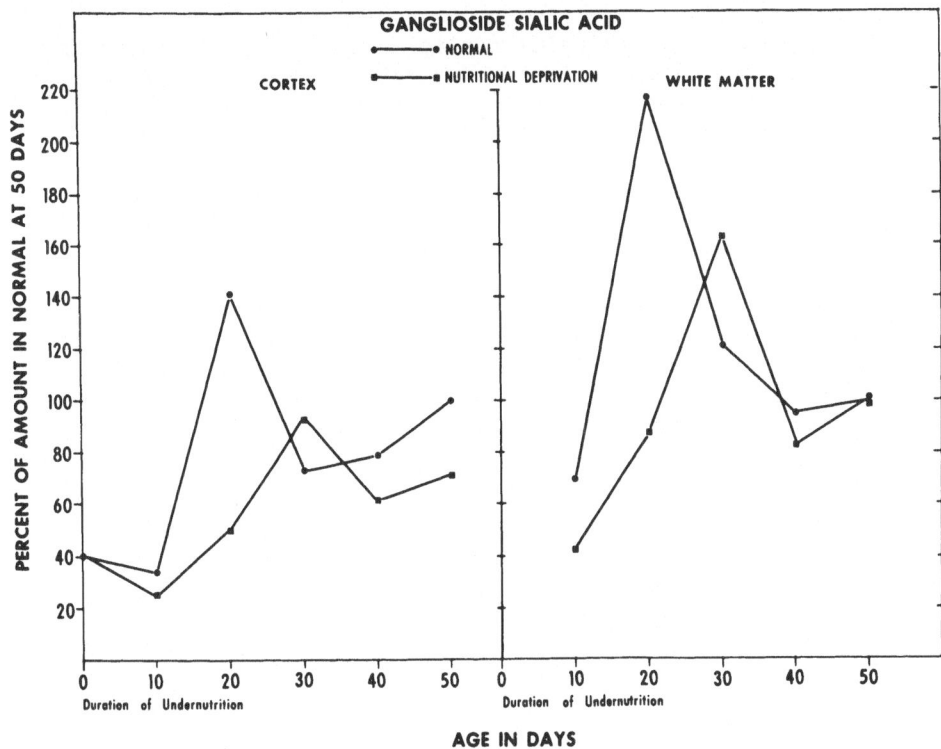

Figure 4. Ganglioside sialic acid in developing cortex and white
matter of normal rats, and in those subjected to nutritional
deprivation. All values are expressed as percentage of ganglioside
sialic acid in normal specimens at 50 postnatal days.

        In animals subjected to nutritional deprivation, weaning cannot
be accomplished until the 35th postnatal day without incurring a
mortality of greater than 50%. In these animals, peak values of
ganglioside sialic acid in cortex are decreased by 50% and are
delayed in appearance until 30 postnatal days. This abnormality
results in a 20% decrease in cortical ganglioside in the neonatally
malnourished but rehabilitated 50 day old rat, and can be correlated
with histologic observations of a lesser mass of fibers impregnated
by silver. In the white matter of neonatally malnourished rats,
the ganglioside peak also is delayed until 30 postnatal days and is
again markedly decreased. Those observations suggest close
relationships between the abnormal peak values in ganglioside sialic
acid after neonatal malnutrition, the delay in behavioral maturation,
and decreased proliferation of axodendritic fibers and formation of
synapses.

Between birth and 10 postnatal days of normal development, a
3.5-fold increase of ganglioside sialic acid per cell is found in
cortex correlating with the process whereby undifferentiated cells
produce dendritic fibers and cortical thickness increases rapidly
(Figure 5). In contrast, ganglioside per cell does not increase
during the first 10 postnatal days of neonatal malnutrition;
undifferentiated cells fail to form dendritic processes, thereby
retarding growth of the cerebral cortex. Similarly, RNA per cell
shows peak levels in cortex of normal animals at 10 days of age,
the values increasing 1.7-fold, followed by return to newborn levels
at 40 and 50 days. By contrast, in the cerebral cortex of neonatally malnourished animals, RNA per cell decreased by 20% during the
first 10 postnatal days. It is postulated that this defect in
cortical growth during the first 10 days of postnatal life has set
the stage for subsequent pathologic alterations of development:
i.e., decreased migration of glial cells, diminished formation of
synapses, and defective myelinogenesis.

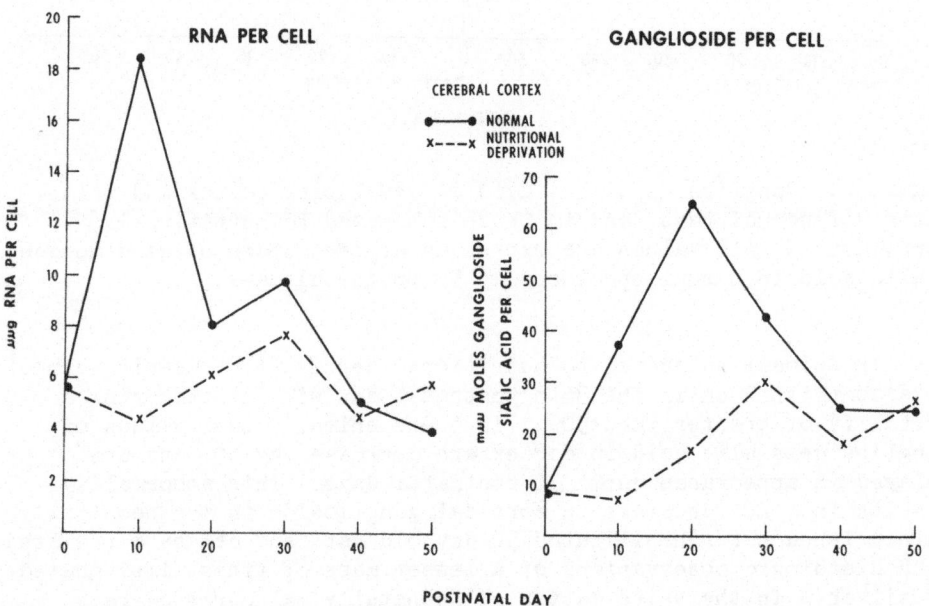

Figure 5. RNA and ganglioside sialic acid expressed as function of
DNA (µµg RNA per cell and mµµ moles ganglioside sialic acid per
cell) in developing cerebral cortex of normal rats, compared to
nutritionally deprived animals.

In the subcortical white matter of normally developing albino rats, myelination as seen in electron micrographs begins about 2 days postpartum (20). Although most axons are unmyelinated and compact repeating myelin periods are only rarely seen, a few fibers with a loosely organized lamellar structure are found. Between 10 and 20 days of age, light microscopy shows an increasing number of immature glial cells and axons impregnated by silver, but stainable myelin is absent. Thirty days of age marks the first appearance of myelin stainable by Luxol-fast blue accompanied by an increase in number of mature oligodendroglia. Between 30 and 40 days of age, myelinated fibers and their associated glial cells assume an adult appearance (2).

In normal rat cerebrum, myelination is continuous between 10 and 50 postnatal days. At 10 and 20 days of age, poorly different- iated glial cells are present before the light microscopic appear- ance of myelin stained by luxol fast blue. During this period, the accumulation of total lipids, cerebrosides, cholesterol and pro- teolipid proteins to about one-third of adult values is correlated with electron microscopic observations of early glial ensheathment of axons. Between 30 and 50 days of age, biochemical components double as oligodendrocytes progressively myelinate increasing numbers of axons to produce fibers with complex lamellation. (Figure 6).

In rats subjected to neonatal undernutrition, although the concentrations of total lipids and cerebrosides were relatively similar in both malnourished and control animals during the first 20 days of postnatal development, total cholesterol and proteolipid proteins were abnormally increased. Between 20 and 50 postnatal days, however, the concentrations of all 4 biochemical components in subcortical white matter were greatly abnormal, values ultimately reaching only 50% of normal adult levels by 50 days of age. Microscopic examination of 50 day old adult subcortical white matter shows decreased staining of myelinated fibers, diminished numbers of mature oligodendroglia, and many undifferentiated hyperchromatic cells deep in subcortical white matter. This pathologic picture, accompanied by a striking reduction in the concentration of myelin components, occurred during rehabilitation of body weight, after the period of undernutrition. It is postulated that decreased formation of myelin resulting from neonatal malnutrition is produced by damage to many glial cell precursors which fail to undergo the differentiation necessary for the formation of normal myelin.

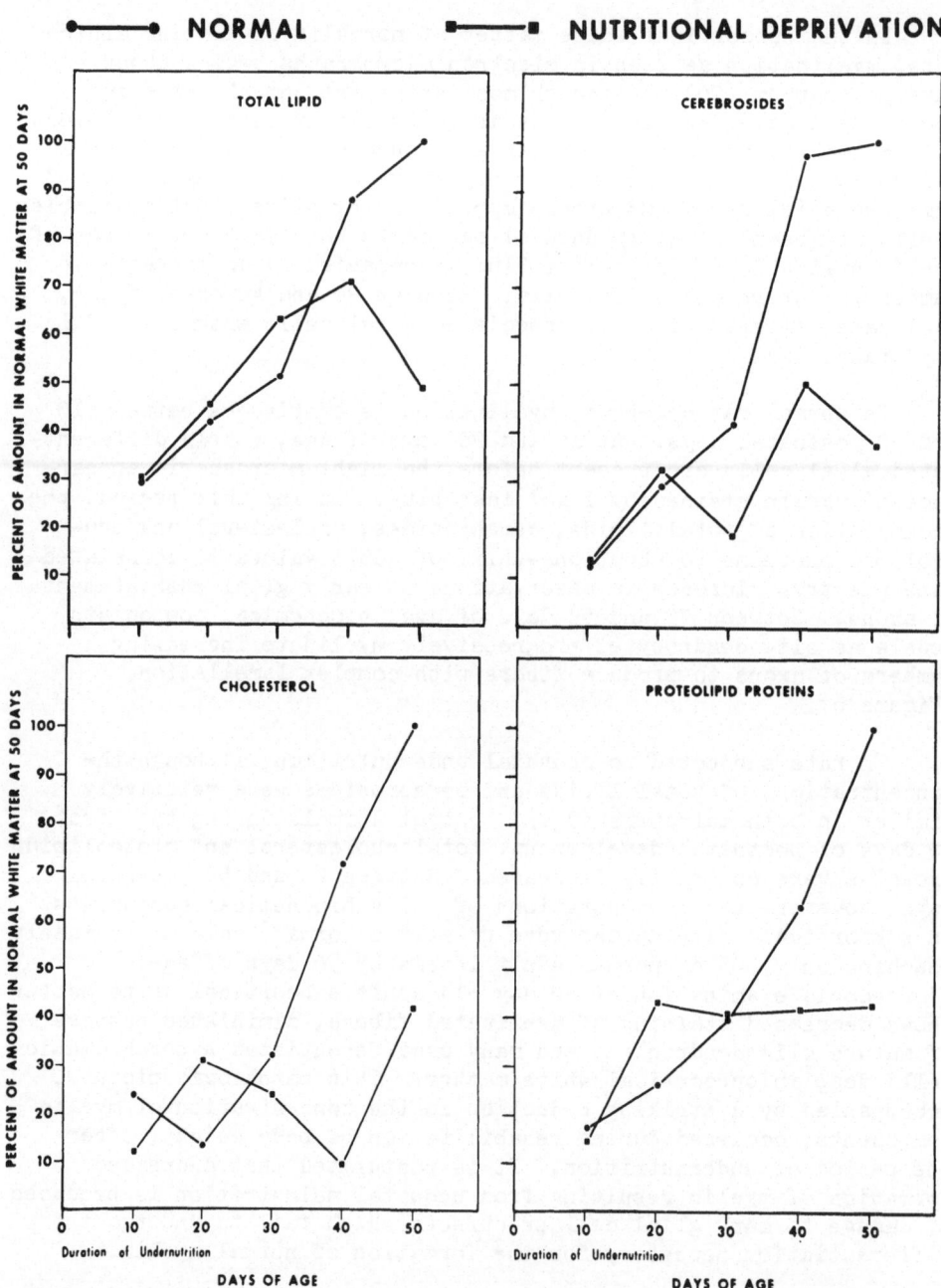

Figure 6. Biochemical profiles of subcortical white matter during postnatal development in normal rats and in animals nutritionally deprived during first 21 days of postnatal life. On ordinate, amount of total lipids, cerebrosides, cholesterol, and proteolipid proteins is expressed as percentage of amount in normal specimens at 50 postnatal days. Postnatal age in days is shown on abscissa.

In conclusion, our observations indicate that the somatosensory area of rat cerebrum is highly vulnerable to pathologic change resulting from undernutrition during the phase of rapid growth prior to the body growth spurt. A 21 day period of undernutrition begun immediately after birth is accompanied by persistent and progressive chemical and histologic alterations despite subsequent recovery of body weight. Changes during the first 10 days resulting from a marked decrease in dendritic proliferation and cytoplasmic differentiation of ribosomal organelles, reduces cortical growth and sets the stage for further derangement of cortical organization between 10 and 50 postnatal days. As a result, migration of glial cells is retarded, proliferation of neuronal fibers decreased, and formation of synapses and myelin diminished.

References

1. Bass, N.H., Netsky, M.G., and Young, E.   Neurology 19:258 (1969).

2. Bass, N.H., Netsky, M.G., and Young, E.   Neurology 19:405 (1969).

3. Flexner, L.B.   Harvey Lect. 47:156 (1952).

4. Hatai, S.   Amer. J. Physiol. 12:116 (1904).

5. Donaldson, H.H.   J. Comp. Neurol. 21:139 (1911).

6. Dobbing, J.   Biol. Neonat. 9:132 (1965).

7. Eichenwald, H.F., and Fry, P.C.   Science 163:644 (1969).

8. Winick, M.   J. Pediat. 74:667 (1969).

9. Bass, N.H., Netsky, M.G., and Young, E.   Archiv. Neurol. 23:289 (1970).

10. Bass, N.H., Netsky, M.G., and Young, E.   Archiv. Neurol. 23:303 (1970).

11. Sugita, N.   J. Comp. Neurol. 29:177 (1918).

12. Widdowson, E.M., and McCance, R.A.   Proc. Roy. Soc. (Biol.) 152:188 (1960).

13. Culley, W.J., and Mertz, E.T.   Proc. Soc. Exp. Biol. Med. 118:233 (1965).

14. Caley, D.W., and Maxwell, D.S. J. Comp. Neurol. 133:17 (1968).

15. Altman, J. Exp. Neurol. 16:263 (1966).

16. Aghajanian, G.K., and Bloom, F.E. Brain Res. 6:716 (1967).

17. Schade, J.P., and Baxter, C.F. In Inhibition in the Nervous
     System and GABA, p. 207 (Ed. E. Roberts) (1960).

18. Deza, L., and Eidelberg, E. Exp. Neurol. 17:425 (1967).

19. Tilney, F. Bull. Neurol. Inst. N.Y. 3:352 (1933).

20. Luse, S.A. J. Biophys. Biochem. Cytol. 2:777 (1956).

ESSENTIAL FATTY ACID DEFICIENCY AND ITS EFFECTS ON THE CENTRAL NERVOUS SYSTEM

C. Galli, H.B. White, Jr.* and R. Paoletti

Institute of Pharmacology, University of Milan

20129 Milan, Italy

Dietary polyunsaturated fatty acids (PUFA) belonging to the linoleic and linolenic acid families were re-cognized several years ago as being essential for balanced body growth and the proper function of many tissues and organs and have since been extensively investigated. This field has recently been reviewed by Holman (1) and Alfin-Slater and Aftergood (2), while the metabolism of PUFA has been discussed by Mead (3).

The symptoms of essential fatty acid (EFA) deficiency are considered to be the consequence of functional altera-tions of cellular and intracellular membranes deriving from changes in the metabolism of polyunsaturated EFA under these conditions. The major biochemical changes detected in the tissues of EFA deficient animals consist of a decrease in PUFA of the linoleate (18:2 w6) and linolenate (18:3 w3) families and a corresponding in-crease in trienes, especially 20:3, derived from oleic acid (18:1 w9). These changes have been interpreted as a consequence of a reduced inhibition in the conversion of oleic acid to more unsaturated, longer chain fatty acids, due to the dietary deficiency of linoleate and

---

* On leave of absence from the Department of Biochemistry, University of Mississippi, Jackson, Mississippi, U.S.A.

linolenate (4).

Fatty acid changes in EFA deficient animals are generally less pronounced in the brain than in other tissues, presumably because of the lower turnover of brain fatty acids, which are part of the structural components of the perennial nervous cells. However, while nutritional deficiencies have a limited effect on the adult brain, they have been shown to influence growth and maturation of the central nervous system (CNS) in experimental animals (5) and in man (6) when induced during the period of brain maturation. Thus, maternal diets were shown to influence brain fatty acid composition of newborn rats (7). Furthermore, pregnant rats, fed an EFA deficient diet from weaning, give birth to offspring with reduced body and brain weights and with changes in brain fatty acid composition (8).

In consideration of these observations, a study was carried out with the aim of following the effects of EFA deficiency on brain lipid composition and fatty acid distribution at various stages of development. This was accomplished by feeding a deficient diet to rats during pregnancy and lactation and then to the offspring after weaning. Brain lipid and phospholipid concentrations and the fatty acid composition of ethanolamine phosphoglyceride (EPG), which is contained in high concentrations in the brain and is particularly rich in PUFA, have been measured after various periods of deficiency. The reversibility of brain EPG fatty acid changes induced by EFA deficiency has also been investigated.

MATERIALS AND METHODS

Pregnant rats of the Sprague-Dawley strain were fed ad libitum with either a fat free or control diet with 2% corn oil added, starting approximately one week before delivery. Gas chromatographic (GLC) analyses revealed that the control diet contained 0.84% linoleate and 0.023% linolenate, while in the deficient diet the values were 0.004% and 0.001% respectively.

Groups of control and deficient rats were sacrificed at 3, 10, 30, 60, 90, 180 and 365 days after birth. Additional groups of rats maintained on an EFA deficient diet for 30 and 90 days after birth were returned to the control diet for an equal period of time and then sacrificed at 60 and 180 days respectively. A few rats from both the control and deficient groups were sacrificed at 1, 2, 5, 10 and 15 days of age and the stomach contents were analysed for fatty acid composition. Body and brain weights were measured and brains from all groups were pooled, with the exception of the 60 day old control and deficient groups, which were examined individually to evaluate the individual variability of the data.

Lipids were extracted (9) and purified (10). Total lipid and phospholipid concentrations were measured on the purified lipid extract, while the phospholipid composition was determined by two dimensional thin layer chromatography (2D-TLC) (11). EPG was purified by chromatography (12) and fatty acid methyl esters were prepared, purified and analysed by GLC (12). Fatty acid identification was achieved by use of reference compounds and by the combination of gas chromatography-mass spectrometry (model LKB 9000). Myelin was purified by ultracentrifugation according to the procedure described by Autilio et al. (13). Myelin EPG was isolated, the fatty acid methyl esters prepared, purified and analysed as described above.

RESULTS AND CONCLUSIONS

Intake of EFA

The analysis of the fatty acid composition of the stomach contents obtained from the suckling rats showed a considerable decrease in 18:2 after only a few days of lactation in the EFA deficient group. 18:2 made up 14.2%, 9.3% and 11.9% of total fatty acids at 5, 10 and 15 days in the control animals and 4.2%, 1.3% and 1.2% respectively, in the deficient ones. Hence, the intake of EFA in the sucklings of mothers fed the EFA deficient diet during pregnancy was considerably decreased.

C. GALLI, H. B. WHITE, JR., AND R. PAOLETTI

TABLE I

BODY AND BRAIN WEIGHTS AND BRAIN PHOSPHOLIPID (PL) CONCENTRATIONS IN CONTROL (C) AND EFA DEFICIENT (D) RATS.

Days/ Group	No. in Group	Body Weight[*]	Brain Weight	Brain PL Concentration[‡] (mg/g FW)
10 C	11	21 ± 0.4	955 ± 14	27.2
10 D	9	18 ± 0.4***	858 ± 8***	27.6
30 C	4	46 ± 1.1	1388 ± 7	48.2
30 D	4	45 ± 1.8	1346 ± 30	51.6
60 C	8	177 ± 10	1702 ± 35	not determined
60 D	5	108 ± 7***	1655 ± 21	not determined
90 C	4	375 ± 9	1911 ± 46	45.6
90 D	3	128 ± 2***	1517 ± 39**	40.1
180 C	2	400 ± 4	1838 ± 1	51.9
180 D	2	214 ± 6**	1726 ± 56	45.8
1 yr C	4	586 ± 40	2230 ± 90	46.0
1 yr D	6	417 ± 32**	2000 ± 30*	40.7

[*] Values are mean ± standard error. Values with asterisk are significantly different from control values at the P level, as follows:
  *** = $P < 0.001$;   ** = $P < 0.01$;   * = $P < 0.05$

[‡] PL concentrations expressed as mg/g fresh weight. The values are the mean of three determinations made on the lipid extracts from the pooled brains.

Brain and Body Weights

Under these dietary conditions body growth was significantly reduced, but also brain size and phospholipid concentration were significantly affected (Table I). Total phospholipid composition, however, was not modified by EFA deficiency.

Brain EPG Fatty Acid Changes

Analyses of brain EPG fatty acid composition showed considerable changes in the deficient animals. These variations mainly affected polyunsaturated acids, while saturates were virtually unchanged and monoenes, which increase considerably in the developing brain, remained slightly below control values in deficient rats after six months. Polyenes from EPG normally tend to decrease from approximately 50% of EPG fatty acids at birth to approximately 30% in the adult rat. After a long period of EFA deficiency polyene values remained above those of control. Examination of the individual components of the polyene group revealed an increase of trienes (20:3 and 22:3 of the w9 family), a slow accumulation of pentaenes (22:5 w6) and a corresponding decrease of tetraenes (20:4 and 22:4 w6) and, also, later on, of hexaenes (22:6 w3). The overall elevation of polyenes in EFA deficiency was due to the increase of trienes and pentaenes exceeding the reduction of the other components.

The extent of brain EPG fatty acid changes in growing EFA deficient rats was indicated by the triene/tetraene ratio, which is a biochemical index of EFA deficiency (14). This index remained at approximately 0.05 in control animals while, in an almost linear progression, it reached the value of 1.95 in the deficient animals after one year on the deficient diet (Fig. 1).

The overall fatty acid changes in brain EPG in EFA deficiency can be summarized by considering the percentages of the various fatty acid families (w9, w6 and w3) at various ages in both control and deficient animals (Table 2). In the deficient rats, w9 fatty acids, derived from oleic acid (18:1 w9), reached a value approximately 30% higher than the control value, while w6 and w3 fatty acids remained approximately 30% below that of the control. In spite of these changes the unsaturation index (summation of percentage of individual unsaturated fatty acids multiplied by the number of double bonds) of brain EPG fatty acids remained virtually unchanged in both groups of rats. This indicates the ability of tissue phospholipids to maintain a certain

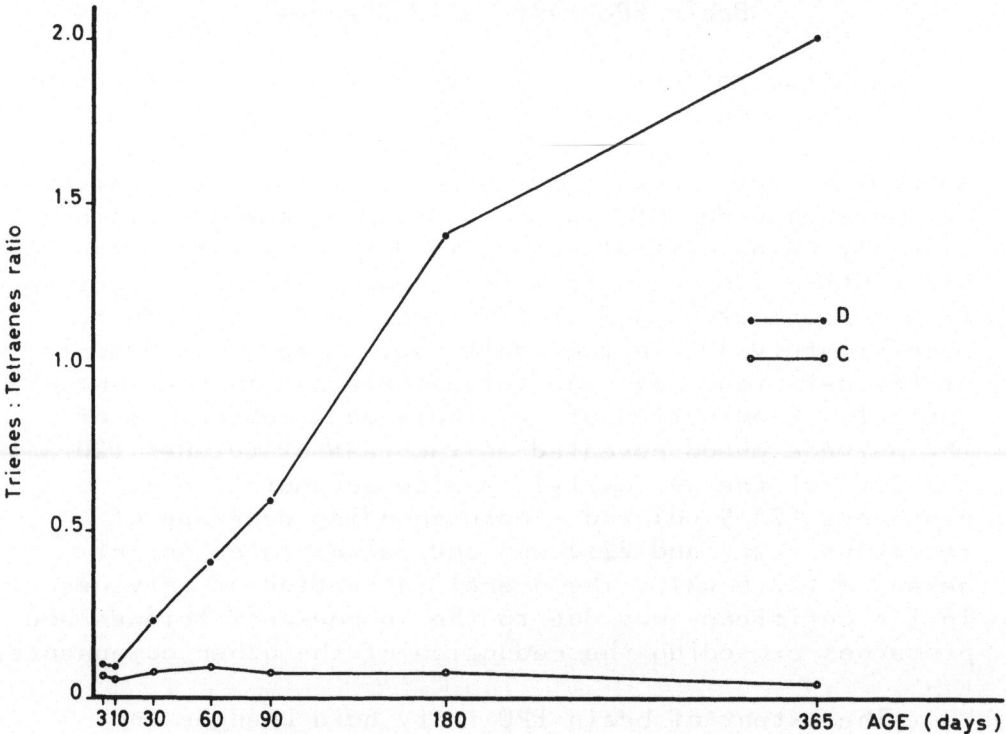

Fig. I.   Triene/tetraene ratio of brain EPG fatty acids
in growing control and EFA deficient rats.

degree of unsaturation, regardless of the dietary intake
of EFA.

Recovery from EFA Deficiency

A return to the control diet after a period
of EFA deficiency (groups D + C = animals kept for 30
and 90 days on the deficient diet before being returned
to the control diet for an equal length of time) was
able to restore brain and body growth and, not only
reverse the brain EPG fatty acid changes but also induce
an overcompensatory effect. In fact, w6 acids exceeded
the values of control animals of the same age, while
w9 acids were reduced to below control values (Table 3).
Analysis of the polyenoic group of acids revealed that
tetraenes (belonging to the w6 family) increased above
control values, while monoenes (mainly 18:1 w9) were

TABLE 2.

BRAIN EPG FATTY ACID FAMILIES AS WEIGHT PERCENTAGE OF TOTAL FATTY ACIDS IN CONTROL (C) AND EFA DEFICIENT (D) RATS.

Days	w9 acids		w6 acids		w3 acids		U.I.	
	C	D	C	D	C	D	C	D
3	10.6	11.2	30.2	28.3	16.2	18.1	241	259
10	11.2	10.0	29.2	27.6	16.3	17.9	237	238
30	22.0	26.6	28.4	22.5	14.2	14.6	230	218
60	30.7	37.2	23.9	20.2	11.3	10.7	(6)* 205 ± 16	(5)* 202 ± 7
90	33.2	40.4	25.7	19.7	11.3	13.4	212	222
180	39.5	48.9	20.5	17.3	15.7	12.0	218	225
365	37.8	49.0	18.5	12.6	11.2	5.8	178	172

U.I. Unsaturation Index : summation of percentage of individual unsaturated fatty acids multiplied by the number of double bonds.

* These values represent the mean + standard error of determinations performed on individual animals. The number of animals is given in brackets.

TABLE 3.

BRAIN EPG FATTY ACID FAMILIES IN CONTROL (C) AND

EFA DEFICIENT (D) RATS AND EFA DEFICIENT RATS

RETURNED TO CONTROL DIET (D+C)

	60 DAYS			180 DAYS		
	C	30D+30C	D	C	90D+90C	D
w9 acids	30.7	28.8	37.2	39.5	32.7	48.9
w6 acids	23.9	26.1	20.2	20.5	24.4	17.3
w3 acids	11.3	11.6	10.7	15.7	15.8	14.5

Values are expressed as weight percentages of total EPG
fatty acid.

reduced below controls and trienes returned virtually to
normal values.

Fatty Acids of Myelin EPG

Fatty acid analysis of myelin EPG in the three
groups of 180 day old rats (control, deficient and
deficient returned to control diet after 90 days) revealed that EFA deficiency induced in myelin changes
similar to those observed in the whole brain (Table 4).

Trienes were increased and tetraenes decreased.
As a consequence, w9 acids were elevated and w6 reduced.
These modifications, however, were somewhat less pronounced than those occurring in the whole brain. Monoenes, by far the major fatty acid group in myelin EPG,
were slightly reduced while saturated acids were virtually
unaffected.

Return to the control diet after a 90 day long
period of deficiency was found to normalize the fatty
acid pattern even in myelin, although the overcompensatory effect seen in the whole brain does not appear.
The unsaturation index of myelin EPG in the three groups
of animals was not affected by the different dietary

TABLE 4.

FATTY ACID DISTRIBUTION (WEIGHT PERCENTAGE) OF EPG IN MYELIN OF 180 DAY OLD MALE RATS.

FATTY ACIDS	6 Months CONTROL	3 Months + 3 Months DEFICIENT + CONTROL	6 Months DEFICIENT
SATURATES	13.5	13.9	13.5
MONOENES	64.0	63.9	60.1
POLYENES	22.2	21.8	25.6
Dienes	1.8	1.8	1.9
Trienes	2.5	4.2	13.0
Tetraenes	13.9	12.8	7.3
Pentaenes	0.5	1.1	1.6
Hexaenes	3.5	2.0	1.8
w9 acids	63.4	65.7	70.6
w6 acids	16.7	15.3	9.9
w3 acids	3.7	2.2	2.0
Unsaturation Index	154	148	151
Triene/ Tetraene	0.18	0.33	1.74

treatments.

These results indicate that the brain is affected by EFA deficiency and that considerable changes in brain fatty acids can be induced during the early periods of life. Investigations now in progress to study the effects of EFA deficiency on animal behaviour provide evidence that the performance of deficient rats in an avoidance/escape test is also affected. It is thus possible that brain functional changes as well as bio-chemical modifications are induced by the EFA deficient diet.

It is interesting to observe that, in spite of the

alterations in brain fatty acid composition, the unsaturation index of the fatty acids of brain EPG is
retained. This suggests the existence of regulatory
mechanisms for the retention of precise physico-chemical
characteristics of membrane constituents.

The rate of fatty acid modification (as indicated
in the triene/tetraene ratio shown in Fig. 1) is maintained for a long period of time even after the rate
of new lipid deposition is considered to be greatly
reduced. This observation, together with the almost
complete recovery of the brain from fatty acid changes
induced by EFA deficiency, indicates that a continuous
replacement of fatty acids takes place in the brain and
suggests that the constituents of nervous membranes
undergo dynamic processes.

REFERENCES

1. Holman, R.T. in Progress in the Chemistry of Fats
    and other Lipids (Ed. R.T. Holman) Vol. IX, Pergamon
    Press (1968)

2. Alfin-Slater, R.B., and Aftergood, L. Physiol. Rev.
    48:758 (1968)

3. Mead, J.F. in Progress in the Chemistry of Fats and
    other Lipids (Ed. R.T. Holman) Vol. IX, Pergamon
    Press (1968)

4. Holman, R.T. and Mohrhauer, H. Acta chim. Scand.
    17:584 (1963)

5. Howard, E., and Granoff, D.M. J. Nutrition 95:111
    (1965)

6. Graham, G.G. Fed. Proc. 26:139 (1967)

7. Walker, B.L. Lipids 2:497 (1967)

8. Steinberg, A.B., Clarke, G.B., and Ramwell, P.W.
    Developmental Psychobiol. 1(4):225 (1968)

9. Rouser, G., Kritchevsky, G., Galli, C., Yamamoto, A.
    and Knudson, Jr., A.G. in Inborn Disorders of Sphingo-
    lipid Metabolism (Eds. Aronson and Volk) Pergamon
    Press (1966)

10. Siakotos, A.N., and Rouser, G. J. Am. Oil Chem. Soc. $\underline{42}$:913 (1965)

11. Rouser, G., Siakotos, A.N., and Fleischer, S. Lipids $\underline{1}$:86 (1966)

12. White, H.B., Jr., Galli, C., and Paoletti, R., J. Neurochem. (in press)

13. Autilio, L.A., Norton, W.T., and Terry, R.T. J. Neurochem. $\underline{11}$:17 (1968)

14. Holman, R.T. J. Nutrition $\underline{70}$:405 (1960)

CLINICAL OBSERVATIONS ON LATE EFFECTS OF EARLY MAL-NUTRITION

Zdzisław Rydzyński

Młynarska 2, Głowno, Poland

In relation to the papers of Drs. Dobbing, Bass and Galli, I would like to report some clinical observations made in Poland after the Second World War on Jewish and Polish children who, unfortunately, were born or brought up in German ghettos or concentration camps.

These children were underfed in their early child-hood. Their diet was not only low in calories but also poor in protein and practically without fats. The psychological and psychiatric state of the children was similar. All of them showed great disturbances of both recent and reproductive memory, and had great difficulty in concentration. Their mood was rather apathetic and anxious with a tendency to irritability, explosiveness and fear reactions. Their intelligence was frequently normal, or a little below the standard. The children said they felt as if they were permanent convalescents after the severe somatic disease.

Maladaptation to family and school life appeared to be the result of their character and intellectual disturbances. Some of them were psychotic but their psychopathological symptoms, not being characteristic for any psychiatric classical nosological picture, were difficult to classify.

Nowadays, in psychiatric clinics we may see analogous clinical pictures in children with minimal brain
damage after birth trauma and with long lasting infant
diarrhoea. This seems to be the important argument for
the thesis that the main reason for the psychic defects
is undernutrition and not the psychological stress
which the children underwent in early childhood.